AF900711

Digital Innovations in Architecture, Engineering and Construction

Series Editors

Diogo Ribeiro, Department of Civil Engineering, Polytechnic Institute of Porto, Porto, Portugal

M. Z. Naser, Glenn Department of Civil Engineering, Clemson University, Clemson, SC, USA

Rudi Stouffs, Department of Architecture, National University of Singapore, Singapore, Singapore

Marzia Bolpagni, Northumbria University, Newcastle-upon-Tyne, UK

The Architecture, Engineering and Construction (AEC) industry is experiencing an unprecedented transformation from conventional labor-intensive activities to automation using innovative digital technologies and processes. This new paradigm also requires systemic changes focused on social, economic and sustainability aspects. Within the scope of Industry 4.0, digital technologies are a key factor in interconnecting information between the physical built environment and the digital virtual ecosystem. The most advanced virtual ecosystems allow to simulate the built to enable a real-time data-driven decision-making. This Book Series promotes and expedites the dissemination of recent research, advances, and applications in the field of digital innovations in the AEC industry. Topics of interest include but are not limited to:

- Industrialization: digital fabrication, modularization, cobotics, lean.
- Material innovations: bio-inspired, nano and recycled materials.
- Reality capture: computer vision, photogrammetry, laser scanning, drones.
- Extended reality: augmented, virtual and mixed reality.
- Sustainability and circular building economy.
- Interoperability: building/city information modeling.
- Interactive and adaptive architecture.
- Computational design: data-driven, generative and performance-based design.
- Simulation and analysis: digital twins, virtual cities.
- Data analytics: artificial intelligence, machine/deep learning.
- Health and safety: mobile and wearable devices, QR codes, RFID.
- Big data: GIS, IoT, sensors, cloud computing.
- Smart transactions, cybersecurity, gamification, blockchain.
- Quality and project management, business models, legal prospective.
- Risk and disaster management.

Ilaria Trizio · Emanuel Demetrescu ·
Daniele Ferdani
Editors

Digital Restoration and Virtual Reconstructions

Case Studies and Compared Experiences for Cultural Heritage

Springer

Editors
Ilaria Trizio
ITC-CNR
L'Aquila, Italy

Emanuel Demetrescu
ISPC-CNR
Montelibretti, Italy

Daniele Ferdani
ISPC-CNR
Montelibretti, Italy

ISSN 2731-7269 ISSN 2731-7277 (electronic)
Digital Innovations in Architecture, Engineering and Construction
ISBN 978-3-031-15320-4 ISBN 978-3-031-15321-1 (eBook)
https://doi.org/10.1007/978-3-031-15321-1

© The Editor(s) (if applicable) and The Author(s), under exclusive license to Springer Nature Switzerland AG 2023
This work is subject to copyright. All rights are solely and exclusively licensed by the Publisher, whether the whole or part of the material is concerned, specifically the rights of translation, reprinting, reuse of illustrations, recitation, broadcasting, reproduction on microfilms or in any other physical way, and transmission or information storage and retrieval, electronic adaptation, computer software, or by similar or dissimilar methodology now known or hereafter developed.
The use of general descriptive names, registered names, trademarks, service marks, etc. in this publication does not imply, even in the absence of a specific statement, that such names are exempt from the relevant protective laws and regulations and therefore free for general use.
The publisher, the authors, and the editors are safe to assume that the advice and information in this book are believed to be true and accurate at the date of publication. Neither the publisher nor the authors or the editors give a warranty, expressed or implied, with respect to the material contained herein or for any errors or omissions that may have been made. The publisher remains neutral with regard to jurisdictional claims in published maps and institutional affiliations.

This Springer imprint is published by the registered company Springer Nature Switzerland AG
The registered company address is: Gewerbestrasse 11, 6330 Cham, Switzerland

Foreword

Virtual reconstruction and restoration is gaining momentum at a surprising pace and this is only partially due to the availability of powerful information and computational technologies. Indeed, virtualization has proved to be a multidisciplinary subject whose applications can go beyond what initially expected. Virtual representation of archeological sites and architectural artefacts, as well as digital exhibitions and installations involving virtual, augmented, or mixed reality have already made their path in the framework of cultural heritage and new techniques and solutions appear at daily frequency.

From the perspective of the preservation and use of cultural built heritage, however, the digital restoration has still some exciting steps to take. In countries like Italy, where the built heritage is extraordinarily large and in the vast majority of cases still used for public and private purposes, one of major challenge is to manage the "trade off balance" between philological restoration and modernization of ancient, but still "alive" constructions.

Aside from museums and other cultural buildings, there are in Italy thousands of constructed facilities belonging to the monumental built heritage hosting various crucial activities like hospitals, schools, universities, banks, and public offices. Furthermore, a significant part of the road and railways bridges are centuries old and built in masonry, therefore in many case it has the "right" to belong to our built heritage. In such cases, in order to keep these facilities running for their functions, a specific maintenance is needed to reach satisfactory safety levels in terms of structural behavior and sufficient service levels in terms of comfort performance. This, in turn, means the need for the adoption of structural measures and for the installations of plants (i.e., HVAC systems, energy production or transformation elements, etc.) in a complex and tight path where unavailable constraints of restoration should meet the need for new or upgraded structural and plant elements.

From this point of view, thus, the exploitation of digital restoration techniques and technologies can be looked at as a tremendously powerful tool for evaluating advantages and drawbacks of any single maintenance operation on the built heritage. The articles included in this issue yield the seeds of this powerful application of Virtual reconstruction and restoration.

Antonio Occhiuzzi
Head of ITC-CNR
San Giuliano Milanese, Italy
occhiuzzi@itc.cnr.it

Preface

The digital technologies currently in use for the virtual representation of archaeological sites, and architectural artefacts offer researchers and scholars a wider range of possibilities than a few decades ago. The rapid evolution of ICT applied to the Cultural Heritage field has greatly advantaged the archaeological interpretation process; the latter, thanks to the development of three-dimensional acquisition, analysis, and visualization methodologies, is now able to extract previously unthinkable information and advance reconstructive hypotheses for landscapes, sites, and artefacts. Simultaneously, in the architectural domain, ICTs have made clear the interpretation process by integrating data resulting from the field survey with those relating to the state of surface degradation, finally making them readable directly on the virtual models, thanks to accurate ontologies. This has made it possible to create virtual restoration and simulations or, when possible, three-dimensional reconstructions based on analytical interpretation obtained by crossing the documentary sources with the material evidence that can be read directly on the artefacts.

Within this framework, archaeology, architecture, and conservation have often operated independently by referring to their theoretical background, operational needs, and fields of application. Although with extensive cross-contamination, each discipline has adapted or developed its methodologies, processes, and terminologies based on its own operational needs and aims.

According to scientific literature, Digital (or Virtual) Restoration (DR) consists in applying digital techniques in the field of restoration. Given that, DR is limited to the digital domain without any intervention on the physical artefacts. This definition is rather fuzzy, vague and it is used with different connotations according to the contexts and field of applications (architecture, sculpture, artworks, paintings, etc). Some of the most widespread meanings in academia are presented below.

DR can be intended as a digital intervention to simulate the result of physical restoration. Most often this is a digital reassembling or anastylosis where digitised fragments of artworks are digitally manipulated to find matches. In this case, digital restoration tools allow restorers to perform actions, in a virtual environment, that would be difficult or impossible to do in a physical context. Indeed, in many cases, the size or weight of fragments or their fragility limits the possibility of physical intervention. An example is the case of the restoration of Madonna di Pietranico (L'Aquila, Italy). An earthquake struck L'Aquila in 2009, damaged the city and the terracotta masterpiece was destroyed. The recomposition of the fragments was assisted by computer simulation: after complete digitisation of the fragile fragments, the researchers analyzed the matches among the fragments in a virtual environment, preventing damages, and allowing to simulate different reassembly alternatives.

In other cases, the term refers to digital actions to support the physical restoration, such as the recreation of missing parts of a broken artefact using automatic algorithms or manual photo retouching and 3D modelling techniques. An example is the case of the ancient funerary busts, dated between the 2nd and the 3rd century A.D., rescued from Palmyra (Syria). Firstly, the researchers designed the missing portions using 3D modelling and subsequently printed them using synthetic nylon powder and rapid prototyping technologies.

DR is also used to describe the digital rehabilitation of lost heritage, destroyed by natural or human-caused catastrophic events, in its former state of preservation and beauty. The recent cases of destruction of cultural heritage caused by war, as in the case of the ancient city of Palmyra (2016) and the Buddha of Bamiyan in Afghanistan (2011), or by accidental events, as in the case of the fire which caused severe damage to the Notre Dame cathedral (2019), has led international communities to an unprecedented need for digital preservation through projects of virtual restoration. In the first two cases, physical restoration is not always possible and the only way to restore these sites to their former beauty and make them accessible is only through 3D modelling and virtual reality technologies.

In some areas, like music, photography, or cinematography—related to visual or acoustic assets—DR is, de facto, the only restoration possible especially when the matter of the work of art cannot be restored or when the restoration to the tangible support (films, vinyl records) is limited. For instance, in film restoration, there are procedures of colour correction for recovering and enhancing the detail, look, and tone of the films that can be done only on the digital copies without endangering the original materials. When the restoration was performed only on analogue films, the operations were limited by the state of conservation of the product, but nowadays, this kind of digital procedure makes it possible to scan, edit and reconstruct images for which the advanced level of degradation precludes any physical interventions.

Finally, DR is also defined as the process used to reconstruct the unity of style or, in other words, the hypothetical original aspect of an artefact. The main goal of this digital edition is to provide an undisturbed reading of the whole artefacts in its integrity and improve better legibility for interpretation and dissemination purposes.

DR is then used to remove alteration from the digital copy of a painting and bridge the gaps in a mimetic way gathering the missing information from analogous elements present on the surface of the artwork. This stylistic restoration is carried on many fields, from paintings and mosaics restoration to written work and sculpture, especially when the gaps are small and easy to fill through unassailable evidence. When the evidence is not enough to complete the reconstruction, then "traditional" approaches from the physical restoration are commonly used to complete the work (neutral retouching, chromatic dampening, etc).

A complete stylistic restoration of large gaps, on the other hand, can be carried out under certain conditions. However, in this case, we are dealing with Virtual Reconstruction (VRC) rather than restoration. According to literature, stylistic restoration is sometimes used as synonymous with virtual reconstruction, especially in the field of historical architecture and building archaeology where the artefactsts are often preserved as ruins. Both terms refer to processes of simulation of the past aimed at restoring the unity of style of an artefact, however, in VRC the concepts of "hypothesis" and "conjecture" play an important role and specific precautions must be evaluated to ensure the reliability and consistency of the work.

According to the Principles of Seville, VRC is a digital process that uses "a virtual model to visually recover a building or object made by humans at a given moment in the past from available physical evidence of these buildings or objects, scientifically reasonable comparative inferences, and in general, all studies carried out by archaeologists and other experts in relation to archaeological and historical science". Given that, when the gaps exceed what is preserved and the evidence is not sufficient to complete the reconstructive model and guarantee the legibility of the artefact, it is necessary to push the critical hypothesis beyond the context and rely on sources and comparisons.

The most criticized issue regarding VRC stands on authenticity. Advantages and drawbacks of the simulation models of the past have been widely discussed in academia, leading to the development of guidelines and best practices, particularly for what concerns the issues related to the philologic study, authenticity, and scientific transparency which are the fundamental background to guarantee the reliability of the work and avoid arbitrary interpretations and reconstructions. Especially in the field of archaeology and ancient architecture, the debates have led to the creation of international documents such as the London Charter and the Principles of Seville.

As noted above, the case studies presented within this volume cover a variety of chronological contexts with methodologies specific to archaeology, architecture, and conservation. Thus, there are some elements that unify the various approaches. In almost all cases the authors present, albeit sometimes in a sketchy way, the use of the reconstructive model for valorization purposes. It is worth noting that the first wave of virtual reconstructions in the 1990s and 2000s was strongly aimed at a visualization for enhancement while in the last decade, thanks to increased methodological awareness, the visualization approach is becoming more widespread to increase scientific understanding of the cultural context. Given this premise, the addition of a valorization step, even where it is merely a sketch or a forecast of future development, seems to indicate a still-living link between reconstruction and valorization. This link, although on the one hand, it remains legitimate for the natural development of project analysis and synthesis, on the other hand, it seems to demonstrate a widespread expectation of the public (including the "academic" public) towards "compulsory" reuse of reconstruction for valorization and educational purposes. This, in our opinion, can be seen as a sign of not complete autonomy of the scientific reconstructive process from the needs of musealization and valorization, which, although absolutely legitimate and important, represent only a part of the purposes of the reconstructive process and should not be considered "obligatory" within the study project. In other words, virtual reconstruction is still struggling to be distinguished from valorization, probably because it is still not fully considered an integral part of the analytical and study aspect of the monument or archaeological context.

Given this premise, in this volume, we would like to focus on the current application of virtual restoration and reconstruction in different Cultural Heritage domains by comparing and discussing several case studies. The book can provide a representative state of the art for archaeologists, architects, restorers and experts in the representation, enhancement and protection of cultural heritage.

L'Aquila, Italy Ilaria Trizio
Montelibretti, Italy Emanuel Demetrescu
Montelibretti, Italy Daniele Ferdani

Contents

Virtual Reconstructions

Interpreting and Visualizing the Past Through Virtual Archaeology: From Site to Museum Experience

Daniele Ferdani, Emanuel Demetrescu, and Marco Cavalieri

Abstract This article illustrates the reconstructive hypothesis of the trefoil hall found in the roman Villa of Aiano (Tuscany, Italy) built at the beginning of the fourth century AD and characterised by monumental architecture and decorations. The core of the article focuses in detail on the research questions and issues faced during interpretation describing, step by step, all the sources and the reasonings that have led to the scientific visualization. Furthermore, a breakdown of the use of 3D modelling as a research tool to simulate and debate the proposed hypothesis, its findings, and its implications is discussed. The article ends by presenting a project for the exploitation of the reconstructive work through a museum exhibition in San Gimignano. It includes not only the musealization of archaeological finds but, above all, the presentation of a short movie narrating the evolution of the Roman villa and the trefoil hall and the display of a physical replica of the site created using rapid prototyping techniques.

Keywords Digital archaeology · Virtual archaeology · Virtual reconstruction · Late antiquity · Building archaeology · Extended matrix · Villa of Aiano

1 Introduction

The archaeological site of Aiano is placed in Val d'Elsa (Tuscany, IT). Since 2005, an Italian-Belgian mission has been coordinated by the UCLouvain as part of the international project *VII Regio. The Elsa Valley during Roman Age and Late Antiquity*, has carried out several excavation campaigns discovering the vestiges of an ancient Roman villa, dated between the end of the third and the seventh century AD (Fig. 1). The results of the investigations allowed us to partially understand and reconstruct

D. Ferdani (✉) · E. Demetrescu
Institute of Heritage Science, National Research Council, Rome, Italy
e-mail: daniele.ferdani@cnr.it

M. Cavalieri
Institut des civilisations, arts et lettres, Université Catholique de Louvain, Louvain-La-Neuve, Belgium

© The Author(s), under exclusive license to Springer Nature Switzerland AG 2023

I. Trizio et al. (eds.), *Digital Restoration and Virtual Reconstructions*, Digital Innovations in Architecture, Engineering and Construction,
https://doi.org/10.1007/978-3-031-15321-1_1

Fig. 1 The villa of Aiano in 2018

the history of the villa whose architectures cover a large area, and the rich and varied findings are proof of the ancient magnificence of its architecture.

Only a small part of the villa has been brought to light and the so-called "trefoil hall" is certainly the most complex and fascinating architectural element. It was built between the end of the fourth and the beginning of the fifth century AD, during a monumental reorganization of the villa.

The first project of the hall, some decades before the trefoil hall, provided a central room with six apses surrounded by a corresponding *ambulatio*. This project was probably never completed and at the end of the fourth century three of the six apses were replaced by rectangular rooms [1]. In the sixth century, the hall lost its residential function and became the centre of several handcraft activities. These activities are documented by some pits and small furnaces used for the processing of recycled metals, glass, and ceramics. In that period the site became a quarry of raw materials such as marbles, mosaics, and other building materials. The spolia were recycled in workshops built within the ruins of the villa and finally, in the seventh century the site was abandoned [2, 3].

The archaeological site was discovered at the beginning of the twentieth century when ancient artifacts were found, but the remains of villa were brought back to light in the 1960s thanks to the Associazione Archeologica Sangimignanse and to the investigations carried out from the Soprintendenza Archeologica della Toscana.

In 2017, restoration activities began to guarantee the conservation and enhancement of the site and finally in 2020 a project of valorisation has been supported and funded by the Municipality of San Gimignano which curated the setting-up of a permanent exhibition in the museum located in the ex Conservatorio of Santa Chiara.

The interpretation, reconstruction, and dissemination of the archaeological site of Aiano present many critical issues due to the state of conservation and a peculiar architectural configuration which turned it into a *unicum*. In this project aimed at re-enacting the late antique shape of the villa, we used virtual archaeology as a research tool to support the interpretation and dissemination of the site.

The methodology used for the virtual reconstruction has already been published in [4, 5]. In these articles, after a summary, we will focus on the interpretative issues that have led to the scientific visualization of the reconstructive hypothesis and its communication to the public through a project of museum valorisation.

2 Material and Methods

Virtual archaeology was born in the 1990s. Over the years, issues related to the scientific approach in the creation of simulative models of the past and their communication have been discussed to ensure that the methods were applied with academic rigor. In 2006, the London Charter, which regulates the principles of visualization in Cultural Heritage, was drafted, while in 2009, its contents were put into practice by the Seville Principles. Reconstruction in archaeology today is understood not only as a mere visualization of an artifact as it should have appeared in the past, but as part of the interpretive process for data analysis and synthesis [6]. If carried out with a scientific approach, it allows researchers to achieve a higher level in understanding, processing and communication of complex information coming from archaeological sites [7, 8].

The reconstruction of the trefoil hall is grounded on a theoretical and methodological approach that refers to these principles to ensure the intellectual and technical accuracy of the work and the transparency of the data. Dealing with the interpretation of the past is a very complex path and full of unpredictable issues. The approach used in addressing the interpretation and virtual reconstruction of this archaeological site involves 5 steps (Fig. 2), which range from fieldwork to the publication of data, and it was published in [9]:

1. sources collection
2. data management and analysis
3. interpretation and virtual reconstruction
4. design of a representation model
5. publication and digital dissemination.

The reliability of the interpretation is strongly influenced by the amount and quality of data collected, therefore, before designing the reconstruction, a complete 3D survey of the site was carried out [4] and all archaeological and historical sources were analysed and discussed by the research team. Once collected, all the sources

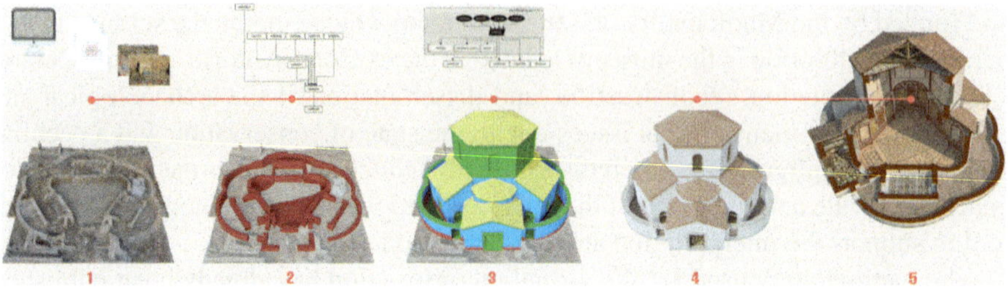

Fig. 2 5 steps workflow used for virtual reconstruction projects of archaeological sites

were organized within the database in which the reconstructive hypothesis, using the Extended Matrix (EM), was developed [10].

EM is a formal language and method for managing, reconstructing, and publishing an archaeological site and it is based on the stratigraphic approach widely used in archaeology: the Matrix of Harris.

Therefore, in this step the stratigraphic analysis and annotation of the physical remains and 3D visualization within a virtual environment were performed by means of a semantic model called "proxy". A proxy model is a simplified geometric representation of the stratigraphy modelled with Blender, using a "digital replica" [11] as reference to sketch the volume, and connected to the EM graph database to query relevant information (Fig. 2, 2).

The term "extended" indicates that the Matrix includes and defines not only the archaeological stratigraphy but also its hypothetical reconstruction, and therefore called Virtual Stratigraphic Units (USV). This is the core of the work. Using the EM, it was possible to account, through a graph database connected to a 3D model, all the relevant sources and the logical-interpretative processes of analysis and synthesis that have led from archaeological evidence to the reconstructive hypothesis. In this step, another proxy model that simulates the reconstructive hypothesis was modelled (Fig. 3). It allows a conceptual visualization using semantic queries that exploit a colour coding related to different levels of reliability.

The colour coding of reliability allows the scientific community to visualize, evaluate, and track reconstruction work and it is organized as follows:

- Red: extant structures.
- Blue: structural reconstruction of existing structures.
- Yellow: reconstruction, by anastylosis, of the fragments completed and repositioned in their hypothetical location.
- Green: non-structural reconstruction of all those parts for which we have no archaeological evidence and for which the reconstruction is entrusted to comparisons or interpreted sources.

The 3D modelling was integrated in the interpretation processes and supported them through simulations. In this phase, it was possible to design the missing parts of the building using computer graphics.

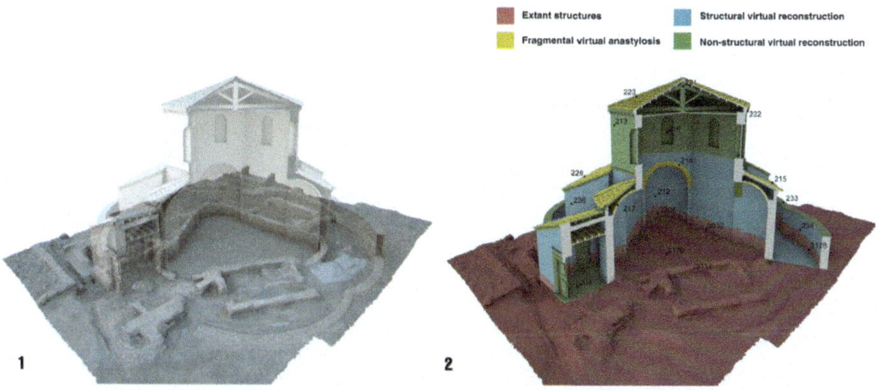

Fig. 3 Model of the trefoil hall. **1** Polygonal modelling of the reconstructive volumes. **2** Reconstructive proxy model with USV and colour coding related to level of reliability of the structures based on archaeological evidences

3 Discussion: The Shape of the Trefoil Hall

To formulate a plausible reconstruction, we had to deal with many factors such as structural issues, height and dimension of the walls and the construction materials used. However, the analytic studies allowed us to hypothesize the shape of the trefoil hall, which probably was one of the main rooms of the seigniorial area, in the fifth century, and simulate it.

3.1 The Masonry

According to the stratigraphic study and the analysis of the preserved masonry, the hall is the result of at least two different phases (see Fig. 4.1). The proposed virtual reconstruction refers to the second building phase. In a first phase, in the fourth century A.D., the hall had six exedras. The masonry technique consisted of regular travertine ashlars, arranged in horizontal rows. Between the end of the fourth and the beginning of the fifth century A.D. the volumes were modified: the south-east, north and south-west exedras were demolished and replaced with rectangular rooms, H, I and L.

The design scheme identified in the plan is characterized by a formal rigor, symmetry, and harmony in the proportions due to mathematical relationships between the parts (see Fig. 4.2). The measures found are multiple of the Roman foot (29.6 cm).

The original project of the plan with six exedras is obtained from an intersection of circles that describe the major and minor exedras having, as radius, the side of the inscribed hexagon. By analysing the plan, it is possible to reconstruct the procedure used to design the project which was based on a module of 40 Roman feet and precise

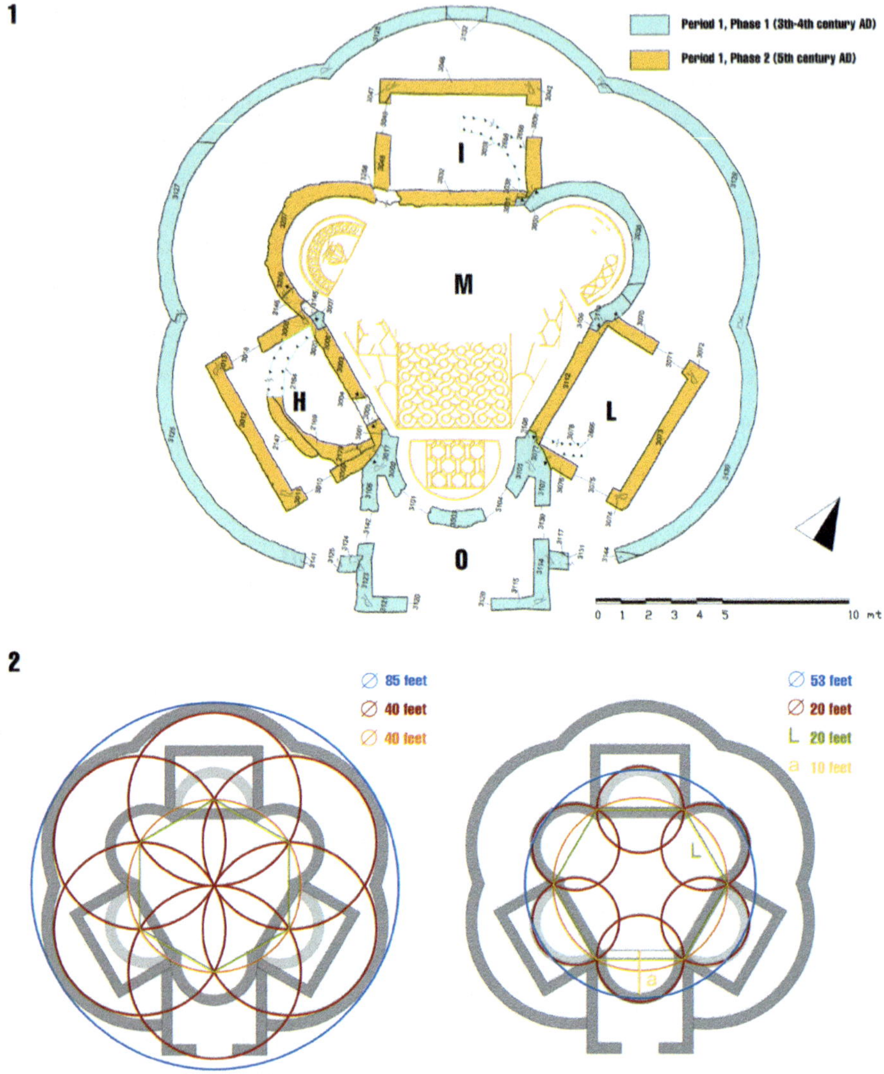

Fig. 4 **1** Plan of the trefoil hall, the two building phases are highlighted with different colors (archaeological drawing made by the dott. Nadia Montevecchi); **2** Proportional ratios and symmetries identified in the plan (dimensions are presented in roman feet)

proportional ratios of symmetry that refers to the principles of Vitruvius [12]. For instance, the internal diameter of the minor exedras is 16 feet while the span of the latter is 10 feet and therefore in a golden ratio of 1:6.

The original design was modified in a second phase but still seems to have considered the same patterns of symmetry. The rectangular rooms H, I, L are set perfectly on the central hexagon and the external walls fall along the perimeter of the hexagon obtained by connecting the intersections of the first six tangent circles which have a diameter of 40 feet.

Similar design scheme can be found mainly in Romanesque or pre-Romanesque hexapetal churches, especially in Croatia [13]. The only example in the Italian territory is the Canonica di San Niccolò in Montieri [14]. For the Late Antique period, we did not find any direct comparisons. Some late imperial structures, such as the mausoleum of Diocletian, and other late-antique buildings in Ravenna, like the baptistery of the Orthodox, the baptistery of the Aryans, the mausoleum of Galla Placidia or the basilica of San Vitale, although with different plans and functions, show some similar structural solutions which helped us to solve some architectural issues.

Some similarities in the floor plans and perhaps in the functions, can be identified in some late antique villas. As stated in [15], reception or representation rooms in the large late antique villas have circular shape or are provided with apses, like the villa of Casale or Cazzanello. Currently the most reliable hypothesis is that the trefoil hall was a representation room or, at least, one of the principal rooms of the noble area. Taking into consideration all the aspects above mentioned, we assumed that the building, in the second phase, was composed by three lateral exedras covered by hemispherical semi-domes and a central area characterized by a hexagonal drum higher than the exedras and the rectangular rooms. During the digging, the remains of collapsed ceilings were found and were sufficient to hypothesise its reconstruction. It was a suspended ceiling composed of reeds then covered by white plaster. Finally, the exedras were covered by a semi-dome ceiling. Some fragments of arches in *opus latericium* were found in the ruins of the northeast and northwest apsidal rooms. To hypothesise the height of the rooms, we based our calculations on the modules, symmetries and the ratios identified in the design scheme of the plan. For example, the circle "C" that circumscribes the *ambulatio* (85 feet) and the circle "A" which circumscribes the exedras (53 feet) are approximately in a golden ratio of 1.618 (Fig. 5). The same ratio has been used to suppose the elevations so that the base of the entire complex (C) and the total height (A) are in golden ratio of 1:6.

Regarding the coating, the external walls must have been covered by a thin layer of plaster. During the excavation some fragments of whitish plaster were found. The travertine used as a building material is very porous and subject to weathering and therefore the plaster was used to protect the wall surfaces as well as for aesthetic reasons. The trefoil hall and its rooms were surrounded by an *ambulatio* with five apses. The excavations have not highlighted any evidence of the floor and therefore it is not certain whether this space was used as a garden or was paved.

However, considering the small space and the fact that this circuit was the only passage for the access to rooms I, H and L, the latter was preferred. Since this was an open space, we assumed that the floor was made of bricks. The *opus spicatum*, an alternating arrangement of bricks with an inclination of 45°, was widely used especially in foundations and floors (see some examples in the Trajan Markets or in the Great Baths of Hadrian's Villa) [16].

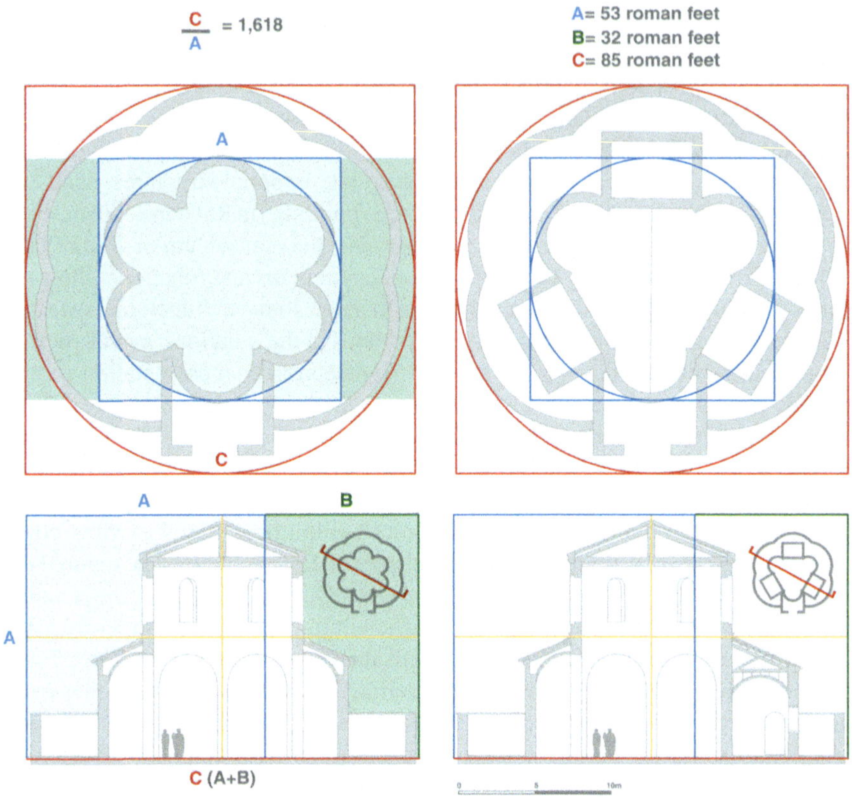

Fig. 5 Reconstructive plans and sections of the Hall in the first (left) and in the second (right) phase. The hight of the building units has been hypothesised using symmetries and golden ratio found in the plan

3.2 *Roofs and Wooden Components*

Nails, beams and collapsed plaster with reed footprints are the only evidence of the wooden infrastructures. Based on these findings and the study of Roman construction, the following hypotheses have been proposed (Fig. 6).

As already mentioned above, the discoveries suggest that the ceilings and apsidal basins were composed by reeds anchored to wooden beams and covered with plaster [16]. This system is called *camorcanna*. This system, made of lightweight material, does not weigh on supporting walls, but it looks like a real vault or ceiling. The coverage of the apsidal basins, given the small size, was probably completed by a fan truss, and covered with terracotta tiles (*tegulae and imbrices*).

The roofing system of the central room was a pavilion roof with six arris. The hypothesis of a vaulted roof was rejected considering the wide span of more than 10 m between the walls. The truss was the most probable solution. In this case, given

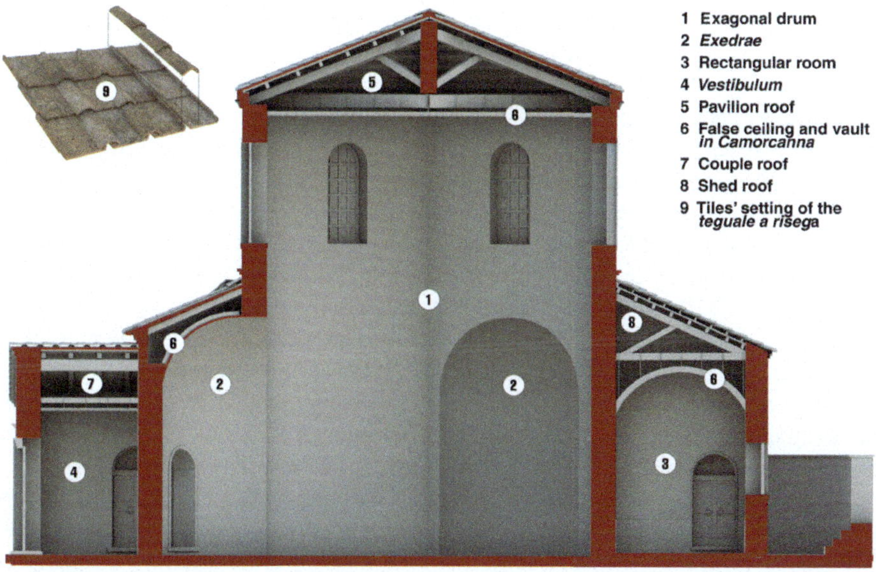

Fig. 6 Section of the hall and wooden infrastructure

the hexagonal shape of the drum, the most suitable framing system was a circular one, where the six hip rafters radiate from the centre to the corners of the hexagon. The system has a common king post on which the radial rafters and tie beams are connected in form of triangles [16–18].

A similar static system can be found in some historical buildings such as, for example, the baptistery of the episcopal complex of the Euphrasian Basilica in Poreč in Istria or the baptistery of Albenga. The wooden structures in this case are relatively recent restorations (the covering of the baptistery of Albenga was restored at the beginning of the twentieth century) [19], however, the carpenter who built them followed a know-how that has remained unchanged over the centuries. The oldest preserved truss, equipped with king post and rafters belongs the Basilica of St. Catherine in the fortress of Mount Sinai dating back to the sixth century AD [20–22].

The shed roof in Roman times is used mainly to cover porches or parts of buildings with a small span. It was composed of a series of rafters fixed to walls at either end of the roof span. It was used for small spans (less than 7 m.), indeed, this kind of roof does not have intermediate support and has a tendency, if overloaded, to push the supporting walls outwards at the top causing structural damage [23]. It is possible that such systems were adopted on the roofs of the rectangular rooms H, I and L, whose span measures 3.7 and 5.8 m respectively. Finally, the vestibule must have been covered with a couple roof with a classical system of rafters fixed at the ridge and at the wall plate. One of the few late antique examples of trusses that can be used as comparison can be seen in a Rondelet's drawing, the Basilica of St. Paul Outside the Walls. The drawing depicts the basilica a few years before the great fire of 1823 that destroyed the late antique carpentry [24].

For the reconstruction of the roofing tiles, we relied on the large quantities of tiles found during excavation. Some *tegulae* and *imbrices* were found intact and allow us to reconstruct the roofing scheme. During the excavation two typology of *tegula* were found: *tegula a incasso* and *a risega* [25].

The second type is among the most widely spread and was taken as a reference model to calculate the pattern and the number of elements needed to cover the pitches. The *tegulae* found in the excavation measure 55–57 × 42–46 × 2.5/3 cm, and weigh about 5 kg while the *imbrices* measure approximately 46–46.4 × 19–21 cm. Their thickness is between 1.5 and 2.5 cm. Given this dimension, the estimated number of tiles to cover the roof of the hall is between 3300 and 3350 (Fig. 6).

Given the absence of material traces, the representation of doors and windows has been mostly based on the available iconographic documentation, with reference to the mosaics of the fourth-fifth century [26]. To reconstruct the doors, we use, as reference, a typology widely used in Roman times and late antiquity: a double door topped by an arch, whose space is compartmentalized by a series of frames covered with sheets of glass. To ensure sufficient illumination of the interiors we provided the upper part of the hexagonal drum with large arched windows. The openings were compartmentalized by a series of frames cover by glass panels according to the above-mentioned reference.

3.3 Decorative Apparatus

The decorative apparatus of the trefoil room is the element of greatest uncertainty. According to archaeological data, between the end of the fifth and the middle of the sixth century, the villa was abandoned, and the architectural decoration was dismantled, re-used, or recycled as raw materials in workshops.

The best-preserved part of the site is the floor in *opus signinum* of the trefoil room [27] and the remains are sufficient to carry out a virtual restoration of the missing parts. Given the geometric pattern of the decoration, all the information necessary to fill the gaps were taken directly from analogous and recurring elements present on the surface (Fig. 7).

On the other hand, the decorative apparatus of the *vestibulum* and rectangular rooms is completely lost. Some *sectilia*, glass tesserae, marble *crustae* and fragments of painted plaster found out of context, can help us to imagine the decorative apparatus [28]. Unfortunately, the remaining materials are insufficient to reconstruct accurately the decorations.

Therefore, the simulation of these elements, although based on archaeological evidence, does not have the same level of reliability of the structural construction material mentioned above. In this case, the simulation of the decorative apparatus is just an evocative approximation, but it is useful to understand the luxury of the villa whose interiors were characterized by relations among colours, light, and shadow.

Fig. 7 Virtual restoration of the *opus signinum*

While we do not know exactly how the original decoration was, we do know for certain that it must have been rich and must have followed typical late roman decorative schema. While materials, colours and decorative types can be deduced from archaeological material, decorative patterns and styles have been completely assumed by comparison with coeval evidence.

The floor of the vestibule O was probably an *opus sectile* with geometric decoration. Indeed, *sectilia pavimenta* were quite common in the aristocratic residences between the end of the third and fourth century: examples are the domus under Palazzo Valentini [29], Vittoriano, Domus Valeriorum (Rome), *domus* of Cupid and Psyche and *domus* outside Porta Marina (Ostia Antica) [30]. The decorative schemas are very varied. In our case, considering the marble fragments found, we opted for simple geometric pattern with rhomboidal, triangular, and rectangular shapes (see the Q, Q2, Q3 and Q4 motifs in [31]) in Giallo Antico, Pavonazzetto, Serpentino, Porfido Verde Antico, Cipollino, Carrara and Siena marbles [2].

During the excavations, travertine wall blocks were found that still preserve the original painted plaster. The backgrounds were painted in red, yellow, and black, sometimes decorated with vegetation patterns (leaves and racemes) or with irregular red stripes that simulate marble veins (Fig. 8).

Almost all these materials, however, were found out of context, buried in pits where they were accumulated, perhaps to be reused. Only in a few cases the remains were found in the collapses and therefore attributable to a specific room. According to archaeological data, most of the fragments found were part of wall painted decoration simulating marble panels with narrow red bands (e.g., Giallo Antico marble). The lower part of the wall was decorated with a red colour while the upper part of the wall paintings should have had a white background.

Based on the available data, comparisons have been made with other villas and *domus* of the late antique period to identify similar decorative solutions.

The most direct comparisons have been identified in the wall paintings of the roman houses of the Celio hill in Rome [32] dated back between the second and the

Fig. 8 Archaeological finds belonging to the architectural, pictorial and floor decoration of the villa of Aiano: 1. painted plaster simulating marble panels; 2. painted plaster with vegetal decoration; 3. stucco relief used in the architectural decoration; 4. fragment of mosaic; 5. marble elements used for the *sectilia pavimenta*

fourth century AD. These houses are preserved in excellent condition and feature frescoes which simulate panelled marbles. These references were used to organize all the scattered decorative material found in the excavation and hypothesise the reconstruction (Fig. 9).

4 Dissemination of the Interpretative Studies

As often happens in archaeology, the visitor is not able to understand the original appearance of an artifact because of its state of preservation. The virtual reconstruction is an important support for the communication of the past to wide public. However, the scientific illustration is not able to tell the whole interpretative process and the evolutionary history of the artifact itself. For this reason, a short-movie, and a 3D printing of the archaeological site in its actual state of conservation have been produced. These, once included in the museum itinerary, will contribute to tell the story of the Villa and explain to the public all the interpretative work above described through an immediate narrative and visual language (see Fig. 10).

The short movie, lasting about five minutes, has been produced to be displayed in one of the museum's rooms. The movie retraced the evolutionary history of the site, from its origins to its discovery, also narrating the most recent events related to the archaeological excavations and restoration combining real footage with computer graphic animation (Fig. 9). The images are supported by a voice-over dubbing that engages the visitor using a documentary-style to convey the cultural message and provide an emotional impact.

Fig. 9 Representation model of the trefoil hall

A physical replica of the site in its actual state of conservation has been produced using rapid prototyping technologies. The 3D photogrammetric model of the entire archaeological site, carried out in 2018, has been post-processed through digital sculpting in Blender and optimized (retopology) to obtain a "waterproof" mesh suitable for 3D printing (Fig. 11). The physical replica was printed by UnoCAD S.r.l. using stereolithographic 3D printing process and coloured manually using water-based colours (dimension: 1 × 0,50 m).

These products, in addition to narrating the story of the archaeological site, also have the function of attracting visitors' attention to a minor cultural site in the town of San Gimignano, redirecting part of the large tourist flows in the surrounding area.

Fig. 10 Some shots from the short movie. Top: camera matching between real footage and computer graphic animation; bottom: presentation of excavation and restoration activities

Fig. 11 Physical replica of the archaeological site obtained with 3D printing technologies (Curtesy of dott. Ivano Ambrosini, UnoCAD S.r.l.)

Thus, the research work, between conservation, comprehension, and promotion of an archaeological past, can have an important impact on tourism and sustainability of the territory by promoting Aiano as an important archaeological site.

5 Conclusion

The analysis of the archaeological evidence, an accurate collection and management of the sources, the use of 3D models in the analytical study, have allowed us to discuss and compare the interpretative hypotheses and verify them through virtual simulations, thus culminating in a thorough reconstructive proposal of the trefoil hall.

The value of this virtual reconstruction, however, lies not only in the design of a visual synthesis, but above all in the creation of a semantic model that organizes all the sources and makes transparent the reasoning and the interpretations involved in the reconstructive hypothesis. In fact, not all parts of the structure have the same level of reliability. This hermeneutic limit is stated (Fig. 3), mitigated, when possible, using sources and comparisons, and represented by means of graphic conventions. Finally, a fundamental and often overlooked matter: the final goal of our research process also included communication activity by means of a museum installation which trasform complex data into information through a narrative and visual language.

References

1. Cavalieri, M., Gloriana P., & Lenzi S. (2019). Aiano-Torraccia di Chiusi (San Gimignano, Siena): A Roman Villa in Central Italy during Late Antiquity. In: Drijvers, J. W. & Lensky, N. (Eds.), *The Fifth Century: Age of Transformation. Proceedings of the 12th Biennial Shifting Frontiers in Late Antiquity Conference* (pp. 93–103). Edipuglia, Bari.
2. Cavalieri, M., Lenzi, S., & Crisanti E. (2011). *Disiecta membra*: i sectilia Della villa tardoantica di Aiano-Torraccia di Chiusi (San Gimignano, Siena). Primi dati su litotipi, sistemi decorativi e reimpiego. In: *Atti Del Xvii Colloquio Dell'associazione Italiana Per Lo Studio E La Conservazione Del Mosaico - Aiscom (Teramo, 2011)* (pp. 119–131). Tivoli.
3. Deltenre, F. D., & Orlandi, L. (2016). «Rien ne se perd, rien ne se crée, tout se transforme» Transformation and Manufacturing in the Late Roman Villa of Aiano-Torraccia di Chiusi (5th-7th cent. AD). *Post-Classical Archaeologies, 6*, 71–90.
4. Ferdani, D., Demetrescu, E., Cavalieri, M., Pace, G., & Lenzi, S. (2020). 3D modelling and visualization in field archaeology. From survey to interpretation of the past using digital technologies. *Groma* 4.
5. Cavalieri, M., Lenzi, S., Pace, G., Ferdani, D., & Demetrescu, E.: Le ricerche alla villa romana di Aiano (San Gimignano-Siena): dall'interpretazione stratigrafica alla rielaborazione 3D. In: Baldini, I. & Sfameni, C. (Eds.), *Abitare Nel Mediterraneo Tardoantico, Atti del III Convegno Internazionale del Centro Interuniversitario di Studi sull'Edilizia abitativa tardoantica nel Mediterraneo (CISEM) (Bologna 28–31 ottobre 2019)* (pp. 273–284). Edipuglia, Bari (2021).
6. Limoncelli, M. (2011). Applicazioni digitali per l'archeologia: il restauro Virtuale. *DigItalia, 1*, 42–59.
7. Ferdani, D., Fanini, B., Piccioli, M. C., Carboni, F., & Vigliarolo, P. (2020). 3D reconstruction and validation of historical background for immersive VR applications and games: The case study of the Forum of Augustus in Rome. *Journal of Cultural Heritage, 43*, 129–143.
8. Gabellone, F. (2020). Archeologia virtuale. Teoria, tecniche e casi di studio. Grifo, Lecce.
9. Demetrescu, E., & Ferdani, D. (2021). From field archaeology to virtual reconstruction: A five steps method using the extended matrix. *Applied Sciences, 11*(11), 5206.
10. Demetrescu, E. (2015). Archaeological stratigraphy as a formal language for virtual reconstruction. Theory and practice. *Journal of Archaeological Science, 57*, 42–55.

11. Demetrescu, E., d'Annibale, E., Ferdani, D., & Fanini, B. (2020). Digital replica of cultural landscapes: An experimental reality-based workflow to create realistic, interactive open world experiences. *Journal of Cultural Heritage, 41*, 125–141.
12. Gros, P. (Ed.). (1997). *Vitruvio. De architettura*. Einaudi, Torino.
13. Vežić, P. (2012). Dalmatinski šesterolisti–sličnosti i razlike". *Ars adriatica, 2*, 41–74.
14. Bianchi, G., Bruttini, J., & Grassi, F. (2012). Lo scavo della Canonica di San Niccolò a Montieri (Gr). *NSBAT, 8*, 564–567.
15. Sfameni, C. (2008). Le ville in età tardo-antica. Il contesto storico-archeologico. In: Russo, A. & Di Giuseppe H., (Eds.), *Felicitas Temporum. Dalla terra alle genti: la Basilicata settentrionale tra archeologia e storia, Lavello (PZ)*. Rome (pp. 471–487).
16. Adam, J. P., Guidobaldi, M. P., Torelli, M., & Torelli, M. (1988). *L'arte di costruire presso i Romani: materiali e tecniche*. Longanesi & C., Milano.
17. Giuliani, C. F. (1990). *L'edilizia nell'antichità, Nuova Italia Scientifica*. Roma.
18. Laner, F. (2005). *Diagnostica delle strutture lignee*. Flap, Mestre.
19. Mannoni, A. T., Cagnana, A., & Brandt, O. (2018). Il Battistero di S. Giovanni ad Albenga (SV). Le travagliate vicende di un cantiere tardoantico di lunga durata. *Archeologia dell'Architettura, 23*, 157–182.
20. Tampone, G. (1996). *Il restauro delle strutture di legno*. Hoepli, Trento.
21. Barbisan, U., & Laner, F. (2000). *Capriate e tetti in legno: progetto e recupero, tipologie, esempi di dimensionamento, particolari costruttivi, criteri e tecnologie per il recupero, manti di copertura*. Franco Angeli, Milano.
22. Valeriani, S. (2005). *Monaci, dardi e colonnelli. Genesi e caratteristiche delle capriate italiane. In: Actas del Cuarto Congreso Nacional de Historia de la Construcción (Cádiz 2005)*. Madrid (pp. 1039–1049).
23. Centola, V. (2018). I sistemi di copertura nelle domus di età romana, Tesi di dottorato in Storia, Critica e Conservazione dei Beni Culturali, 30° ciclo, Università degli Studi di Padova, Dipartimento dei Beni Culturali: Archeologia, Storia dell'Arte, del Cinema e della Musica, Padova.
24. Bianchini, M. (2010). *Le tecniche edilizie nel mondo antico*. Dedalo, Roma.
25. Sheperd, E. J. (2007). Considerazioni sulla tipologia e diffusione dei laterizi da copertura nell'Italia tardo-repubblicana. *Bullettino della Commissione Archeologica Comunale di Roma, 108*, 55–88.
26. Gala, A. V., & Mata, J. A. G. (2017). Roman window glass: An approach to its study through iconography. *Lucentum, 36*, 159–176.
27. Cavalieri, M. (2010). Il pavimento in cementizio della villa tardoantica di Aiano-Torraccia di Chiusi (Siena). Primi dati su decorazione musiva, tecnica esecutiva e orizzonte cronologico. In: *Atti del XV colloquio dell'associazione italiana per lo studio e la conservazione del mosaico - aiscom (Aquileia 2009)*. Tivoli (pp. 365–376).
28. Cavalieri, M., Camin, L., & Paolucci, F. (2016). I sectilia vitrei dagli scavi Della villa Romana di Aiano-Torraccia di Chiusi (Siena, Toscana). *Journal of Glass Studies, 58*, 286–291.
29. Napoli, L., & Baldassarri, P. (2015). Palazzo Valentini: Archaeological discoveries and redevelopment projects. *Frontiers of Architectural Research, 4*(2), 91–99.
30. Guidobaldi, F. (2009). Sectilia pavimenta tardoantichi e paleocristiani a piccolo modulo dell'Italia Settentrionale. *Rivista di Archeologia Cristiana, 85*, 355–410.
31. Guidobaldi, F. (2003). Sectilia pavimenta e incrustationes. I rivestimenti policromi pavimentali e parietali in marmo o materiali litici e litoidi dell'antichità romana. In: Giusti, A. (Ed.), *Eternità e nobiltà di materia: itinerario artistico fra le pietre policrome*. Firenze (pp.15–69).
32. Englen, A., Filetici, M. G., Palazzo, P., Pavolini, C., & Santolini R. (Eds.). (2014). *Caelius II: tomo 1: pars inferior. Le case romane sotto la basilica dei santi Giovanni e Paolo*. L'Erma di Bretschneider, Roma.

"Sources of Research" for the Virtual Reconstruction of Ancient Monuments: The Case of Architectural Models

Massimo Limoncelli

Abstract The use of architectural scale models, also known by the French word maquette or the Italian names plastico or modello, is an ancient practice, which probably began when architecture emerged from its purely practical and utilitarian phase to become a formal category of knowledge with specifically sought-after and deliberately selected characteristics, linked to the increased dimensions and the enrichment of forms [1]. The model constitutes a representation, generally on a smaller scale and made of different materials but similar in effect, of a work to be created. Models are made of various materials, in accordance with the tastes of Scholars and with need, such as wood, wax, terracotta, stucco and so on. Precisely due to the highly perishable nature of the materials with which models in antiquity were made, this type of indirect source is very rarely found in archaeological contexts. This contribution concerns a case study in which an architectural model found in an archaeological excavation was used as the main of research source for the 3D reconstruction of an ancient monument.

Keywords Architectural models · Reconstructive study · Research sources

1 Introduction

In the field of Virtual Archaeology, the reconstructive study of an ancient monument by means of virtual reconstruction "involves using a virtual model to visually recover a building or object made by humans at a given moment in the past from available physical evidence of these buildings or objects, scientifically-reasonable comparative inferences and in general all studies carried out by archaeologists and other experts in relation to archaeological and historical science" [2, p. 3].

The methodology to be applied was set out in the preamble to the *London Charter for the Use of 3D Visualisation in the Research and Communication of Cultural*

M. Limoncelli (✉)
Dipartimento Culture e Società, Università degli Studi di Palermo, Palermo, Italy
e-mail: massimo.limoncelli@unipa.it

© The Author(s), under exclusive license to Springer Nature Switzerland AG 2023
I. Trizio et al. (eds.), *Digital Restoration and Virtual Reconstructions*, Digital Innovations in Architecture, Engineering and Construction,
https://doi.org/10.1007/978-3-031-15321-1_2

Heritage, conceived in 2006 as a means of ensuring the methodological rigour of computer-based visualisation in this field.[1]

The Charter's aim was to set out the rigorous methodological principles "that will ensure that digital heritage visualisation is, and is seen to be, at least as intellectually and technically rigorous as longer established cultural heritage research and communication methods. At the same time, such principles must reflect the distinctive properties of computer-based visualisation technologies and methods" [4, p. 2]. Specifically, principle n°3, 'Research Sources', stresses the notion that "In order to ensure the intellectual integrity of computer-based visualisation methods and outcomes, relevant research sources should be identified and evaluated in a structured and documented way" [4, p. 7]. "Research Sources" refers to the retrieval of "All information, digital and non-digital, considered during, or directly influencing, the creation of the computer-based visualisation outcomes" [4, p. 7]. Indeed, this information is required in order to ensure the intellectual integrity of the methods and results of the digital visualisation and it must be identified and assessed in a documented and structured way.

Concerning Principle n°4,[2] pertaining to 'Documentation', paragraph 4.5, on the Documentation of Research Sources, declares that "A complete list of research sources used and their provenance should be disseminated" [4, p. 8], while paragraph 4.6, regarding Documentation of Process (Paradata), states that "Documentation of the evaluative, analytical, deductive, interpretative and creative decisions made in the course of computer-based visualisation should be disseminated in such a way that the relationship between research sources, implicit knowledge, explicit reasoning, and visualisation-based outcomes can be understood" [4, pp. 8–9].

Lastly, the glossary specifies in greater detail the meaning of 'paradata' [5], i.e. "Information about human processes of understanding and interpretation of data objects. Examples of paradata include descriptions stored within a structured dataset of how evidence was used to interpret an artefact, or a comment on methodological premises within a research publication. It is closely related, but somewhat different in emphasis, to 'contextual metadata', which tend to communicate interpretations of an artefact or collection, rather than the process through which one or more artefacts were processed or interpreted" [4, p. 12].

[1] The London Charter was created following a Symposium and Expert Seminar at the British Academy, London and the Centre for Computing in the Humanities, King's College London, from 23 to 5 February 2006, jointly sponsored by the AHRC ICT Methods Network and EPOCH. During the two-day symposium, 50 delegates debated approaches to the issue of transparency, and on the third day, a smaller group of experts discussed the first 'discussion document' phase of the draft London Charter [3, p. 2].

[2] Principle 4: Documentation. Sufficient information should be documented and disseminated to allow computer-based visualisation methods and outcomes to be understood and evaluated in relation to the contexts and purposes for which they are deployed [4, p. 8].

2 'Research Sources'

On the basis of these considerations, the information to be gathered before the start of a reconstructive study of an ancient building can be divided into three types: primary direct, primary indirect and secondary [6, pp. 186–187; 7]. The first of these indicates all the information that is still "available on the body of the monument or is immediately connected to it, especially physically, or can be deduced from it" [8]. It is therefore information that can be found and verified by direct analysis of the architectural artefact, by means of a survey, the stratigraphic reading of the elevations and the data from the archaeological excavation. Indirect primary sources on the other hand consist of all the information found in historical research, which cannot therefore be directly traced in the living body of the monument in question [8, E20]. This can be divided into two main categories: written and iconographic evidence. The former corresponds to archive data that can provide specific information on a building, such as diaries and excavation reports carried out on the site in the past, while iconographic evidence includes drawings, surveys and graphic reconstructions of buildings, as well as engravings, paintings, prints, depictions on coins and seals or, more rarely, scale models (architectural and otherwise) that may contain information useful for reconstructive purposes. Finally, secondary sources concern all the information that can be derived by analogy via the comparative analysis of buildings of the same type and epoch and, where possible, of the same geographical area [9, p. 68; 10, p. 205]. We are therefore dealing here with data that have emerged from precise architectural comparisons in accordance with the criteria of 'analogy' and 'style'. Style is a "historical and formal reality, unified and consistent, limited in time and clearly defined in terms of its figurative modes" [11, coll. 344, ff]. In contrast, analogy consists of the juxtaposition and comparison of two or more buildings linked by shared formal and functional features (type, use of materials, construction techniques, etc.).

Having a large number of direct, indirect and secondary sources available makes it possible to obtain a high degree of historical rigour in the proposed reconstructions, whose scientific visualisation is supported by solid historical and archaeological research and documentation. The retrieval of all the sources then translates into the creation of 3D models, whose hyper-realistic rendering is the fruit of three processes: reconstruction of the volumes, restitution of colours and simulation of light.

Unfortunately, it is not always possible to draw on all of the above-described types of information, but sometimes even just one of those sources can be enough to virtually recreate an archaeological monument with a high level of reliability.

An emblematic case of the use of models as the only source available for the reconstructive study of a building is the scale model discovered in fragments inside the so-called Contra-temple of Soknopaios and Isis Nepherses in the settlement of Soknopaiou Nesos, today known as Dime es-Seba, on the northern edge of the pseudo-oasis of Fayyum in Egypt [12], where a Virtual Archaeology project aiming at the reconstruction of the city in the Ptolemaic and Roman periods has been active since 2013 [13–15].

3 Architectural Scale Models in Archaeology

The use of architectural scale models, also known by the French word *maquette* or the Italian names *plastico* or *modello*, is an ancient practice, which probably began when architecture emerged from its purely practical and utilitarian phase to become a formal category of knowledge with specifically sought-after and deliberately selected characteristics, linked to the increased dimensions and the enrichment of forms [1]. The model constitutes a representation, generally on a smaller scale and made of different materials but similar in effect, of a work to be created [16]. The earliest definition of *modello* is found in Baldinucci's *Vocabolario toscano dell'arte del disegno* of 1681, in which it is described as "the thing that the sculptor or architect makes in order to exemplify or show what must be created, in varying proportion to the final artefact, since the model is sometimes smaller, sometimes of the same size. Models are made of various materials, in accordance with the tastes of Scholars and with need, such as wood, wax, terracotta, stucco and so on. The model is the first and principal task of the entire undertaking, by which, through trial and error, the Craftsman arrives at the most beautiful and the most perfect. It enables architects to establish the lengths, widths, heights and thicknesses: the number, dimensions, type and quality of all things, as they must be, in order for the workmanship to be perfect; and to facilitate deliberation on the various skills that must be employed to create the building, and to calculate the expense that it requires" [17, p. 15, our translation].

There are two types of model: the 'scale type', with all dimensions exactly proportional to the work to be created, and the 'illusory type', intended to make a visual impression, which presents a single frontal view. It is therefore important to further distinguish between 'design' (i.e. architectural) models—linked to a specific monument, generally made to a pre-determined scale, in which the proportional ratios are exactly calculated and from which it is possible to derive the measurements and dimensions of the monument—and models of buildings created for other purposes, for example votive reasons, which have no claimed value in terms of design, aiming rather to provide a general idea of the shape of the building.

Design models thus represent a scale reproduction of a building, made of terracotta, wood or other perishable material, which serve to visualise and verify the formal hypotheses and the constructional, structural and functional choices associated with the agreed design.

Precisely due to the highly perishable nature of the materials with which models in antiquity were made, this type of indirect source is very rarely found in archaeological contexts. Indeed, for the period preceding the late Middle Ages, there is an almost total absence of surviving models and only a small number of historical references to their use in architectural practice, while the earliest examples are dated to the Renaissance [18], in most cases associated with prestige buildings (Fig. 1).

The first literary reference to architectural models is found in the Book of Chronicles: "Then David gave to Solomon his son the pattern of the porch, and of the houses thereof, and of the treasuries thereof, and of the upper chambers thereof, and

Fig. 1 Architectural models of the votive type: the defensive towers of Hattusa (above) and a portico discovered in the votive deposit of the North Gate in Vulci (below)

of the inner parlours thereof, and of the place of the mercy seat".[3] The practice is also attested in the ancient Greek world by Aristotle, who states that the duties of the Council of Five Hundred in 4th-century Athens included the adjudication of tenders for public works: "The Council also used to assess the models and the *peplon*; but now this work is done by a tribunal chosen by lot, since the Council appeared to be biased in its judgements".[4] In addition, in an episode of the life of Pompey, Suetonius narrates that he had a model of the theatre of Mytilene brought to Rome in order to build one of a similar shape but larger dimensions [20, p. 117; 21, p. 59].

[3] Chronicles I,28,16 (King James Version); note the use of 'pattern' for 'model' [19].

[4] Aristotle, *Athenian Constitution* (XLIX, 3), our translation. The reference to models (παραδείγματα) in this passage is taken by some to refer to the temple of Delphi.

The use of architectural models is also documented in the subsequent centuries, in both literary texts and the numerous scenes of donation present in medieval iconography. In his work *Le vite de' più eccellenti pittori, scultori et architettori*, Vasari mentions many examples of the use of models, including the one created by Jacopo Tedesco, who sent "the model of a tomb in Sicily to the abbey of Monreale, for the Emperor Frederick" [22, p. 7, our translation]. In addition, Vasari states that Arnolfo di Lapo created one of the models for the church of Santa Maria del Fiore in Florence, which was then destroyed in 1367 following the approval of the definitive model by Benci di Cione and Neri di Fioravante [22, p. 13, our translation]. Lastly, Vasari himself prepared for Marco da Faenza the sketch of the fresco painted in 1556–1558 in the stateroom of Cosimo de' Medici in the Palazzo Vecchio in Florence showing the architects Brunelleschi and Ghiberti in the act of presenting to Cosimo the model of the church of San Lorenzo.

Another reference to models is to be found in *De Re Aedificatoria* by Leon Battista Alberti, where the following appears: "By means of models, then, buildings must be designed. But the design cannot limit itself to what must be built; it is also necessary, on the basis of the same models, to establish and thus to procure that which will be of use in the course of construction" [23, (*IX*, *IX*)]. A further reference appears in the treatise by Vincenzo Scamozzi, according to which "models must be prepared for public or private works of great importance in order to have before one's eyes a correctly proportioned body corresponding to the form of the structure to be built; it must show the lengths and heights of all of its parts, including the walls, vaults and roofs, and lastly, in the case of the architect not being available, it must be possible to clearly see and understand with reference to the model what their intentions were, at least in general" [24, p. I, XV].

Among the very few ancient examples of "design" models still conserved is that of the temple of Seti I in Heliopolis, found in Tell el Yahudiya in the Nile delta, today kept in the Brooklyn Museum, datable to 1290-1279 BC [25, pp. 1–23]. Then there is the marble maquette, on a scale of 1:24, of the *adyton* of Temple A in Niha in Baalbek, Lebanon, dated to the second century AD [26]. Furthermore, we must mention the architectural model, found in fragments in 1957, referable to the temple of Contrada Mango in Segesta and datable to the fifth century BC. Reference [27] and the base of a temple found in the Collegio degli Augustali in Ostia [28] (Fig. 2).

Lastly, we could hardly fail to mention the models intended to represent archaeological excavations or ancient monuments. The most famous in this case is the imposing model on a scale of 1:100 of the ruins of Pompeii, conserved in the National Archaeological Museum of Naples, commissioned by Giuseppe Fiorelli with the objective of facilitating a more immediate understanding of the entire site of Pompeii. It also served to meticulously and expertly document the state of conservation, reflecting the phases of the rediscovery [29]. The model was created in a number of stages from 1861 to 1908 by Felice Padiglione, son of Domenico Padiglione, who had already created other works of this type, including models of the temples of Paestum and the *macellum* in Pozzuoli. The model of Pompeii, which is composed of 8 parts, represents an extraordinary piece of evidence, which shows the original

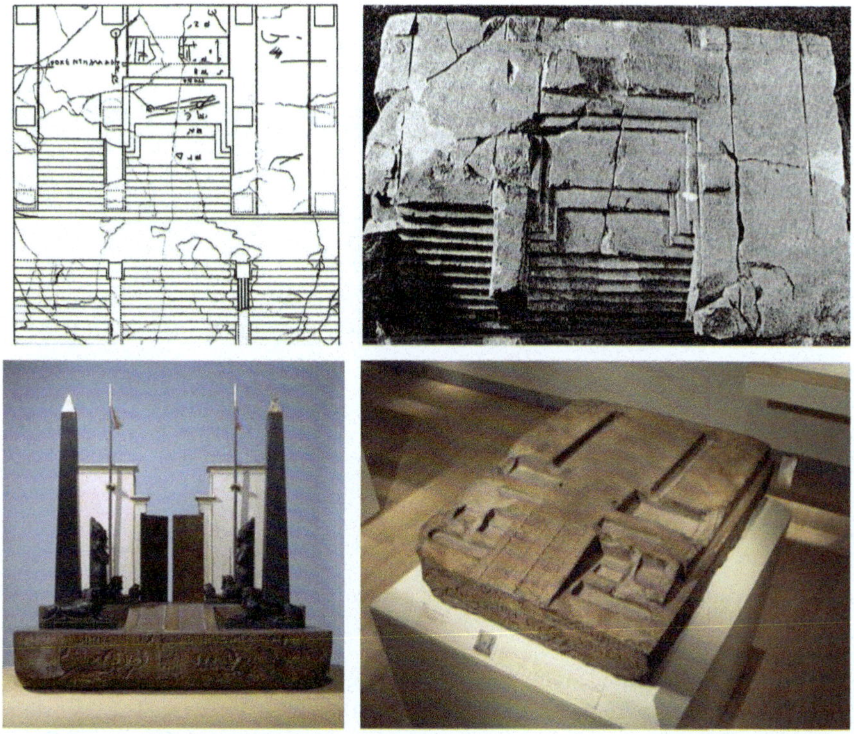

Fig. 2 Architectural design models: the *adyton* of Temple A in Niha, Baalbek (above) and the model of the temple of Seti I in Heliopolis (below)

decoration and arrangement in terms of the floor plans and volumes of the monumental remains. It was created on a plywood base, with the constructions made of cork, stucco or gypsum plaster. The frescoes are reproduced in every detail using temperas or watercolours and the floors are made of paper, incised in the case of mosaics. The frescoed ceilings are not glued to the rest of the structure, so that they can be lifted up in order to observe them.

The value of models as an indirect source needs to be assessed by means of a careful examination of the data, since they might be associated with building projects that underwent major modifications in the course of construction. Where possible therefore, it is advisable to verify the veracity of the measurements seen on the models by comparing them with those that can still be observed on the artefact. Conversely, they can be useful in the reconstruction of a building's early phases of life before any subsequent interventions that modified its original physiognomy with extensions, demolitions and conversions.

As mentioned earlier, a model constituting the only available source for the reconstructive study of a building (a unique case) was discovered in Soknopaiou Nesos.

Fig. 3 Soknopaiou Nesos, aerial view of the sacred area of the city (photograph by Bruno Bazzani)

The settlement, located along the northern edge of the pseudo-oasis of Fayyum in Egypt, is about 600 m from north to south and 320 from east to west. The excavations conducted to date have focused on the area bounded by the wall surrounding the temple compound, which covers about 5% of the archaeological site, within which 28 buildings are still visible. Of these, four have been recognised as temples (ST 6-18-19-20) and one as a contra-temple (ST 203). There are also two chapels, one of which is in the classical style with columns and intercolumnial walls (ST 7) and the other has decorative elements in the Egyptian style (ST 5), built against the north side of the *temenos*. The area enclosed by the *temenos* had an irregular shape and covered about 10,300m^2. The north side measures 86 m, the south 88 m, the east 114.5 m and the west 124.5 m (Fig. 3).

4 From Architectural Scale Models to 3D Reconstruction Model: An Example

In 2003, the archaeological excavations to the north of the temple still visible on the surface (ST18) brought to light a second temple building called ST20, a monumental building dated to the late Ptolemaic period (first century BC) [30, 31]. Only the ground floor, measuring 27.40 m × 19 m, is fully conserved, to a maximum height of 1.40 m. It has 17 rooms arranged symmetrically with respect to the building's axis, oriented N-S, with three rooms preceding the cella, linked by ramps; two staircases, one to the east and another to the west, leading to the upper floors; and four small staircases hidden in the walls that led down to subterranean crypts [30, 31]. Together with the monumental doorway to the south, a single entrance on the western side gave access to the building. The blocks of the conserved walls, in *opus isodomum*, are made of yellow limestone. In some cases, on the inside of the building they are polished and ready for decoration, while on their external faces they had highly protruding truncated-pyramid bossage decoration. The floors are conserved only in the central rooms, where they consist of slabs of grey limestone, while in the lateral rooms and corridors slabs of yellow limestone were used: restored in antiquity, they are only partially conserved.

The layout of Temple ST20 finds parallels with the architectural schemes of Greco-Roman temples in Fayyum and the larger temples of Upper Egypt [32]. A precise match in terms of 'style' and 'analogy' is the temple dedicated to the god Sobek in Qasr Qarun, the ancient Dionysias, also situated in the pseudo-oasis of Fayyum about 3.5 km from Lake Qarun, to the west of the modern Bahr Qarun channel [33–37] (Fig. 4).

Dated to the period from the late Ptolemaic to the first century AD, this is the only temple in Fayyum that is conserved almost intact, up to the roof, although the numerous attempts at restoring the monument are clearly visible [38]. Making use of comparative analyses (secondary sources) based on other buildings that are similar in terms of construction techniques, stereometry and the stylistic features of architectural and decorative elements is a widely used approach in the virtual reconstruction of ancient monuments, making it possible to reconstruct the volumes of the missing parts, in some cases with a high degree of reliability. Therefore, by filling in the missing metric data of the temple of Soknopaios and Isis Nepherses with reference to those of the temple of Sobek in Dionysias, it was possible to proceed with the virtual reconstruction of the building (Fig. 5).

In the course of the 2016–2019 campaigns, the structures of the so-called contra-temple, also referred to as ST203, were brought to light. This chapel, built against the north wall of the temple, was also dedicated to the gods *Soknopaios* and *Isis Nepherses* [39, 40]. Oriented N-S, it was of considerable size, 14.66 m × 12.30 m, with a layout composed of a hypostyle hall surrounded by columns connected by intercolumnial walls. Four internal columns subdivided the hall into 9 bays. Along the southern edge of the contra-temple, the columns are replaced by four pillars protruding inwards by more than 2 m, with semi-columns on their inner faces. The entrance was on the north side and was composed of a portal with an architrave made of yellow limestone decorated with friezes of uraei painted in red and black and with the winged sun motif, part of which was discovered in a collapsed position. Lastly the roof of the building consisted of limestone beams positioned on top of the columns.

Of the contra-temple the entire layout is conserved, to a maximum height of 1.85 m, with walls composed of yellow limestone blocks in *opus isodomum* whose internal faces are completely smooth and ready for decoration, while on the outside they have truncated-pyramid bossage decoration. The floors, consisting of yellow limestone slabs, are conserved almost throughout the interior (Figs. 6, and 7).

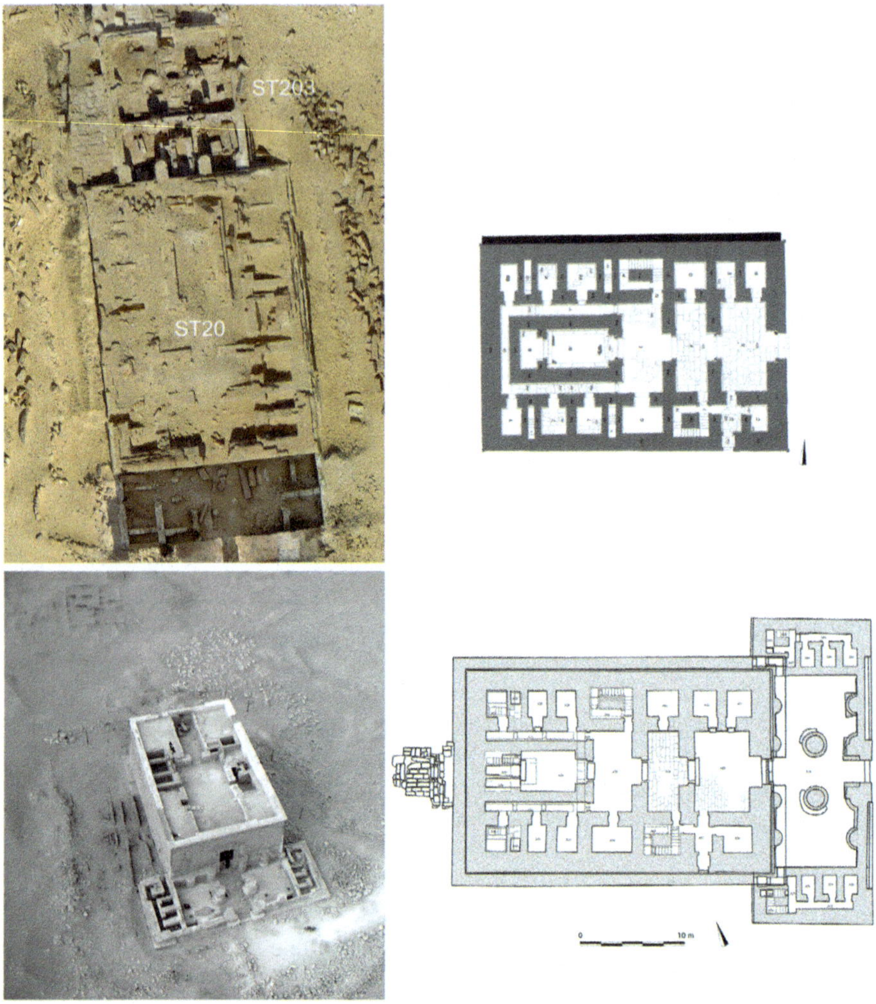

Fig. 4 Temple ST20 in Soknopaiou Nesos (above) and the temple dedicated to the god Sobek in Dionysias (below)

The stratigraphic investigations indicate two constructive phases: the first, datable to the first century AD, saw the construction of the contra-temple itself, while the second is associated with the modification of the functions of the internal spaces, which took place in the second century AD [41]. To this second phase may be attributed the installation of a central pavement composed of black-and-white bichrome tiles which leads from the north entrance to a cella corresponding to the three southern bays.

During the archaeological excavations conducted inside the building, specifically in the south-east bay of the hall, numerous pieces of yellow limestone belonging to

Fig. 5 Virtual reconstruction of Temple ST20 in Soknopaiou Nesos (presentation by Massimo Limoncelli)

the architectural model of the contra-temple were discovered. The discovery in an archaeological context of an architectural design model corresponding to the building being investigated is unique in the field of archaeology.

In total, 31 fragments were identified, some of which had been discovered in 2003, during the excavations of temple ST20, a dozen or so metres from the building in question. Of these, 8 belong to the base, 1 to the entrance portal, 5 to the columns, 2 to the pillars with semi-columns, 5 to the intercolumnial walls, 3 to a single capital, 4 to the architraves and lastly 3 to the roof. The recomposition of the fragments of the architectural model indicated a ratio of about 12:1 between the building and the maquette, which measured 1.40 m × 1.28 m × 0.80 m (Fig. 8).

The data from the survey of the structures in situ (plans, elevations and cross sections) and the analysis of the architectural elements discovered during the excavation were compared with the measurements derived from the virtual recomposition of the fragments of the architectural model. This made it possible to propose a reconstruction of the monument's original appearance, reconstructing the vertical

Fig. 6 Contra-Temple (ST203) in Soknopaiou Nesos (photograph by Massimo Limoncelli)

structures, such as the columns and intercolumnial walls and the roof, with relative precision (Fig. 9).

The reconstruction of the building's floor plan was based purely on the metric data from the survey. Indeed, although the base of the maquette, with the exclusion of the central part, was almost completely conserved, it displayed divergences with respect to the building as it was actually constructed. Specifically, the greatest discrepancies are seen in the arrangement of the columns inside the hall, which in the model are more widely spaced in the centre than at the sides, while in reality the spacing is uniform. In addition, on the southern wall of the maquette there is an entrance leading into Temple ST20, of which there is no trace in the surviving masonry.

Regarding the reconstruction of the volumes of the columns, which are currently conserved to a maximum height of 1.85 m, for the base, reference was made to the metric data from the survey, and for the columns, to the height indicated in the maquette, i.e. 5.50 m. The number of rows of blocks (each 0.25 m high) in each column was calculated on this basis to be 22.

According to the maquette, the column capitals were almost 1/4 of the height of the columns, and were thus about 1.3 m high. They were of the type with three orders of flowers, as the fragment of a lower part discovered in 2017 seems to show. On the basis of these calculations it can be proposed that each column, from the base to the capital, was about 7.40 m high, to which must be added the echinus, composed of

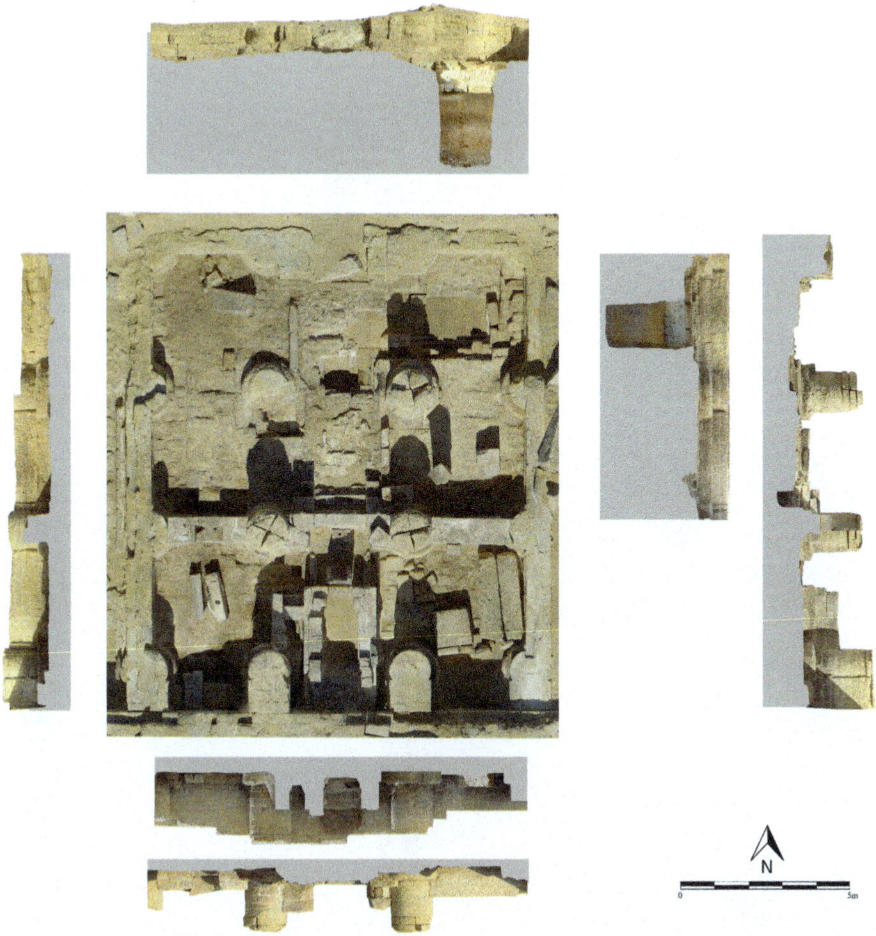

Fig. 7 Photogrammetric survey of the Contra-Temple (ST203) in Soknopaiou Nesos (presentation by Massimo Limoncelli)

a single parallelepiped block with a square base, measuring about 1 m on each side and 0.4 m in height.

Like the columns, the intercolumnial walls, characterised by *tori* on both faces and flaring at the top, were composed of rows of limestone blocks 0.25 m high. Again on the basis of the architectural model, they are believed to have had 15 rows and a height of 3.75 m, with a ratio of 1:2 with respect to the total height of the columns.

The roof of the contra-temple was composed of an initial order of parallelepiped stone architraves, square in cross section, laid on top of the echini. Above them was a second order of stone beams, which were laid adjacent to each other and constituted the actual roof of the building. This hypothesis seems to be supported by the maquette, in which the arrangement of the stone beams is indicated by parallel incisions applied

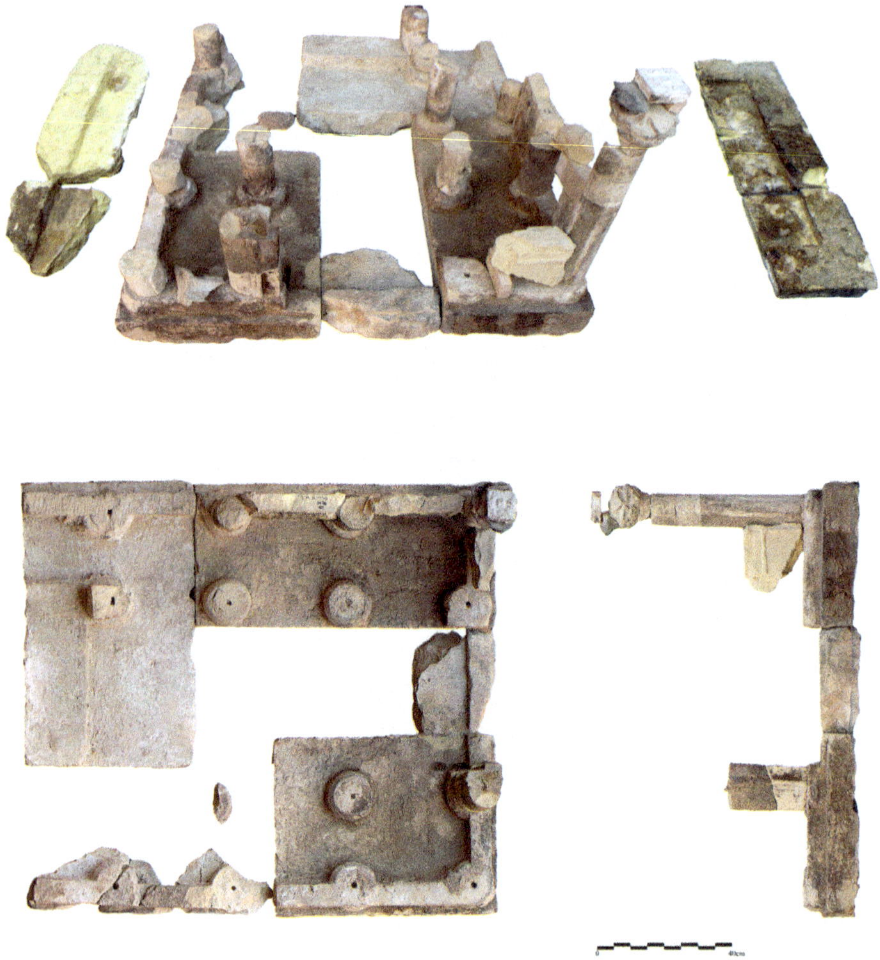

Fig. 8 Model on a scale of 1:12 of the Contra-Temple (ST203) in Soknopaiou Nesos (presentation by Massimo Limoncelli)

to the lower part of the fragment of the roofing. From the calculations, the building is 10.25 m high (Fig. 10).

The interaction between direct sources (survey and excavation data) and indirect sources (the maquette) made it possible to virtually restore the architectural unity of the building, assisting in the reconstruction of the missing parts and the stylistic recreation of the decoration. It also helped our understanding of the ratios between volumes, spaces and routes, knowledge of which had been lost.

In conclusion it can be stated that, just as in architecture the model enables the verification of the construction solutions adopted in the project, in archaeology, particularly virtual archaeology, the digital model should not be understood merely

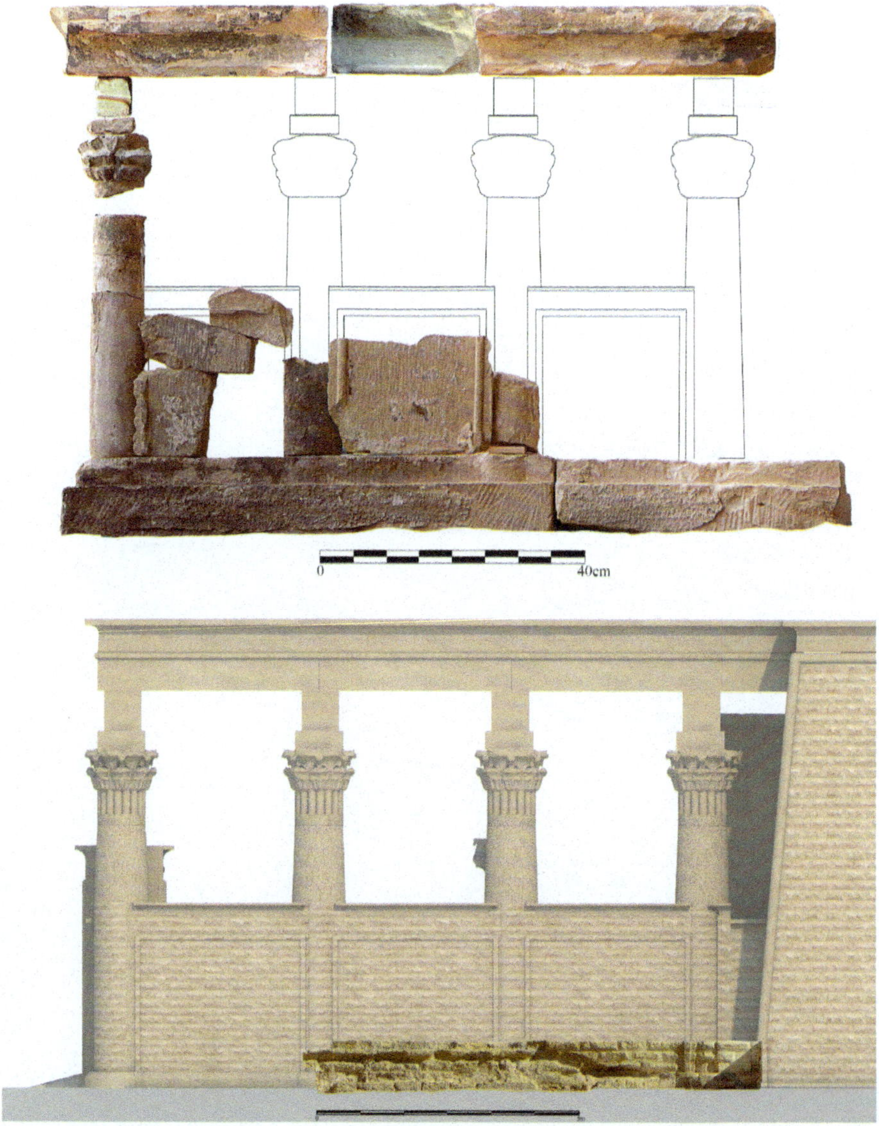

Fig. 9 From the model to the reconstructive study of the structures of the Contra-Temple (ST203) in Soknopaiou Nesos (presentation by Massimo Limoncelli)

as an ideal representation of a building, but rather as a tool for the verification of archaeological data (Fig. 11).

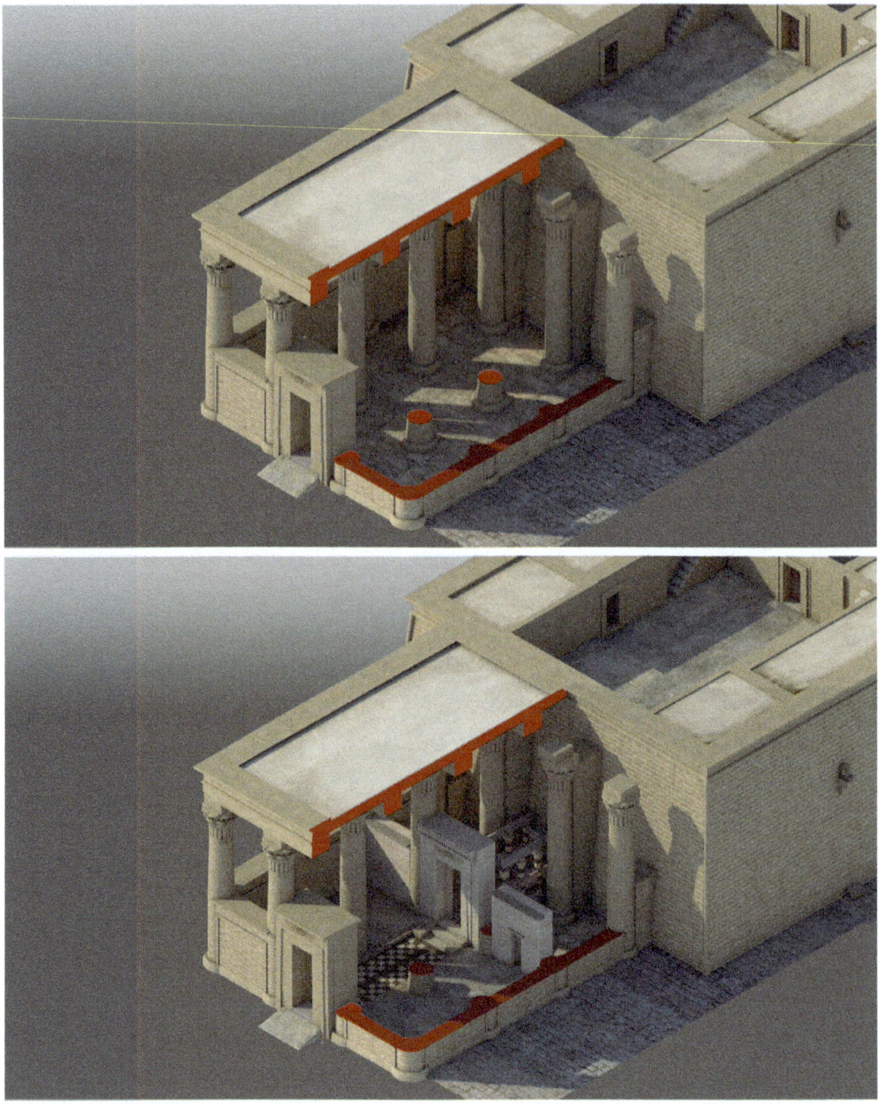

Fig. 10 Virtual reconstruction of the Contra-Temple (ST203) in Soknopaiou Nesos (presentation by Massimo Limoncelli)

Fig. 11 Virtual reconstruction of the Contra-Temple (ST203) in Soknopaiou Nesos (presentation by Massimo Limoncelli)

References

1. Curcio, G., & Manieri Elia, M. (1992). *Storia e uso dei modelli architettonici*, Roma-Bari.
2. The Seville Principles, 2013.
3. Beacham, R., Denard, H., & Niccolucci, F. (2006). *An Introduction to the London Charter*. The E-volution of ICTechnology in Cultural Heritage, Papers from the Joint Event CIPA/VAST/EG/EuroMed Event.
4. London Charter, 2009.
5. Bentkowska-Kafel, A., Denard, H., & Baker, D. (2012). *Paradata and transparency in visual heritage*. London: Ashgate.
6. Limoncelli, M. (2012). *Il restauro virtuale in archeologia*, Roma.
7. Limoncelli, M., Forthcoming.
8. Fancelli, P. (2001). *Indagini preliminari e diagnostica. Rilievo storico-critico delle fasi costruttive*. In Zevi, L. (Ed.), *Il Manuale del Restauro Architettonico*, Roma.
9. Ceschi, C. (1970). *Teoria e storia del restauro*, Roma.
10. Léon 1951, p. 205.
11. Bonelli, R. (1963).*Il restauro architettonico*. In C. Brandi *et alii* (Eds.), *Restauro*, «Enciclopedia Universale dell'Arte», vol. XI, Venezia-Roma, coll 344–351.
12. Capasso, M., & Davoli, P. (Eds.). (2012). *Soknopaiou Nesos Project I (2003–2009)*, Pisa-Roma.
13. Limoncelli, M. (2016). *Un progetto di Archeologia Virtuale a Soknopaiou Nesos nella* regione del Fayyum: stato dell'arte e prospettive di ricerca. In «Studi di Egittologia e Papirologia» (Vol. 13, pp. 45–64), Pisa.
14. Limoncelli, M. (2018). Atlante di Soknopaiou Nesos: piano di conoscenza e di conservazione digitale. In «Papyrologica Lupiensia» (Vol. 27, pp. 47–67), Lecce.

15. Limoncelli, M., & Schepis, L. (2020). Atlante topografico di Soknopaiou Nesos 3D surveying della città. Metodologie di acquisizione dei dati e sistemi di rilevamento per la documentazione archeologica (Campagna 2019). In «Studi di Egittologia e Papirologia» (Vol. 17, pp. 113–124), Pisa.
16. Salerno, L. (1958). *Modello*, in «Enciclopedia Universale dell'Arte» (Vol. V, p. 121). Roma.
17. Baldinucci, F. (1681). Vocabolario toscano dell'arte del disegno: opera di Filippo Baldinucci agli illustri e virtosissimi signori accademici della Crusca, Firenze.
18. Millon, H. A., & Magnano Lampugnani, V. (Eds.). (1994). *Rinascimento da Brunelleschi a Michelangelo. La rappresentazione dell'architettura*, Roma.
19. Virgulin, S. (1975).*Libri delle Cronache*, Roma.
20. Di Pasquale, S. (1996). *L'arte di costruire. Tra conoscenza e scienza*, Venezia.
21. Gros 1976, p. 59.
22. Vasari, G. (1647). *Le vite de' più eccellenti pittori, scultori et architettori/scritte da Giorgio Vasari pittore et architetto aretino*, Bologna.
23. Orlandi, G., Portoghesi, P., & Alberti, L. B. (Eds.). (1966). *L'architettura (De re aedificatoria)*, Milano.
24. Scamozzi, V. (1838). *L'idea dell'architettura universale*, Milano.
25. Badawy, A., Riefstahl, E. T. (1973). *A Monumental Gateway for a Temple of Sety I: An Ancient Model Restored*, Miscellanea Wilbouriana, Brooklyn.
26. Kalayan, H. (1985). *Le Dessin d'Architetcture Dans Les Societes Antique*, Brill, Strasbourg.
27. De Cesare, M., & Montali, G. (2022). *Elementi di un modellino architettonico in pietra dal Santuario di Contrada Mango a Segesta*. In «THIASOS» (Vol. 11, pp. 3–18).
28. Pensabene, P. (1996–1997). Modello templare ostiense in marmo lunense dal Collegio *degli Augustali*, «Rivista dell'Istituto Nazionale d'Archeologia e Storia dell'Arte», III Ser. XIX-XX (pp. 148–151).
29. Amara, G., Malfitana, D., & Mazzaglia, A. (2020). *La grandiosa Imitazione: Il Plastico Di Pompei. Dal Modello Materico Al Modello Digitale*, Roma.
30. Davoli, P. (2013). Il tempio di Soknopaios e Iside Nepherses a Soknopaios Nesos/Dime (El-Fayyum). In L. Giardino & G. Tagliamonte (Eds.), *Archeologia e luoghi delle pratiche di culto. Atti del convegno (Cavallino, 26–27 gennaio 2012)*, Bari (pp. 45–56).
31. Davoli, P. (2015). The temple as Spatial and Architectural Reality. In M. Capasso & P. Davoli (Eds.), *Soknopaios. The Temple and Worship. Proceedings of the First Round table of the Centro di Studi Papirologici of Università del Salento, Lecce, October 9th 2013* (pp. 119–154). Lecce.
32. Zignani, P. (2010). *Le Temple d'Hator à Dendara. Relevés et étude architecturale*. Le Caire.
33. Carpentiero, G. (2016). Continuity and change in Hellenistic town planning in Fayum (Egypt): Between tradition and innovation. In N. Mugnai, J. Nikolaus, & N. Ray (Eds.), *De Africa Romaque: MergingCultures across North Africa: 73–84*, London.
34. Cestari, I. (2010). *Dionysias*, Imola.
35. Jomard, E. F. (1809). Description des Ruines d'Antinöé, dans Description de l'Egypte. Recueil des observations et des recherches qui ont été faites en Egypte pendant l'expédition de l'Armée Française, publié par les ordres de S.M. l'Empereur Napoléon le Grand, Tomo II, chap. XV, pp. Paris.
36. Papi, E., Bigi, L., Camporeale, S., Carpentiero, G., D'Aco, D., Kenawi, M., Mariotti, E., & Passalacqua, L. (2010). *La Missione dell'Università di Siena a Qasr Qaroun - Dionysias (2009–2010)*. In «Ricerche Italiane e Scavi in Egitto» (Vol. 4, pp. 239–250).
37. Römer, C. (2019). *The Fayoum Survey Project: The Themistou Meris. Volume A: The Archaeological and Papyrological Survey*, Leuven-Paris-Bristol.
38. Carrié, J. M. (1974). *Les Castra Dionysiados et l'évolution de l'architecture militaire romaine tardive*. In «Mélanges de l'École française de Rome Antiquité» (Vol. 86, pp. 819–850), Paris.
39. Davoli, P. (2018). Il tempio e il suo modello: nuovi rinvenimenti a Soknopaiou Nesos. In M. Betrò, S. De Martino, G. Miniaci, & F. Pinnock (Eds.),*Egitto e Vicino Oriente antichi: tra passato e futuro: studi e ricerche sull'Egitto e il Vicino Oriente in Italia* (Vol. 1. pp. 135–148), Pisa.

40. Davoli, P. (2019). The contra-temple of Soknopaios and its architectural model. In «Egyptian Archaeology» (Vol. 55, pp. 41–43).
41. Davoli, P., Capasso, M., Ikram, S., & Bertini, L. (2016). Soknopaiou Nesos Project. Missione Archeologica del Centro di Studi Papirologici dell'Università degli Studi del Salento, Lecce, a Soknopaiou Nesos/Dime (El-Fayyum - Egitto) Tredicesima Campagna, Ottobre-Dicembre 2016. In G. Capriotti Vitozzi (Ed.), «RISE, Ricerche italiane e scavi in Egitto» (Vol. 7, pp. 181–196), Cairo.

The Virtual Reconstruction of the *Cine-teatro Olympia* in Catania for the Documentation and Memory of Places

Carmela Rizzo, Mariateresa Galizia, and Cettina Santagati

Abstract The research here presented is aimed at the virtual reconstruction of the *Cine-teatro Olympia* in Catania, designed by architect Francesco Fichera in 1913, and has been inspired by considerations mentioned above, keeping trace of the metadata and the reconstructive assumptions during all the process. Indeed, the 3D modeling followed the cognitive phase and was based on the principles of transparency and ethical reconstruction, in order to ensure a clear distinction between original surviving parts and fully reconstructed parts (no longer existing in reality). Therefore, an in depth research has been carried out to harvest additional sources (books, relations, newspapers, historical images and catalogues). The several sources have been reordered, catalogued and compared. From this comparison have emerged several reconstructive hypotheses that have led to the formulation of reconstructive choices. Such choices have been made explicit, in the respect of the scientific transparency, in the model through visual codes.

Keywords Digital cultural heritage · Virtual reality · Virtual reconstruction · 3D modeling

1 Introduction

The cognitive process underlying the interpretation of Architectural Heritage is very complex. We have to deal with artefacts that are subject to processes of transformation induced by the social context, the cultural and artistic trends belonging to different eras, as well as changes in intended use. We have the need to read and interpret the

---

C. Rizzo · M. Galizia · C. Santagati (✉)
Department of Civil Engineering and Architecture, University of Catania, Catania, Italy
e-mail: cettina.santagati@dau.unict.it

C. Rizzo
e-mail: mely.rizzo96@gmail.com

M. Galizia
e-mail: mgalizia@dau.unict.it

© The Author(s), under exclusive license to Springer Nature Switzerland AG 2023
I. Trizio et al. (eds.), *Digital Restoration and Virtual Reconstructions*, Digital Innovations in Architecture, Engineering and Construction,
https://doi.org/10.1007/978-3-031-15321-1_3

transformations of individual buildings in order to grasp their history and representative features. These operations are possible through historical-documentary research and the relative synthesis of the data in our possession, in this way we are able to faithfully interpret the object under examination and analyze it in an overall view.

Virtual reconstructions are an extraordinary way to visualize and return to the community episodes from the past which are ruined and/or modified from the original configuration. From this point of view, digital technologies offer the advantage of being able to simulate and hypothesise scenarios and solutions [1, 2]. Undoubtedly, the high realism of digital reconstructions raises several ethical issues related to the authenticity of replicas, and implies the need to integrate documentary sources and interpret them as stated in the London Charter and Seville Principles [3–5]. For this reason, it is essential to encourage innovative procedures for the visualization and validation of the reconstructive process of a monument or archaeological site which is no longer existing or transformed. These approaches make explicit the link between the reconstructed elements and the information underlying the reconstruction (thus showing the gap between the interpretation and the original data), as well as the different levels of plausibility and uncertainty of the 3D modeled parts [6].

The research here presented is aimed at the virtual reconstruction of the *Cine-teatro Olympia* in Catania, designed by architect Francesco Fichera in 1913, and has been inspired by considerations mentioned above, with the objective to keep trace of all the steps during the reconstruction process in terms of metadata and assumptions. Indeed, the 3D modeling followed the cognitive phase (collection of all the historical, bibliographic and archive sources) and was based on the principles of transparency and ethical reconstruction, in order to ensure a clear distinction between original surviving parts and fully reconstructed parts (no longer existing in reality).

The obtained digital content will be part of the exhibition of the *Museo della Rappresentazione* (MuRa) in Catania, a University Museum which preserves and exhibits the drawings of Francesco Fichera Fund.

Therefore, the virtual reconstruction of *Cine-teatro Olympia* has these objectives:

- to give back, albeit virtually, to the citizens a part of the historical heritage of Catania now lost;
- to extend the boundaries of the *Museo della Rappresentazione* creating a connection between the drawings on display and the city;
- to use graphic and visual languages to narrate the complexity of the transformations and bring the visitor closer to the museum's heritage, facilitating understanding.

The paper is a structured as follows: after an in-depth study on the related work (Sect. 2) the methodology is shown (Sect. 3), then the *Cine-teatro Olympia* timeline reconstruction (Sect. 4), the documentary sources research (Sect. 5) and the digital survey (Sect. 6) operations are illustrated; to follow the reconstructive choices (Sect. 7) and the degree of accuracy of the reconstruction (Sect. 8) are detailed; finally the immersive experience in a virtual environment (Sect. 9) designed for the exhibition of MuRa is illustrated as where as the conclusion.

2 Related Works

Digital reconstruction is a comprehensive approach, used as a means of verifying and synthesizing analytical data. Its veracity is therefore function of the analysis and interpretation of collected data (direct and indirect). The virtual environment makes possible to carry out simulations and optimize the legibility of the architecture as well as to conduct interventions that often cannot be made on the original artefact, both in architectural and archeological domain [7–10].

It also allows to reconstruct the multiple morphologies assumed by the building during different eras, becoming itself a historical testimony and a tool for data processing and study. Through digital technologies it is possible to achieve new ways of storing and using information, one example being the Extended Matrix [11], a formal language through which to document the processes of virtual reconstruction, both in the archaeological field and in any other area of cultural heritage. For instance, for the Necropolis of Banditaccia in Cerveteri [12] the overall reconstruction of the site allows the management of scientific documentation through a single archive. Another example is given by the classification schemes for the visualization of Uncertainty proposed by Apollonio [13] and applied in the reconstructive hypothesis of Palladio Projects.

Virtual anastylosis is another reconstructive approach, an evolution of the traditional definition of anastylosis reported in the Venice Charter (Venice Charter, 1964), which can be described, according to the Principles of Seville, as a procedure that involves "relocating dismembered parts of the existing structure through a virtual model" (Principles of Seville, 2011). Virtual anastylosis is closely related to virtual restoration and is widely used. Two examples are the virtual reconstruction of the Church of S. M. delle Grazie in Misterbianco [14] and the medieval Church of Monte Sorbo [15]. Both projects in fact, through careful cataloguing and relocation of isolated fragments, aim to reconstruct the decorative apparatus that no longer exists.

3 Methodology

This study is divided into two distinct but complementary phases. The first is the documentation phase, which embeds the acquisition of direct data obtained from observation and survey of the building, and indirect data, obtained through historical research (iconographic sources, archive data, photographs, engravings, prints, paintings and design drawings).

The second phase is the study and analysis of the data previously collected, through which it is possible to reconstruct the timeline of the events. In this way it is possible to date each architectural element across a time range, thus reconstructing the life and transformations of all the building's components. However, this step is very critical due to the uncertainty of reconstructive hypotheses and requires that all the choices must be stored based on scientific transparency. In this phase the 3D model becomes an instrument of study and investigation for the verification of the hypotheses being formulated. The last step involves the creation of the final 3D model where all the data collected can be processed and interpreted.

The workflow consists of the following steps:

- gathering information from archival and bibliographical research;
- analysis and study of the historical context and building typology;
- metric survey;
- production of 2D metric drawings;
- comparison between 2D metric drawings and historical documentation (plans, sections, elevations);
- definition of the reconstructive choices based on the previous comparison and analysis of the degrees of accuracy;
- definition and modelling of the final 3D model;
- definition of the fruition mode.

4 The *Cine-teatro Olympia* Timeline Reconstruction

The *Cine-teatro Olympia* was designed by Francesco Fichera at the beginning of the twentieth century and opened in 1913. It is located in Palazzo Paola in Catania in an area between Piazza Stesicoro and Via Etnea. Having to deal with an existing nineteenth-century building, Fichera used the old shops prospecting the square to realize the waiting rooms and the ticket office, covered by a wooden false ceiling.

The corridor is divided into six bays and leads to the trapezoidal cinema hall. As written by Fichera [16], he made a restyling of the existing hall, with decorations and frescoes in Art Nouveau style, adding a stage, a horseshoe-shaped cantilevered grandstand and a large lowered dome in reinforced concrete.

Fichera also paid a lot of attention to the interior decoration, which, through stucco and frescoes, recalls the lush Mediterranean vegetation. The decorations of the auditorium, by Salvatore Gregorietti and Gaetano D'Emanuele, are characterised by a white background punctuated by pilasters and frescoes depicting the Sicilian countryside. The architrave and the impost of the dome are decorated with stucco bas-reliefs that echo the floral and phytomorphic motifs present in the rest of the room. The intrados of the dome is frescoed with a geometric pattern that converges towards the top.

Today, only a small part of the decoration is still present, as the cinema hall has undergone several transformations that have affected both its appearance and its use (see Fig. 1).

In the 1930s, the stage was removed, and the hall was converted exclusively to a cinema. During the 1950s the room underwent a drastic transformation caused by the significant removal of the decorative apparatus consisting of the rich bas-reliefs and all the frescoes. This operation entailed an inestimable loss in terms of historical and

Fig. 1 Historical photographs from the 1930s of *Cine-teatro Olympia*. (Source: Francesco Fichera Fund)

artistic value, since they testified the socio-cultural expression of the artistic fervor at the beginning of twenty century.

In 1974 an attempt was made to restore the theatre to its former glory, but unfortunately the original frescoes were not restored and were irreversibly lost, and a small stage was added.

Between the end of the 1970s and the beginning of the 1980s there was a crisis in the cinema industry, which led Olympia to convert its programming to the projection of adult films, causing a rapid decline.

The end of the cinema activities came in 1998, when the McDonald's multinational company bought the premises to use as one of its branches. The change of use did not entail any invasive transformations: the floor inclination has been leveled at the ground floor, while in the grandstand the steps have been removed to leave room to restaurant activities. Furthermore, the modest portion of the survived original decorative apparatus has been maintained, reaching this day undamaged. In 2010, the site has been undergone to an extraordinary maintenance and wooden panelling has been inserted, obtaining its current appearance (see Figs. 2 and 3).

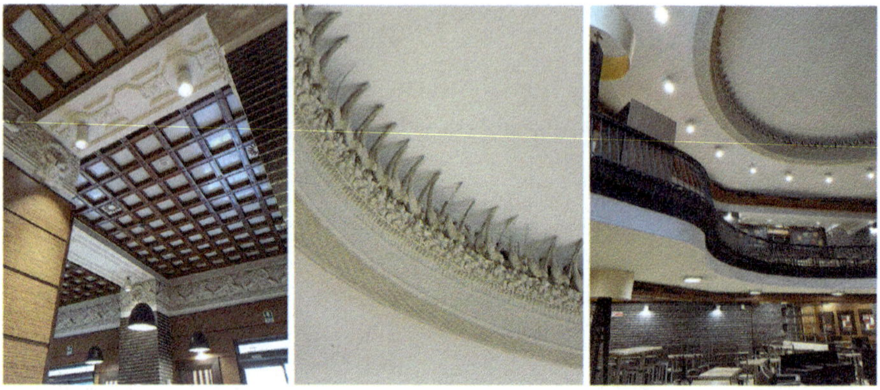

Fig. 2 Photographs of the current appearance of the *Cine-teatro Olympia*

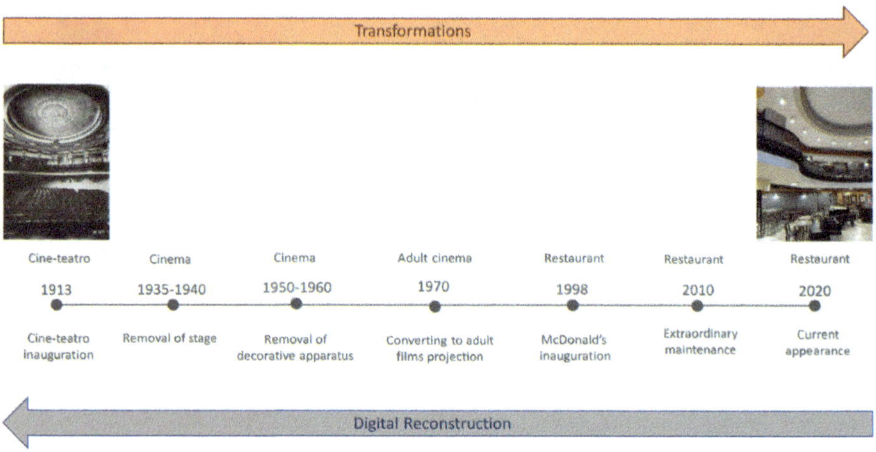

Fig. 3 The timeline of transformations

5 Documentary Sources Research

The first step of our workflow foresaw the collection of historical, archival, iconographic data which was pivotal for the reconstruction. The consultation of multiple sources allowed to trace the transformations undergone by the artefact during its life, identifying the different historical time frames. The analysis and comparison of indirect data enabled the identification of most of the structural and decorative elements, allowing for the recognition of those elements survived to the various transformations and the change of use that took place in 1998. Above all, it was possible to identify and relocate within the building all the components no longer present on site, using an approach similar to virtual restoration. Sources of different nature have been consulted, starting from bibliographic and archival sources (drawings and administrative documents) proceeding with cadastral plans, renovation projects, guidelines for the design of cinemas and furniture catalogs [17]. The various sources were catalogued in chronological order, and each was classified using a code that facilitates the association of the reconstructive information with the 3D model, following the methodology introduced by the Extended Matrix [11].

Indirect data includes:

- Bibliographic sources in text form;
- Bibliographic sources in photographic form;
- Iconographic sources provided by the Francesco Fichera Fund;
- Iconographic sources provided by *Agenzia delle Entrate*;
- Iconographic sources provided by the *Soprintendenza BB.CC.AA* of Catania;
- Iconographic sources relating to works produced within the study courses of the University of Catania (see Figs. 4 and 5).

6 Digital Survey

To acquire the geometry of the building under study, terrestrial laser scanning technique has been used. The shooting was mainly indoor, using a Leica BLK360, a compact and light scanner, equipped with a LiDAR sensor that allows the acquisition of 360° color HDR images, superimposed to a 3D point cloud of millimetric precision. The instrumentation performs a rapid scanning process, taking an average of 6 min to complete each acquisition, with an acquisition rate of 360,000 points per second and a range of 60 m with an accuracy of 6 mm at 10 m. In addition to the LiDAR sensor, it includes infrared sensors for thermal imaging and 360° cameras.

The survey project envisaged the choice of 20 station points appropriately located to allow overlapping between contiguous scans, thus facilitating the registration phase. Using the app Cyclone Field 360 an automatic pre-alignment of the scans has been carried out on site. Then the alignments have been verified through Leica Register 360 software. The link between the pre-aligned scans have been checked using slicing tools and eventually improved with visual alignment tool. Considering that the scans interested an indoor environment we set as alignment error threshold an interval between 1 and 2 mm, in order to keep the average alignment error lower than 2 mm. Therefore, at the end of the verification step, the cloud had an average error of 2 mm. The overall point cloud consists of 303,335,566 points (see Fig. 6).

The information related to the individual scans is shown in the table below (Table 1). Finally, the cloud was exported in .e57 format, for the subsequent processing. The point cloud was imported, filtered, denoised and decimated in CloudCompare. Following this preliminary operation, it was possible to obtain the mesh of the entire model, which was exported in .obj format and inserted into the 3D modelling software Rhinoceros 5 (see Fig. 7).

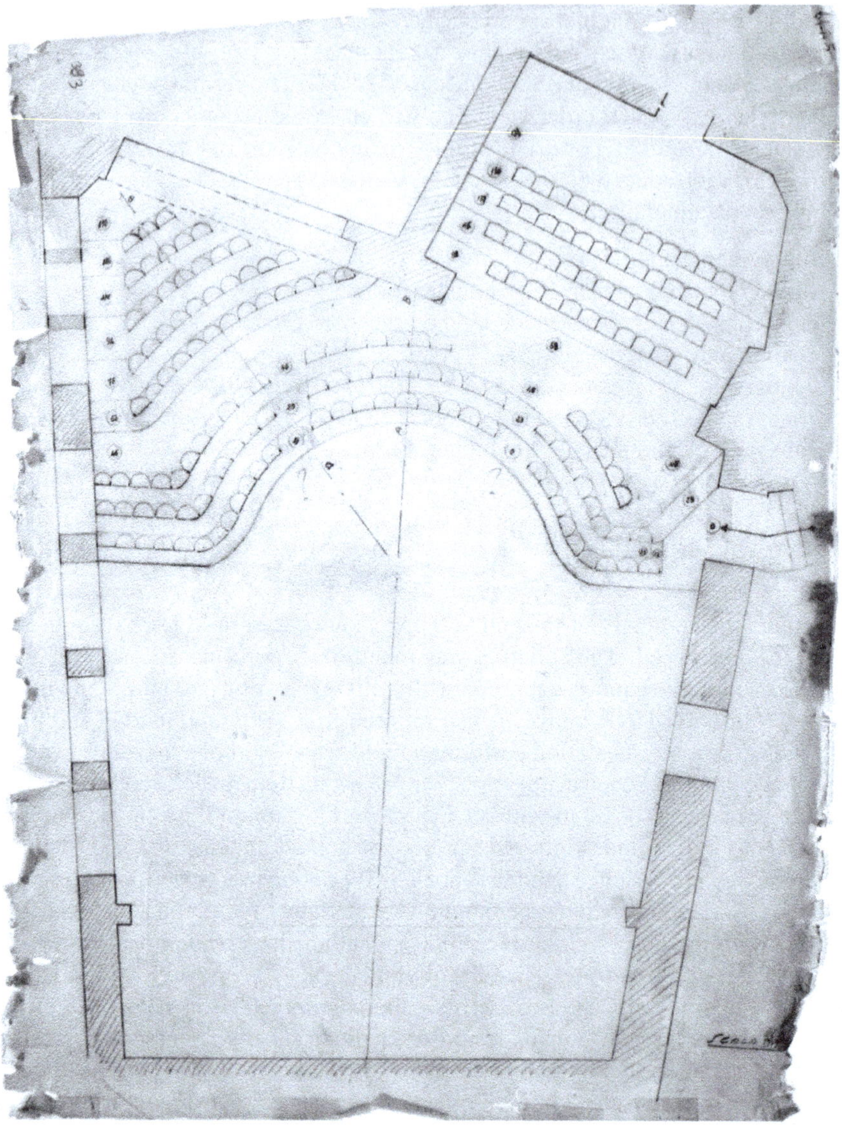

Fig. 4 Fichera's project drawings of the second floor (Source: Francesco Fichera Fund)

7 The Reconstructive Choices

The purpose of the digital reconstruction is to restore the original appearance of the inner environment to 1916, the date of the bibliographic source closest to the inauguration of the *Cine-teatro Olympia*. The modelling of the volumetry was possible through the analysis and comparison of multiple data (point clouds, 2D drawings,

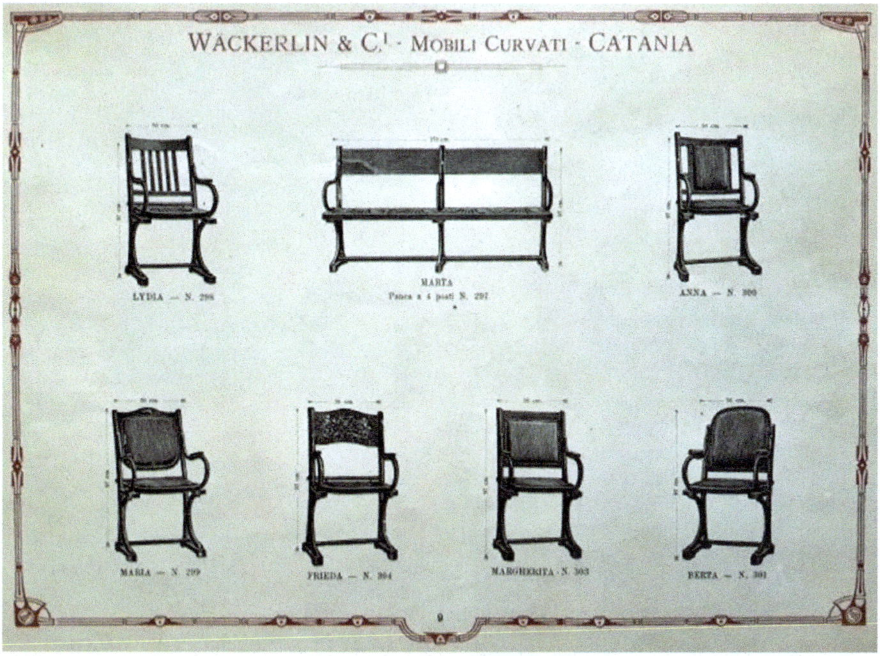

Fig. 5 Catalog of the manufacturer Wäckerlin (Source: Renzi, [17])

cadastral plans). The most difficult reconstructive choice was the one concerning the slope of the stalls, since several reconstructive hypotheses were elaborated; finally, a slope of less than 1% was adopted, which was reconstructed thanks to the consultation of the survey of the Odeon cinema by Carmelo Aloisi.

For each component, various sources, easily identifiable according to a specific identification code, were analysed and compared. These made it possible to formulate various reconstructive hypotheses, which were carefully analyzed to get the final configuration (see Fig. 8).

The photographic sources from the Francesco Fichera Fund and other bibliographic sources, as Francesco Fichera's short essay *L'Olympia nel primo dì di primavera* [12] dealing with architect's design choices, provided a valuable guide for the reconstruction of the decorative apparatus removed in the 1950s. As regards bas-reliefs, the one existing in site they were modeled starting from the mesh and improving details with historical images; the no more existing decorations were modeled using as base the rectified historical photographs (Fig. 9). As for the frescoes, historical photographs were photo-rectified and repositioned in the digital model in their original position. However, they were only reproduced in black and white so as not to create a historical fake, since we had no reliable information on the original colors.

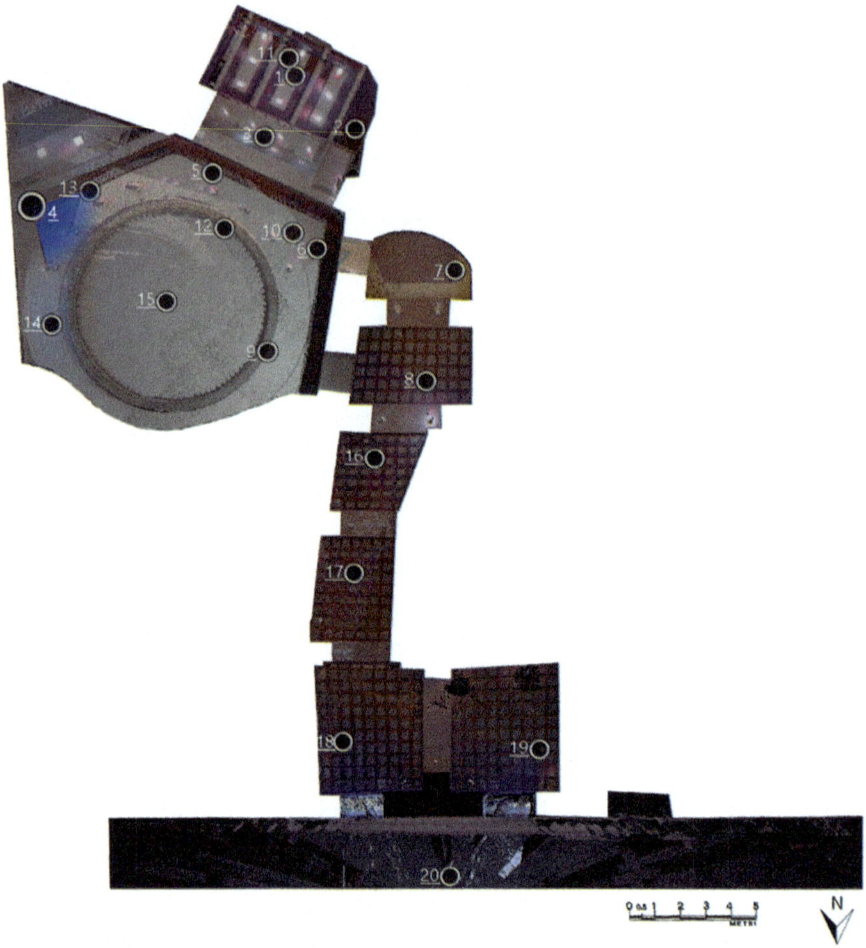

Fig. 6 Plan of the scans taken on site

The issue of colors reproduction was addressed throughout the digital reconstruction process, and it was considered only in presence of reliable written documentation (see Fig. 10).

The furniture in the auditorium, specifically the seats, was modelled on information taken from the catalogue of the manufacturer Wäckerlin, specialized in curved woodwork.

The lighting project drawing by Fichera (Fig. 11) which provides information both on the layout of the lighting bodies and on their use during the various moments of the projection, guided the arrangement of the lighting points.

Table 1 Scanning data

Scans	Number of points	Average error (m)	Minimum overlapping (%)
1	33.926.651	0.003	77
2	36.173.868	0.003	77
3	72.791.440	0.002	63
4	8.388.404	0.002	63
5	6.526.344	0.002	75
6	2.470.352	0.002	46
7	2.888.088	0.002	46
8	2.326.060	0.002	61
9	792.064	0.002	46
10	1.367.088	0.002	54
11	2.236.508	0.002	54
12	72.833.255	0.002	62
13	5.565.756	0.002	65
14	6.215.616	0.002	65
15	1.172.475	0.002	66
16	35.970.004	0.002	66
17	3.741.504	0.003	54
18	3.352.107	0.003	54
19	369.032	0.003	49
20	677.662	0.005	30
Total	303.335.566		

Fig. 7 View of the overall point cloud

Fig. 8 Analysis and comparison between Fichera's project drawing, cadastral plan and digital survey

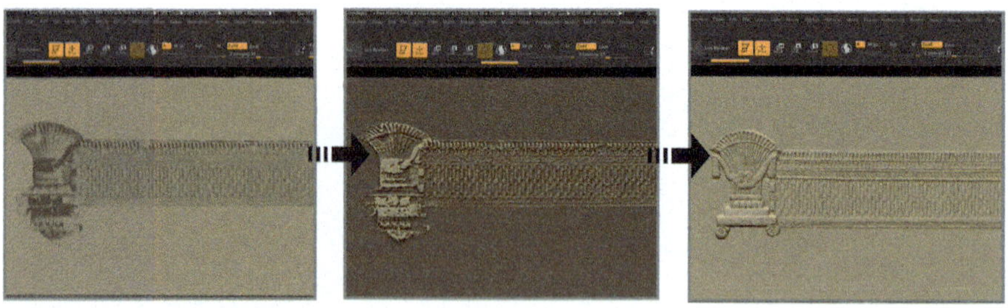

Fig. 9 Reconstruction of bas-reliefs using Zbrush 2020 software

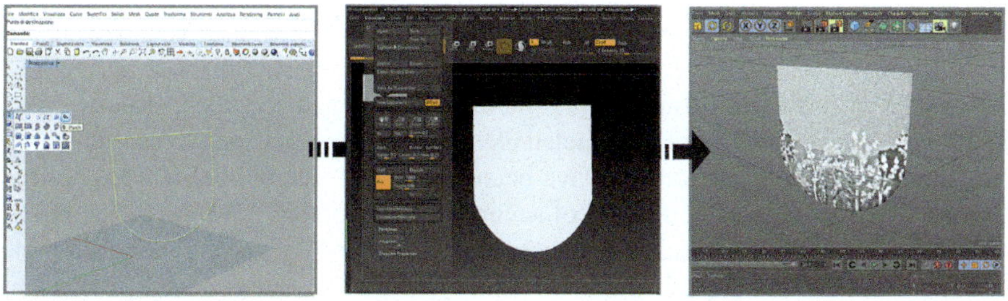

Fig. 10 Modeling and mapping of frescoes using Zbrush 2020 software

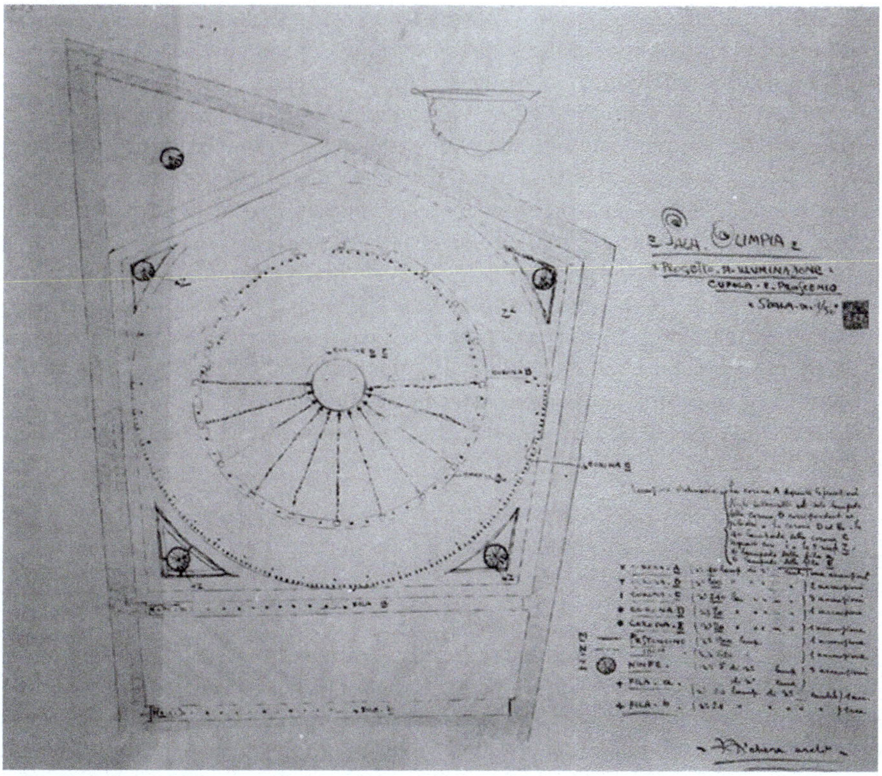

Fig. 11 Fichera's lighting project drawings (Source: Francesco Fichera Fund)

Finally, the last reconstructive choice was the positioning of the projector. Thanks to the consultation of bibliographic sources on early twentieth-century cinemas and the collaboration of the Cinema Museum of Catania, it was assumed that the projection booth was not initially present but was certainly inserted in the years to follow.

8 Degree of Accuracy of the Reconstruction

To ensure the scientific transparence of the process and the reconstructive decisions carried out, for each reconstructed element the reconstruction accuracy and the information on the used sources have been explicated. Among all the possibilities in this work the method of degrees of accuracy developed by [18] has been applied. This method uses a color scale to represent the levels of uncertainty and interpretation of the model. This method represents the different levels of uncertainty related to consistency/pertinence, in relation to the sources used. It should be emphasized that each reconstruction is not a black/white process, but rather generated by several analyses and interpretations, interconnected to specific documentary sources characterized by different degrees of consistency, accuracy/metric quality and subjectivity. The interconnection between these three aspects is represented according to nodes that identify the different levels of interrelation between coherence, accuracy and subjectivity in the definition of each of the reconstructed elements. The coherence index refers to the presence of contradictory sources, the integrity of the original sources (e.g. a fragment of a document), the coherence of the information recorded in the sources, filiation, i.e. sources from other sources, and finally the relevance of each of them. Metric accuracy/quality relates to the degree of fidelity/error resulting from the device used for the survey (e.g. manual survey provides less accurate data than a laser scanner survey); it could also be influenced by the graphic quality of each source. The objectivity/subjectivity is related to the different interpretation capacity of the sources, the credibility and reliability of the data origin, the level of objectivity/subjectivity, i.e. elements assumed through knowledge.

In this work, the colour code assigned to identify the different degrees of accuracy is the following:

- current morphology
- iconographic and/or photographic sources
- interpretation
- assumption.

In particular, the term interpretation means a hypothesis made on the basis of information obtained from some sources but not sufficient to complete the reconstruction. This is the case of the seats in the room visible in the photo, recognized in the Wackelin catalog, but whose disposition in the room has been hypothesized from photographic information. As regards assumption it indicates a reconstruction choice not correlated to any source but formulated on the bases of a typological or a graphical analysis study (see Fig. 12).

The semantic classification of the model gives back the information of the reconstructive process for each element, facilitating the understanding and the dissemination of the reconstruction.

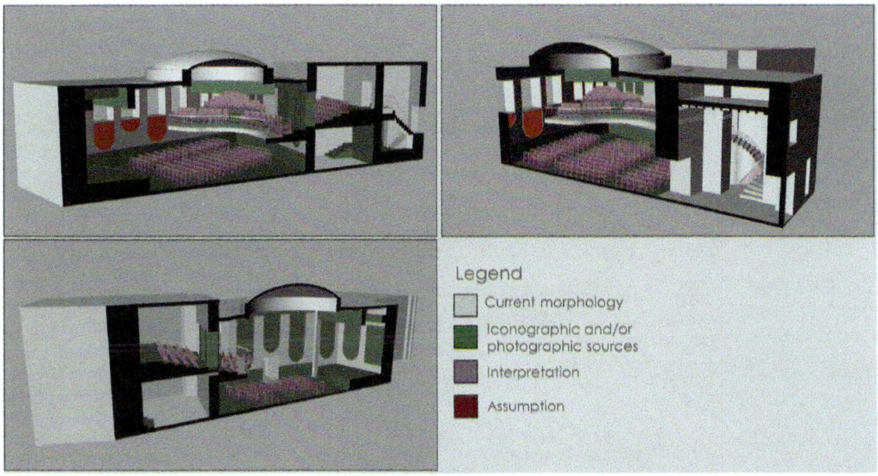

Fig. 12 Axonometric cutaway with degrees of accuracy

9 The Immersive Experience in a Virtual Environment

The reconstructive model provided the groundwork on which designed a possible immersive experience to be installed and exhibited at MuRa as part of the digital section which exploits the potentialities of virtual reality to give the visitors the opportunity to be immersed in the recreated environment enjoining the spatial and visual appearance of Francesco Fichera architectures.

The virtual experience follows the physical exhibitions of the drawings by Francesco Fichera in a room dedicated to Virtual Reality experiences prototypes, which are developed within the didactic and research activities carried out at the Digital Representation, Survey, Reconstruction Laboratory of the Museum.

The model is composed of two macro-categories: the volumetry and the decorative apparatus, the first were modelled respectively in Rhinoceros 5.0 and the second in Zbrush 2020. The VR environment was created using Epic Game's open-source software Unreal Engine, which is a powerful tool for creating immersive and interactive experiences draw on video games and entertainment domain.

For the fruition of the *Cine-teatro Olympia* virtual reconstruction it is envisioned an immersive VR system using Oculus Rift as HMD system which allows the users to become totally immersed in the virtual environment and interact with it. Teleportation was chosen as the system for displacement of the visitor within the virtual environment as it reduces the motion sickness effect.

The experience begins with the user being located in the corridor close to the projection hall, he is free to move around and explore the virtual environment admiring the decoration and the wooden false ceiling; then he/she can chose to enter the hall or walk up the stairs and reach the tribune from which it is possible to

Fig. 13 View of the the prototype for the physical space for visitors

better grasp the details present in the bas-reliefs placed on the architrave and on the spring of dome.

As you enter the theater and reach the front rows, the lights turn out and scenes from the film *Quo Vadis* are played on the projection screen. This movie opened in 1913 the activity of the *Cine-teatro Olympia* and, in our vision, metaphorically inaugurates its virtual rebirth. During the playing, however, it is still possible to admire the environment, the focus being on the dome, which remains illuminated even during the projection (see Fig. 13).

10 Conclusions

The research work here presented falls in the domain of virtual reconstruction/restoration. The applied methodology has taken into account and has been driven by the scientific transparency as defined in the London Charter and Seville Principles. Therefore, the interpretative model keeps trace of the reconstructive choices process. In addition, a virtual interactive experience has been designed to bring the museum visitor closer to this valuable heritage, giving him the opportunity to be immersed in the atmosphere of the *Cine-teatro Olympia* using graphic and visual languages.

Fig. 14 View of the scene, comparison between historical photography and digital reconstruction

Fig. 15 View of the hall, comparison between historical photography and digital reconstruction

The next steps of the research will be addressed to evaluate this first prototype, since during the COVID-19 pandemic the museum has drastically decreased the number of visitors. Furthermore, the research efforts will be focused to explore other ways to give access (for instance remote) or enjoy (AR/XR) this digital cultural content (see Figs. 14 and 15).

Acknowledgements The research has received funding from Programma di Ricerca di Ateneo UNICT 2020-22 linea 2. The authors thanks the *Museo della Rappresentazione* for allowing the consultation and publication of the drawings and images of the Francesco Fichera Fund.

Credits Despites unity intent of this work, the editorial responsibility is as follows: 1,10 (C.S.); 2,4,5,9 (C.R.); 3,7 (C.R., C.S., M.G.); 6,8 (C.R., C.S).

References

1. Pagano, A., Pietroni, E., Ferdani, D., & D'Annibale, E. (2021). User eXperience (UX) evaluation for MR cultural applications: The CEMEC holographic showcases in European museums. *Applied System Innovation, 4*(4), 92.

2. Pietroni, E., & Ferdani, D. (2021). Virtual restoration and virtual reconstruction in cultural heritage: Terminology, methodologies, visual representation techniques and cognitive models. *Information, 12*(4), 167.
3. Hermon, S., & Kalisperis, L. (2011). Between the real and the virtual: 3D visualization in the Cultural Heritage domain-expectations and prospects. *Virtual Archaeology Review, 2*(4), 1989–9947.
4. Frommel, S., & Schlimme, H. (2020). Virtual models and scientific value. *Scientific Research and Information Technology, 10*(1), 1–4.
5. Brusaporci, S., Trizio, I. (2013) La "Carta di Londra" e il patrimonio architettonico: riflessioni circa una possibile implementazione. *Scientific Research and Information Technology, 3*(2), 55–68.
6. Demetrescu, E., & Fanini, B. (2017). A white-box framework to oversee archaeological virtual reconstructions in space and time: Methods and tools. *Journal of Archaeological Science: Reports, 14*(14), 500–514.
7. De Vos, P. J., & De Rijk, M. J. (2019). Virtual reconstruction of the birthplace of Rembrandt Van Rijn: From historical research over 3d modeling towards virtual presentation. *The International Archives of Photogrammetry, Remote Sensing and Spatial Information Science, XLII-2/W15*, 397–404.
8. Wong, L., & Quintero, M. S. (2019). Tutankhamen's two tombs: Replica creation and the preservation of our cultural heritage in the digital age. *The International Archives of Photogrammetry, Remote Sensing and Spatial Information Sciences, XLII-2/W11*, 1145–1150.
9. Zerbi, A., & Mikolajewska, S. (2021). Digital technologies for the virtual reconstruction and projection of lost decorations: The case of the proscenium of the Farnese Theatre in Parma. *Disegnarecon, 14*(27), 5.1–5.11.
10. Agnello, F. (2021). From the Museum to the site and backwards. *Disegnarecon, 14*(27), 16.1–16.12.
11. Demetrescu, E. (2018). Virtual reconstruction as a scientific tool: The extended matrix and source-based modelling approach. *Communications in Computer and Information Science*, 817.
12. Demetrescu, E., & Ferdani, D. (2021). From field archaeology to virtual reconstruction: A five steps method using the extended matrix. *Applied Sciences, 11*, 5206. https://doi.org/10.3390/app11115206
13. Apollonio, F. I. (2016). Classification schemes for visualization of uncertainty in digital hypothetical reconstruction. In S. Münster, M. Pfarr-Harfst, P. Kuroczyński, & M. Ioannides (Eds.), *3D research challenges in cultural heritage II, LNCS, 10025* (pp. 173–197). Cham: Springer.
14. Fallica, S., Garozzo, R., & Santagati, C. (2021). Retraced memories-virtual reconstruction of an architectural landmark. *Virtual Archaeology Review, 12*(25), 124–139.
15. Giovannini, E. C. (2020). WorkFlow for an evidence-based virtual recostruction: The marbles of the ciborium of the early medieval Monte Sorbo Church. *The International Archives of the Photogrammetry, Remote Sensing and Spatial Information Sciences, XLIII-B2*
16. Fichera, F. (1913) Olympia nel primo dì di primavera MCMXIII, Stabilimento di arti grafiche, Catania.
17. Renzi, G. (2018). Sicilia, terra di sedie in legno curvato a vapore, in Legno curvato. Thonet, style, design, Milano. https://legno-curvatodesign.it/sicilia-terra-legno-curvato-a-vapore/.
18. Apollonio, F. I., Grellert, M., Martens, B., & Nußbaum, N. (2018). Working experiences with the reconstruction argumentation method (RAM)-scientific documentation for virtual reconstruction. In: *Proceedings of the 23rd International Conference on Cultural Heritage and New Technologies 2018. CHNT 23*, Vienna.

From Design to Construction: Digital Reconstruction of the Convent's Kitchen Area in the Monastery of El Escorial

Pilar Chías, Tomás Abad, and Lucas Fernández-Trapa

Abstract It is quite common to make decisions during a building process that may have some effect on design guidelines. The proposed case study, that is the old Kitchen wing in the Monastery of El Escorial (Madrid, Spain)—built between April 1564 and May 1571-, shows the various design changes made even during construction. They can be followed in the graphic and written historical documents that are still preserved, to compare them with the built reality by means of an accurate survey. The study of this area in such an interesting World Heritage Site -inscribed in 1984- is justified for two main reasons: first, its formal, geometrical, and constructive complexity shown in the numerous and ingenious ashlar work vaults, arches, and openings; second, the interesting functional solutions displayed on the six-storey wing, from cellars to attics. Both are closely related because the constructive solutions applied gave a clear answer to the original operational needs, and particularly to natural lighting and airing requirements. Moreover, the built artifices are illustrative examples of Mannerist architecture that were fully operational over centuries. Through the digital reconstruction of this area, we aim to recover and interesting part of the Monastery's memory.

Keywords Double façades · Natural lighting · Mannerist architecture · 16th century Spanish architecture · Ashlar vaults · Architectural drawings and tracings

P. Chías (✉) · T. Abad
University of Alcalá, Madrid, Spain
e-mail: pilar.chias@uah.es

T. Abad
e-mail: tomas.abad@uah.es

L. Fernández-Trapa
Hochschule Koblenz, Koblenz, Germany
e-mail: fernandez@hs-koblenz.de

© The Author(s), under exclusive license to Springer Nature Switzerland AG 2023

I. Trizio et al. (eds.), *Digital Restoration and Virtual Reconstructions*, Digital Innovations in Architecture, Engineering and Construction,
https://doi.org/10.1007/978-3-031-15321-1_4

1 Introduction, Precedents, and Innovation

Despite the relevance of the Royal Site and Monastery of El Escorial as a UNESCO World Heritage Site, research on a large part of the enormous building still lacks in-depth study. Been founded by King Philip II in 1564 as a true town -the *Civitas Dei* by Saint Augustin- it contained very different rooms (*cuartos*) as the King's House, the Public Palace, the Convent, the Basilica, the College, and the Seminar. Each one had their own service facilities -to a total of nine kitchens-, but there is still a lack of deeper and more detailed studies on most of their spaces, uses, and constructive solutions.

The original Kitchen area is among them, and for this reason we found it suitable as a case study. As precedents in the field of historic research, studies by Bustamante [1] on the Monastery in the sixteenth century, by Rubio [2], Portabales [3] and Íñiguez [4] must be stressed.

From a graphic perspective some previous surveying have been made. When the kitchen refurbishment and relocation in the Convent's southern cellar in 1963 took place, Andrada [5] made a surveying of the whole building and produced a set of small-scale floor plans. Ortega's [6] handmade surveying in 1988 focused on the formal and dimensional control, as a part of his PhD research work. More recently, in 2009 López Mozo [7] made an accurate survey of the Monastery's ashlar vaults [8] by means of a laser-based total station, regarding stereotomy and its relationship with the historic stonecutting treatises.

Consequently, there was any previous research on the whole Kitchen wing whose aim was to define, on the one hand, the former relationships between rooms, considering their respective uses and functions, and on the other hand, regarding the constructive solutions and shapes of each architectural element. Thus, our main contribution and innovation is to reconstruct the original physiognomy of this area in the Monastery through an in-depth study of historic sources and an accurate survey of the whole wing using a laser scanner device. Our work will complete and complement the preceding perspectives, too.

Fortunately, there is a huge number of historical records about the Monastery's construction, mainly construction contracts and work conditions to build the various architectural elements. Moreover, the many descriptions by witnesses of the construction, by Hieronymite chroniclers, friars, and visitors between the 16th and the nineteenth centuries are available, together with an outstanding collection of historical drawings.

Between descriptions related to the study area that provide information about the original uses of the Convent's kitchen and its rooms, those by Almela [9], Sigüenza [10], Bermejo [11] and Álvarez [12] among other, must be cited.

Construction contracts provide evidence for craftsmen, work conditions for each job, and contractors. These documents are mainly kept in the Real Biblioteca and Archive of the Monastery of El Escorial (from now on AMSLE) [13], the Archive of Simancas in Valladolid (AGS) [14] and the Archive of Instituto Valencia de Don Juan in Madrid (IVDJ) [15].

According to our research objectives, we focused on the period beginning 1564 until 1571, which coincides with the building of the Convent's kitchen and its adjacent spaces.

Construction drawings and tracings produced by architects and craftsmen are kept in the Real Biblioteca and Archive of the Monastery [13]; the original ones by Juan Bautista de Toledo and Juan de Herrera are in the Real Biblioteca de Palacio in Madrid (PRRB) [16], that also custodies the remarkable collection of the sixteenth century plans of the building or *Estampa*s engraved by Pedro Perret based on drawings by Juan de Herrera. Other collections of early printed editions can be found in the Biblioteca Nacional de España in Madrid (BNE) and the Bibliothèque nationale de France in Paris (BnF). Among these drawings, the *Tercer diseño* (Third design) shows a cross section through the smaller cloisters where the Kitchen's volume can be clearly seen (see Fig. 2).

Another interesting collection of drawings of the Monastery is kept in the Canadian Centre for Architecture in Montréal, resulting of the survey made in 1759 by military engineers Bernardo Fillera and Balthazar Bécaud under the direction of José de Hermosilla. The collection consists of four elevations and three sections that provide an updated image of Herrera's *Estampas.*

All these documents were studied to know the evolution of the old Kitchen wing at the foundational period, and to make a comparative analysis of its state in later centuries.

2 Description of the Case Study

The old Kitchen wing is composed by a series of spaces that are organised along a West–East axis. It is located between two minor cloisters at the southwest area of the Monastery. In fact, the old Kitchen separated the Infirmary cloister on the south from the Guest quarters on the northwestern side of the Convent.

The western side of the old Kitchen wing is the Convent's doble façade that opens onto the exterior square or *Lonja*, following the tradition of the German medieval *Westwerke.* The east side has direct access to the bright space of the lantern, that is the well of light of the Convent (see Fig. 1).

The whole Monastery was built using the Castilian foot (about 0,2786 m) as the length unit, to be used for modulating rooms and heights.

We established the zero level at the main floor or *planta al andar de la casa*, that is 5 ft over the *Lonja* level. Cellars are 18 ft below the main floor, and the upper stories rise to 15, 30, 45, and 60 ft respectively (see Fig. 2).

Small cloisters on the south side of the Monastery, together with the Kitchen wing are among the oldest part of the building. They were constructed between the end of 1564 and May 1571, when the first Hieronymite friars came to inhabit the building. Their scale is more 'domestic' and their spatial distribution is more fragmented than those in the Main Cloister [6]. Their constructive solution is also different, as "the

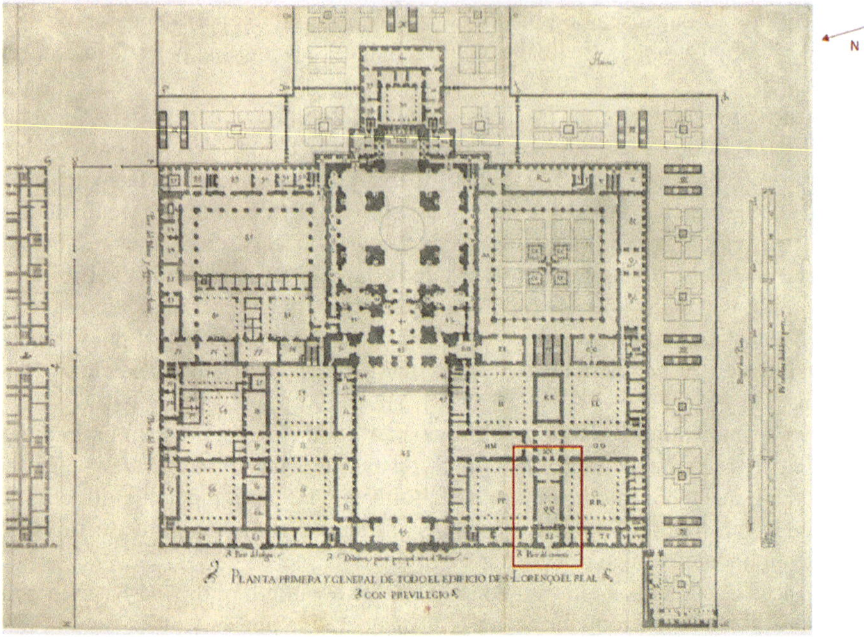

Fig. 1 Juan de Herrera and Pedro Perret (1587): Primer diseño, Planta primera y general de todo el edificio (First design, Main floor, and general plan of the building). Real Biblioteca, Patrimonio Nacional, Madrid. Map legend: NN, Lantern; QQ, Kitchen; SS, Entrance foyer; RR, Cloister of Infirmary; PP, Guest cloister and quarters

roofs on the ground floor are vaulted, while the ceilings on the second and third stories are flat and made of wood" [12].

Spaces of the Kitchen wing are distributed along a West–East axis 115 ½ ft long or 41,5 m, between the exterior surface of the double façade and the transverse plane of symmetry of the lantern. The ordered sequence of contiguous rooms runs from the main entrance in the *zaguán* (hallway) to the old kitchen, with its cellars and upper stories. Behind the kitchen on the main floor are the scullery and the lantern surrounded by a wide ambulatory.

Next is described each space, showing in detail its formal and functional singularities that were brought to light by means of the surveying and digital reconstruction.

3 Surveying Methodology, Analysis, and Main Findings

According to the targets, the first stage of research consisted in carrying out an accurate survey of every indoor space of the old Kitchen wing, from the western double façade to the lantern, and from cellars to attics. In parallel, we analysed the still

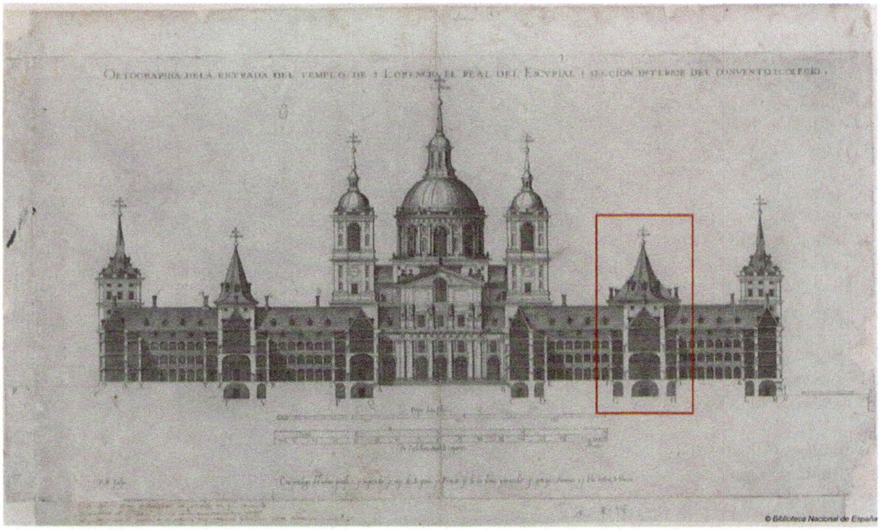

Fig. 2 Juan de Herrera and Pedro Perret (1587): Tercer diseño, Ortographia de la entrada del Templo de S. Lorencio el Real del Escurial i secci6n interior del Convento y Colegio (Third design, Elevation of the Church and cross section of the Convent and the College). Biblioteca Nacional de España, Madrid

extant documentation to check its reliability, focusing on changes and refurbishments over the centuries.

From the methodological perspective, and as a result of previous fieldwork, we found it difficult to interconnect some indoor scan stations (particularly cellars, lantern galleries and attics). To tackle this problem, we increased the number of scan stations, and georeferenced some elements to set reliable referencing points.

Moreover, due to lack of visual and physical accessibility to the double façade, we decided to combine photogrammetric and laser scanning techniques.

3.1 Constructive and Formal Perspectives

Through fieldwork and surveying we proved that the repeated complains about the prolonged absences from works of architect Juan Bautista de Toledo, and the lack of accuracy displayed in his tracings, gave rise to some important construction problems. As an example, errors made during on site setting-out works had direct consequences on the thickness of perimeter bearing walls of the kitchen, as well as in other spaces along the West–East axis. Consequently, the kitchen wing is only apparently symmetrical.

Fig. 3 Tracing for the doorway to the Convent's kitchen wing. Attributed to architect Juan Bautista de Toledo, c1565. Real Biblioteca, Palacio Real, Madrid

In fact, the north wall of the old Kitchen is 7 ft thick (1,92 m) while the south one is just 5 ½ ft (1,55 m). The first consequence is that every room has two planes of longitudinal symmetry that are nearly ½ ft apart (0,15 cm) (see Fig. 12).

Consequences in each space are outlined below along longitudinal axis, from double façade to lantern, and from cellars to attics.

Hallway entrance gives access to the Convent area from the Western *Lonja* through *puerta de las cocinas* (kitchen's doorway). An interesting tracing dated around 1565 (Fig. 3) was compared with the built façade, bringing to light some essential changes affecting fundamentally the two cylindric skew windows. They cast sunlight to hallway and old kitchen spaces where it is needed (Fig. 4).

Hallway layout is a rectangle 35 ft wide (9,73 m) by 20 ft long (5,56 m) that extends the whole width of the kitchen's bay. The upper part of the double façade raises above the *zaguán* (see Figs. 4 and 6).

Five doors give access to the cellars, the small adjacent cloisters, and the kitchen. In the east wall two symmetric doors with a flight of stairs lead to the kitchen, that is 5 ft high from the *Lonja* level. In the central part of the wall, a wider door gives access to the cellars by means of a convenient ramp for pack animals, as we will see later. In the north and south walls two symmetric doors with their corresponding flight of stairs open to the Guest cloister and the Infirmary cloister respectively.

Hallway is domed over with outstanding ashlar vault. It is a domical vault with lunettes in the lower part over the doors, that extends to a flat vault at the highest point by means of straight sections (see Figs. 5 and 12).

The old kitchen adjacent to the hallway to the east is 62 ft long (17,26 m) and 35 ft wide (9,73 m). It has a rectangular ground plan that is only partially domed

Fig. 4 The western façade of the Convent seen from the *Lonja*. (Photograph by the authors)

Fig. 5 The outstanding ashlar vault that domes the Convent's hallway. (Photograph by the authors)

Fig. 6 Skew openings under the Kitchen's sail vault. (Photograph by the authors)

with huge sail vault that was a milestone in Spain at the time [1], with the chimney draught opening in the middle, at a height of 9,46 m above the kitchen's floor.

The sail vault is supported by the perimeter bearing walls of the kitchen, as well as by two barrel vaults at both sides. The west one is 21 ft wide (5,85 m), while the east vault is only 9 ½ ft wide (2,70 m), both with 35 ft span. The supporting arches are stilted, probably because of the need for natural lighting of the kitchen, as stated by the *Accademia* in Florence in its 1567 report on the tracings of the Monastery [17].

Surveying allowed us to verify that the big vault is only apparently a sail vault [8]. Moreover, it does not cover a square shape, but a rectangular one of 35 ft (9,73 m) by nearly 31,37 ft (8,74 m).

As we said before, such unevenness is due to errors during on site setting-out works in the absence of Juan Bautista de Toledo in the autumn of 1564, together with the lack of accuracy of his drawings. As a result, the vault is not perfectly spherical. In such cases, stonework in square courses provides a suitable solution to make each vault sector independent from the rest, what could not be possible if rows were circular and concentric (Fig. 6).

Next to the kitchen lies the former pantry that served the refectory. It is a large rectangular-shaped room that is domed over with barrel vault and pointed lunettes.

On either side there are two interesting rere-arches in ashlar leading to the small cloisters.

On its north side there is a little room domed with small square-shaped sail vault. Andrada's [5] 1964 ground floor plan of the Monastery shows a staircase in this area leading to the room that is 15 ft high over the pantry. Though this staircase is not shown in Herrera's *Primer diseño* (Fig. 1) nor in later drawings as the one by Salcedo de las Heras in 1876, the small sail vault is a clear sign of the missing stair landing (Fig. 7).

Moreover, the room over the pantry opened on to the old kitchen through a wide opening that is now blinded and turned into a vaulted niche (see Fig. 6).

Lantern (Fig. 8) is accessed through a wide ambulatory domed with large surbased barrel vault with pointed lunettes. The character of this space is quite different from three other ambulatories surrounding the lantern, because it was a closed room from the beginning. However, it opens to the adjacent small cloisters. Inside the north wall there is a small staircase to go down to the cellars and the domed space under the lantern. As in the pantry, the stair landing is domed over with sail vault with star-shaped stereotomy (Fig. 7).

Fig. 7 Small sail vaults in the Kitchen area showing various stonecutting solutions. The one on the left is in the office and was at the landing of the missing stair. In the middle, the star-shaped sail vault of the staircase inside a wall in the ambulatory (Photograph by the authors)

Fig. 8 The Convent's lantern. (Photograph by the authors)

Rooms over the old kitchen surround the chimney draught to take advantage of the heat. They are similar in size to the kitchen, but they were severely damaged by fire in the sixteenth century and have been significantly transformed by Andrada in the 1960s.

As we have seen, levels at 15 and 30 ft high are roofed with wooden frames. Rooms on these stories receive indirect sunlight through interesting architectural devices as the double façade, and a varied set of skew and splayed openings to cloisters and lantern.

However, due to a severe termite infestation the original wooden roofs at 45 and 60 ft high stories were partially replaced in the 1960s with lightweight metal structures [18]. Still extant granite supports of the former gabled roof, together with remains of wooden trusses give an idea of the importance of the original roof structure (Fig. 9).

Finally, cellars extend 18 ft below ground floor level. They consist of an interesting series of successive rooms domed with various ashlar vaults, and highly specialised according to their respective functions.

Beneath the kitchen lies a wide room used to store firewood (*bóveda de la leña*), domed with surbased barrel vault 31,23 ft span (8,70 m), that is more than 3 ft shorter than the kitchen barrel vaults. Natural light entered through a set of conical skylights, now blinded. This cellar is accessed through the ramp for pack animals

Fig. 9 Original wooden structure and dormer windows at the 45 ft high level. (Photograph by the authors)

already mentioned and opens to the cellar beneath the lantern domed over with interesting groin vault, although of smaller dimensions than the light well (31,23 ft by 31,23 ft).

Across this cellar and other at the south side of the Convent, monastic orchards can be reached.

3.2 Functional Perspective

From a functional perspective, the diverse solutions designed by Juan Bautista de Toledo to meet the varied needs of the Hieronymite Community show his extraordinary creativity.

As we have seen the ingenious set of architectural devices constructed to illuminate with natural light every room of the building, however humble it was, include skylights in the cellars, skew openings, a double façade, a lantern, and dormer windows (see Figs. 6, 8, 9 and 10). In this respect, historic documents refer to "aposentos a primera y a segunda luz" (rooms at first light and at second light) [11].

Most of the numerous water fountains distributed throughout the Monastery are still in use, and a device for heating water was in the kitchen. Hot water reached the Convent's 30 ft level.

Fig. 10 Upper part of the double façade with the thermal window, over the entrance foyer. (Photograph by the authors)

Fig. 11 Ramp to the cellar for horses and mules. (Photograph by the authors)

Regarding interior circulation, wide luminous corridors, ambulatories, and stairs, together with convenient accesses to every room in the Monastery must be stressed. The ramp for pack animals leading to *bóveda de la leña* is an outstanding example (see Fig. 11).

Specialised storage cellars preserved in the Monastery included an indoor fishery furnished with stone tubs and running water to keep fresh fish, while huge earthenware jars for oil and grain are still in the Convent's cellars.

4 Conclusions

According to our aim, digital reconstruction of the case study evidenced some interesting aspects related to construction and former uses, or in other words, to recover the historic memory of the Monastery.

In relation to construction, different wall thickness in the north and south bearing walls of the Convent's kitchen resulting of on-site setting-out works implied the existence of two longitudinal axis of symmetry affecting many architectural elements in the kitchen wing. Among them, geometry and dimensions of the kitchen vault are just apparent, because the sail vault that does not cover a square shape, but a rectangular one, and the supporting arches are stilted, as López Mozo noticed [7].

Some missing architectural elements as stairs and openings have been found. The small sail vault at the landing of the stair in the pantry permitted us to reconstruct its shape and function, which was to communicate the pantry with the upper room, that could be once used as a rest area for the cooking staff. The central opening in this room, now blinded, could be helpful in surveillance tasks.

In addition to the double façade and the lantern, other interesting lighting and airing devices have been found and stressed. Skylights in the cellars and skew openings in the kitchen and the upper stories evidence the practical vision of architects

Fig. 13 Longitudinal section of the Convent's Kitchen area. (Chias, Abad and Fernández-Trapa 2021)

In the 1960s architect Ramón Andrada provided solutions for a new kitchen and for termite infestation. Both had a deep impact on the Convent's Kitchen wing.

Old kitchen moved to the south side of the Convent, next to *Jardín de los Frailes* (Friar's Garden), while the vacant space was used as a cosy visiting room.

Replacement of the wooden frame of the gabled roof with lightweight metal structures changed the physiognomy of the attics but recovered the original pitch in the roofs that had been deeply altered after a devastating fire in1671.

Through the accurate surveying some interesting areas in the Monastery have been digitally reconstructed as they were in the sixteenth century, displaying their constructive, formal, and functional achievements. It proved to be an essential way to recover the memory of the building.

Acknowledgements The authors wish to thank Patrimonio Nacional and the Augustinian Community of the Royal Monastery of San Lorenzo de El Escorial for the inspiring environment during research work.

References

1. Bustamante, A. (1994). *La octava maravilla del mundo (Estudio histórico sobre El Escorial de Felipe II)*. Ed. Alpuerto, Madrid.
2. Rubio, L. (1948). El Monasterio de El Escorial, sus arquitectos y artífices (observaciones a algunos libros recientes). *La Ciudad de Dios, 160*, 51–108 and 419–474.
3. Portabales, A. (1945). *Los verdaderos artífices de El Escorial y el estilo indebidamente llamado Herreriano*. Gráfica Literaria, Madrid.
4. Íñiguez, F. (1965). *Las Trazas del Monasterio de San Lorenzo de El Escorial*. Real Academia de Bellas Artes de San Fernando, Madrid.
5. Andrada, R. (1964). Las reconstrucciones en El Escorial. In *Monasterio de San Lorenzo El Real. El Escorial en el cuarto centenario de su fundación, 1563–1963* (Vol. 2, pp. 323–349). Patrimonio Nacional, Madrid.
6. Ortega, J. (2000). *El Escorial: dibujo y lenguaje clásico. Sociedad Estatal para la Conmemoración de los Centenarios de Felipe II y Carlos V*, Madrid.
7. López Mozo, A. (2009). *Bóvedas de piedra en el Monasterio de El Escorial*. Unpublished Doctoral Dissertation, Polytechnical University of Madrid.
8. López Mozo, A. (2004). Traza y construcción en la bóveda vaída de la cocina del convento del Monasterio de El Escorial. In: *Actas del X Congreso Internacional de Expresión Gráfica Arquitectónica* (pp. 1021–1031). Universidad de Granada, Granada.
9. Almela, J. A. (1594).*Descripción de la Octava maravilla de el Mundo Que es la excellente y Sancta casa de sant Laurencio el Real Monasterio de frailes hieronimos y Collegio de los mesmos y seminario de letras humanas y sepultura de Reyes y Casa de recogimiento y descanso después de los trabajos de el Gobierno (Manuscript)*. Biblioteca Nacional de España, Ms.1724.
10. Sigüenza, J. (1605). *Tercera parte de la Historia de la Orden de san Jerónimo*. En la Imprenta Real, Madrid.
11. Bermejo, D. (1820). *Descripción artística del Real Monasterio de S. Lorenzo del Escorial y sus preciosidades después de la invasión de los franceses*. Imprenta de Doña Rosa Sanz, Madrid.
12. Álvarez, F. (1843). *Descripción del Monasterio y Palacio de San Lorenzo, Casa del Príncipe, y demás notable que encierra bajo el aspecto histórico, literario y artístico el Real Sitio del Escorial, para uso de los viageros y curiosos que lo visiten*. Imprenta de Vicente Lalama, Madrid.
13. AMSLE, Doc. I-44; Doc. 1-58; Doc. I-64; Doc. I-65; Doc. I-66; Doc. I-83; Doc. I-84; Doc. I-85; Doc. I-94; Doc. II-3; Doc. II-10; Doc. II-21; Doc. II-29; Doc. II-38; Doc. II-47; Doc. II-50; Doc. II-52; Doc. II-57; Doc. II-63; Doc. II-67; Doc. II-78; Doc. II-80; Doc. II-93; Doc. II-101; Doc. II-103; Doc. II-104; Doc. II-106; Doc. II-117; Doc. II-134; Doc. II-155; Doc. IV-20. Archivo de la Procuración de San Lorenzo II/10, II/38.
14. AGS, Obras y Bosques, Escorial, leg. 1; leg. 2, ff. 87, 89, 99, 102, 131, 212, 216, 219; leg. 3; leg. 4; leg. 5; leg. 6; leg. 7. AGS, Casas y Sitios Reales, leg. 258, ap. 1, ff. 7–8; leg. 258, ap. 2, ff. 207–220; leg. 259, ff. 395, 414; leg. 260, f. 422; leg. 280, ff. 342, 480–481. AGS, Consultas y Providencias Generales, leg. 275, ap. 2; AGS, Mapas, Planos y Dibujos 63,036.
15. IVDJ, Envío 61 (1), ff. 173, 239–240; 327–330.
16. PRRB, Planos y Dibujos, IX/M/242/1(24), IX/M/242/1(25), IX/M/242/1(29).
17. Archivio di Stato di Firenze, Accademia del Disegno, f. 157, inserto 2.
18. Andrada, R. (1969). Total renovación de las cubiertas del Monasterio del Escorial. *Reales Sitios, 6*(19), 19–28.

De RE Virtual RES. The Virtual Reconstruction of Rocca Janula in Cassino for a Meaningful "Reading" of the Historical Stratification

Assunta Pelliccio, Marco Saccucci, and Virginia Miele

Abstract The restoration project, according to the Italian ministerial guidelines, gives the photographic campaign a truly significant importance. The photos, which capture the state of things, before, during and after the restoration, have the task of transmitting correct historical information to the observer. For this reason, the ministerial indications suggest affixing the photos inside the restored site with the aim of highlighting above all those parts added to complete the architectural gaps. When this procedure is not performed and when the restoration does not meet the criteria of distinguishability and reversibility, it is necessary to carry out a reverse operation to the photographic campaign. Starting from historical documentary and iconographic sources, it is necessary to create virtual models such as to provide the observer with the correct historical information. At this point the images play a fundamental role. The aim of the work is to define different levels of virtual models based on the accuracy of the acquired historical source. Simplified and monochromatic models, for example, serve to tell what cannot be documented. These models, in fact, are abstracted from reality to the point of transferring to the observer the perception that the virtual reconstruction is hypothetical. More complex and photorealistic models, obtained with the latest digital technologies, virtualize the real object that is perceived by the observer as objective. The methodology presented, and based on different levels of digital models, was applied to the Rocca Janula di Cassino case study, with the aim of clearly and unequivocally communicating the main historical phases that characterize its current state.

Keywords Virtual representation · Digital model · Digital photogrammetry · Historical analysis

A. Pelliccio (✉) · M. Saccucci
University of Cassino and Southern Lazio, Cassino, Italy
e-mail: pelliccio@unicas.it

M. Saccucci
e-mail: m.saccucci@unicas.it

V. Miele
Silesian University of Technology, Gliwice, Poland
e-mail: virginia.miele@polsl.pl

© The Author(s), under exclusive license to Springer Nature Switzerland AG 2023
I. Trizio et al. (eds.), *Digital Restoration and Virtual Reconstructions*, Digital Innovations in Architecture, Engineering and Construction,
https://doi.org/10.1007/978-3-031-15321-1_5

1 Introduction

Images, as the reproduction of reality, are useful for seeing more and understanding more [1]. In recent decades, the use of images as a communication tool has assumed increasing importance, invading almost all sectors of human knowledge. In the architectural field, or more generally of material cultural heritage, images have always played a fundamental role, both in the design and work execution, and in the survey, understood as the global knowledge process, necessary for enhancement activities.

In the restoration, as early as the 1960s, the photos (images) take on the dual role of documenting the "state of things" prior to the intervention as well as all the phases of the intervention. To this end, the Italian Ministry (Central Institute of Cataloguing and Documentation—ICCD) has introduced some guidelines that are still active today. The goal of the detailed photographic campaign is, from one hand, to preserve the heritage description by transferring accurate historical information to future generations. For this reason, it has a strong historical value as frames taken at different times, testify the intrinsic evolution of the life of the cultural asset. On the other hand, the photographic campaign of the pre- and post-restoration intervention helps to recognize the completion elements of the architectural gaps that could derive, for instance, from design choices in disagreement with the indications suggested by the Restoration Charters and the Restoration Theory introduced by Cesare Brandi in 1975. It is in fact crucial precisely in the case of a questionable restoration project due to it underlines the significant change in the historical qualities of the property often obtained with even irreversible processing. The documentary frames, according to ministerial guidelines, must provide information on all the elements that make up the asset, from the relationship between the asset and the landscape, including the definition of volumes, the development of surfaces and ornamental details, to the analysis of construction techniques and the state of conservation. Bearing in mind these assumptions, surely the images used as historical-documentary evidence must be as objective as possible to make the visual communication adequate. To this end, the use of colour in the images is of great help in reading the material characteristics of the asset as well as the historical stratifications that determine its current state. In fact, colour, enhancing the chromatic and tonal relationships between the parts, produces a unitary and harmonic perception of the object in its environmental and landscape context and makes evident the connections and relationships typical of the compositional nature of the object itself.

Today many technologies return a realistic view of the object and display a lot of geometric, metric, and technical information, even dynamically, using images based on different colour models (RGB, CMYK, etc.). Digital photogrammetry, thermal imaging cameras, etc. are just some of the most used. The photogrammetric digital model, obtained by processing many photos according to a standard procedure, returns, for example, photorealistic colour images of the entire object together with all the information requested in the most objective way possible. The realistic correspondence between colour and elements of the object is closely related to natural light: at different times of the day or in different climatic conditions the single element can have different chromatic features. From a technical point of view, photogrammetric virtual models are always legible even when they are displayed discretized as a point cloud. At the same time, the same model requires the use of a texture, also defined as the sensitization (natural or artificial) of a surface by means of signs that do not alter its uniformity [1], capable of simplifying the learning process in the context of historical record, just like colour. The use of texture is essential when the visual communication is aimed at a non-technical audience. Briefly, a photogrammetric digital model fully meets the requirements of historical documentation campaign in a restoration process. In addition, non-textured point clouds are also an excellent basis for creating simplified digital models, i.e. in wireframe/solid visualization, which mathematically reproduce the object through primitives, such as polygons, spheres, cylinders etc. as well as the parametric models i.e. H-BIM.

In the case study of the Rocca Janula in Cassino, the methodology applied is the reverse of the traditional photographic campaign, which characterizes the entire restoration process. The fortress, in fact, was subjected to a massive intervention to fill the architectural gaps that are not easily recognizable. The applied methodology matches multi-epoch images and virtual models obtained by processing the point cloud from the aerial digital frame. The goal is to clearly and unequivocally communicate each historical phase of the Rocca that characterizes its current state, focusing attention on the distinguishability of contemporary elements.

"The mind sees what the eyes believe and convinces itself that it is. What you see is not what exists." (Leonardo).

2 The Case Study of Rocca Janula in Cassino (Italy)

Rocca Janula occupies an offshoot of Montecassino and dates to the period between 949 and 967.[1] It was built as a defensive structure of the famous Benedictine abbey of Montecassino, located a little higher.

[…] The picturesque line of this dark and crenelated castle contrasts with the calm shape of the majestic building that dominates Mount Cassino, and its ruined parts suggest the sacrifice that this modest castle had to undergo, for the good of the Monastery […][2].

From the very beginning, in fact, the events of the Rocca are closely linked to those of the Abbey. Its numerous transformations, destructions and restorations were due to the will of the Abbots that followed one another, or to natural events, such as strong earthquakes, which often involved southern Lazio (Fig. 1).

There are several hypotheses on the origin of the toponym "Rocca Janula". The first, according to *Chronica Sacri Monasterii Casinsensi* [3], identifies "Janulo" with the name of pagan divinity of Giano Bifronte to whom a temple was dedicated in the

[1] The name Rocca Janula is indicated for the first time in a document dated 7 June 967 which lists the castles and towers granted to the abbot of Montecassino by the princes Pandolfo and Landolfo di Capua: the castle of Jannula (E. Gattola, Accessiones note 17 page 11).

Fig. 1 Mabillon engraving of Montecassino and San Germano, 1739; Google Earth image, 2022. Both images show the position of Rocca Janula with respect to the Abbey

area where the fortress was built [3]. The second hypothesis derives the toponym from the Latin word iānŭa (small door) due to the presence of a small passage in the walls of the fortress that led to the Abbey of Montecassino. The third and more articulated, is based on the presence of the same toponym in the area around the city of Rieti. The "castellum quod nominatur Ianule" appears in fact in the list of castra acquired by the abbot Berardo I (1047–1089) of the Benedictine abbey of Farfa. Based on these considerations, Rocca Janula was therefore part of a group of seven castra offered by the last counts of Rieti to the abbey of Farfa [4]. From the early Middle Ages to the WWII, the fortress was the undisputed protagonist of the history of the Italian territory: Carolingian, Saracen, Norman (with Frederick I the Barbarossa and Roger II), Svenian (Frederick II), Angevin (Charles of Anjou), Aragonese (Alfonso of Aragon) and Bourbons (Charles III) marked its lifetime. However, very few studies have been conducted on the fortress, as only the abbey of Montecassino has aroused the interest of scholars from all over the world, even though numerous documentary and iconographic sources allow us to retrace its main historical phases. An accurate historical examination, which collects numerous sources, was conducted by Leonardo Paterna Baldizzi for the restoration of the fortress, carried out between 1907 and 1910 [5]. All the historical data collected by Paterna Baldizzi, with the support of the Benedictine father D. Luigi Tosti, and his accurate reliefs were published in the Memorie della R. Accademia di Archeologia, Lettere e Belle Arti (Vol. II, 1911) with the title "Rocca Janula nell'arte e nella storia". The study highlights that the history of the fortress can be divided into three main phases. The first phase concerns the period from its foundation to its demolition in 1221, following the edict of Capua, promulgated by Frederick II of Swabia, which provided for the demolition of all the castles and fortresses of his kingdom. The pentagonal tower was the only element that survived the demolition due to its construction characteristics. The second phase embraces the time that goes from the reconstruction around 1235, which took place with the Abbot Landolfo[2] by the will of Frederick II himself, to the WWII, which caused

[2] In 1226, with the Peace of San Germano, Frederick II himself started the phase of reconstruction of all the fortresses and castles. In 1230 Frederick II appointed Filippo di Citrò, Contestabile of

Fig. 2 Rocca Janula before (1995), during (1996-?) and after (around 2010) the restoration intervention

considerable damage mainly due to the bombing of the allied troops. The third and last phase starts from the post-war period to today and is characterized by a partial restoration of the fortress. The intervention, by the Ministry of Cultural Heritage, begins on November 20, 1996, and ends on September 25, 2015. The restoration has completed the numerous structural lacks, just on the west part of the fortress, and arranged the external area of the perimeter walls with "the redevelopment of the embankment in front of the main entrance and arranged at the time of Frederick II to allow the use of wagons" [5] (see Fig. 2).

The summary of the main historical and natural events that characterized the most important transformations of the fortress from its origins to today is contained in Fig. 3.

Virtual 3D reconstruction as the reverse of the traditional photographic campaign in the restoration.

3 3D virtual reconstruction as the reverse of the traditional photographic enterprise in restoration

The Rocca still represents for the inhabitants of Cassino an element of strong historical, symbolic and identity value. The damage of the WWII and the more recent restorations unfortunately cannot help in the interpretation of its "real phenomenon" and in the perception of the *genius loci.* Generally, according to ministerial indications, the exhibition of a photographic documentation certifying the "state of the things" before the restoration, brings the visitor closer to the knowledge of the past, by correctly transferring the information. Therefore, images play a fundamental role

Capua, superintendent for the fortification of San Germano and of Rocca Janula itself. In the same year (June? / July?) a strong earthquake caused further damage to the fortress of which there is no description. In 1232 Tommaso di Aquino placed in Castellano della Rocca; Taffuro of Capua. In 1235 the fortification works of San Germano and the Rocca were completed.

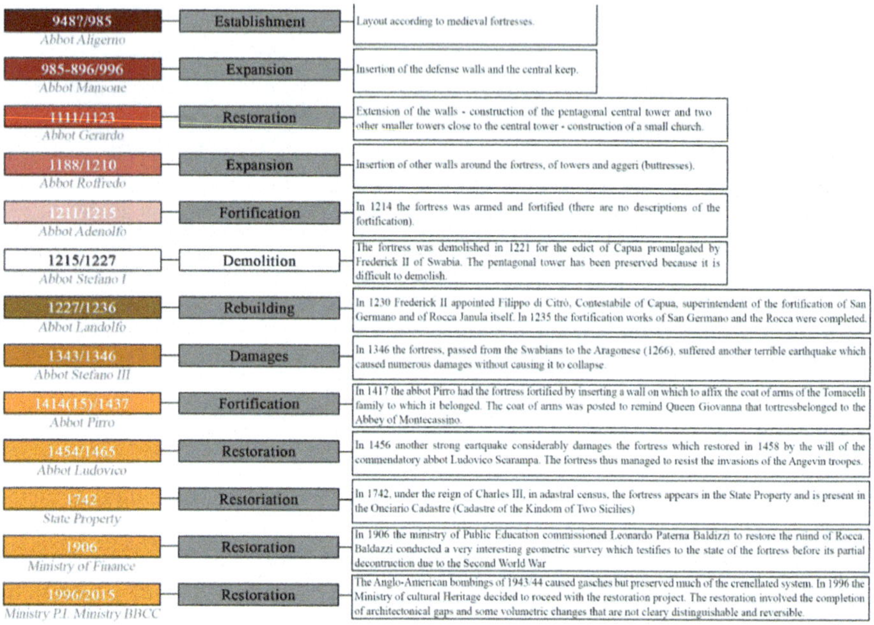

Fig. 3 Synoptic diagram of the main evolutionary phases of the Rocca Janula from the beginning to today

in the process of communicating the value of cultural heritage. When the guidelines are not respected, a process able to retrace historical events, or to narrate the events that have had a decisive impact on the architectural characteristics of the building, starting from the current state, is essential. The process is thought as an inverse process compared to the photographic campaign in the restoration intervention. Obviously, it makes use, once again, of images since visual communication is the most immediate and the most intuitive approach. To this end, digital technologies are fundamental as they virtualize the complexity of the real object. Furthermore, thanks to the application of ad hoc linguistic codes (eg. color) [6], they can represent information in a diversified way in relation to the kinds of the historical data acquired. In the case study, the procedure adopted returns the graphic information of the three main historical phases of the fortress. Monochromatic 3D virtual models return historical information when there is no proven certainty between the data and the "state of things".

In this case, the model is a transposition of a written text and materializes the hypothetical configuration of the object in the volumetric composition, making the description understandable even to less experienced people. Monochromaticity returns an abstract model from reality, avoiding mistaking that model as corresponding to the objective data, while it is a narrated image. Virtual photorealistic 3D models, on the other hand, return the state of things on which historical, documentary, and

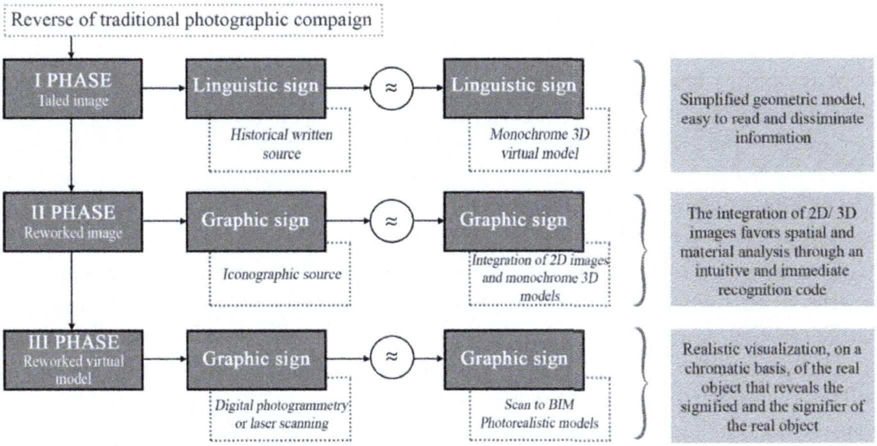

Fig. 4 Articulation of the phases of the reverse process of the photographic campaign. From the first phase, a narrated image is obtained. From the second one a reworked image. From the third a photorealistic virtual model

iconographic data intersect. Photorealism projects the observer into simulated environments, in which the reading and analysis of the spatial relationships between the components are facilitated. Consequently, the historical evolution that has marked the life of the object is readable, so are its material qualities, which often testify to the relationship between the object itself and the place. The photorealistic chromaticity of the images transmits the information as the certainty of the data, among other things acquired from the state of things, and pushes the observer to a more careful critical analysis. In the present case, the process uses digital photogrammetry for the construction of a metrically correct and photorealistic 3D virtual model of the state of things. The same model, with the Scan to Bim procedure, is then imported into a parametric information tool (H-BIM), in order to display both the narrated image and the one reworked from historical iconographic data. The model, deprived of the plot, has been simplified in its geometric conformation and modified on the basis of the acquired historical narrative (Fig. 4).

3.1 First Phase. The Modeling of the Narrated Object

On the first historical phase of the fortress, which, as previously mentioned, goes from its origins to 1221, we have only a written narration. Therefore, the graphic model created has the aim of transforming the linguistic sign into a visual sign. The model in this case is the synthetic reproduction of the narrated object and establishes an iconic relationship between the narrated reality and the image, according to an intuitive and immediate recognition code [7]. The narration is taken from the study of Leonardo Paterna Baldizzi according to which the fortress, located on an area that has a strong overhang to the north due to a topographical gorge, has an original nucleus with a quadrangular plan, (about 3400sm), defined by two sides of about 42mt and 66mt. The nucleus follows the line of the urban development of the underlying city of Cassino. It was born with an architectural structure similar to the early medieval fortification structures [8], and therefore it can be assumed that the system was geometrically regular, characterized by circular towers affixed to the corners, as evidenced by the only circular tower still visible, and rich battlements that run along the perimeter of the walls and on the towers themselves. The first expansion occurs with the abbot Mansone (985/996) who inserted a high central keep (about 30 m) into the structure. Abbot Gerardo (1111–1123) widens the fortress with the insertion of two smaller towers flanking a mighty pentagonal-based tower, whose geometry is very reminiscent of the more recent bastion systems. Inside, a small church was also built, dedicated to the Annunziata, below which there is a cistern, connected to a rainwater collection system, which guaranteed the military garrison's self-sufficiency in the event of a prolonged siege. The fortification activity, mainly due to the numerous sieges for the attribution of abbatial power, pushes Abbot Roffredo (1188/1210) to furnish the fortress with a mighty wall system equipped with buttresses (Fig. 5). The entire structure of the Rocca is in irregular masonry, of local calcarenite rock, with the exception of the pentagonal tower whose masonry consists of small regular ashlars, in Roman travertine, and mortar.

In 1235, after the demolition commissioned by Frederick II of Swabia, the fortress was rebuilt, by the will of the abbot Landolfo (1227–1236), on the structure that has been preserved until today.

3.2 *Second Phase. The Integrated Virtual Modelling*

The virtual modeling of the second phase, which includes the period from the reconstruction of 1235 to 1944, is based on iconographic data such as the relief drawings by Leonardo Paterna Baldizzi, which return a realistic image of the state of things up to 1907–1909. The historical data is however a graphic sign which, associated with the simplified monochromatic model of the narrated object, allows the observer to perceive the real phenomenon with greater awareness. The model is obtained with the integration of the two graphic structures, digital and raster. Background transparency has been applied to raster images:

1. "Tav. I-Rocca Janula presso Cassino—Iconografia—Scala di 1:100";
2. "Tav. II-Rocca Janula presso Cassino—Prospetto della Torre—Scala di 1:100";
3. "Tav. III-Rocca Janula presso Cassino—Sezione longitudinale della Torre—Scala di 1:100";
4. "Tav. IV-Rocca Janula presso Cassino—Prospetto posteriore della Torre- Scala di 1:100";

to allow the overlay reading with the digital model. Imported into the digital model, the raster images are appropriately scaled and positioned in relation to the views

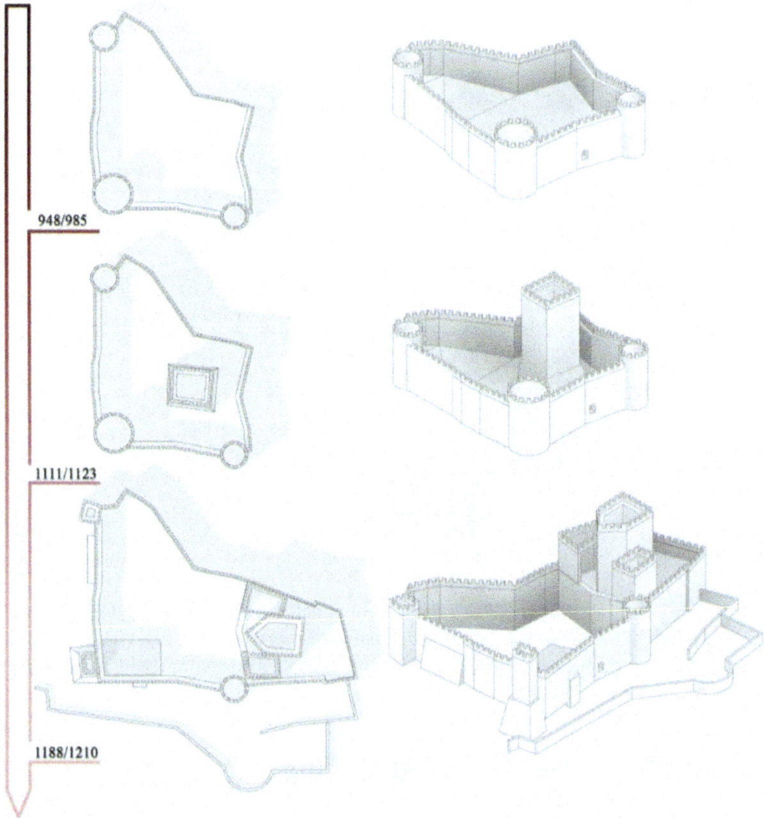

Fig. 5 Phase I. Told or narrated image. The model describes the historical evolution of Rocca Janula according to the narration of historical sources. The simplified model is not realistic so it does not confuse the observer who manages to understand that it is a hypothesis and not the state of things

contained (plan, elevation, section). This operation allows to digitize the architectural objects contained in the raster structure and compare them with those obtained from the narrated object. In this way, it is possible to highlight the architectural transformations undergone by the fortress. The use of an integrated raster/digital model projects the observer into a simulated dimension whose abstraction is partial since the images of the relief drawings by Paterna Baldizzi still give a realistic view of the fortress, by communicating the historical information lost over the subsequent decades (Fig. 6).

Fig. 6 Phase II. Reworked image. The model is obtained from the intersection of iconographic sources, which bring the observer closer to a realistic height, and a 3D virtual model, which, being monochrome, provides the observer with the correct information which is a hypothetical reconstruction

3.3 Third Phase. The Virtual Modelling

The various earthquakes and bombings of the Second World War had reduced the fortress to little more than a ruin, except for the pentagonal tower, which still rises today. The perimeter wall to the north-west has instead partially collapsed. Aiming to restore the fortress' splendor, the recent restoration, ended in 2015 completed some structural gaps in an uncertain work of local material, making the interventions hardly distinguishable, especially to less experienced eyes. It involved the trapezoidal area of the entire plant and the external arrangement of the fortress. The interventions carried out have also modified the geometry of some elements, for example, incorporating the surviving battlements in new walls, which have been raised for this purpose. In this case, the goal of virtual modeling concerns the possibility of correctly communicating historical information by distinguishing historically original elements (photorealistic texture) from those introduced by the restoration (compact gray) with different color choices. For this reason, the choice of modeling fell on photorealism using digital drone photogrammetry. The surveys were carried out according to the standards required by the execution of digital photogrammetry and the acquired images were processed as a cloud of points (Fig. 7).

According to the SCAN to BIM procedure, the point cloud was texturized, thus replicating the real object in its dimensions and in its material characteristics. This

Fig. 7 Phase III. Aerial digital photogrammetry: Flight paths and 3D cloud points reconstruction

virtualization that simulates reality increases awareness in the observer and promotes correct communication of information. This model implements, albeit virtually, the principle of distinguishability and reversibility at the basis of the principles of contemporary restoration, very often disregarded, frustrating the historical essence of the cultural asset.

This virtualization, with the application of the texture, replicates the real object in its dimensions and in its material characteristics and is, therefore, able to increase the awareness of the observer by promoting correct communication of information (Fig. 8).

Fig. 8 Phase III. 3D H-BIM model of the fortress designed using the point cloud obtained from aerial digital photogrammetry: overlapping of the two models

4 Conclusions

Rocca Janula is a fortification created for the Abbey of Montecassino and although it has experienced the most important historical events in our area from the early Middle Ages to today, it is little known except by local historians. This important symbol for the local community requires an enhancement process able to correctly convey the deep historical significance and that is easily communicated. For this purpose, the use of images is of great help. The history of the fortress is very ancient and according to the documentary and iconographic data, it has been divided into three phases. For each phase, a graphic model is identified to communicate, and therefore enhance, the historical information correctly. The first model, or the model of the narrated object, arises from a written narrative. It has no direct references to the state of affairs and therefore requires an abstraction from reality to avoid communicating inferred but untestable information. The second model, the Integrated virtual model, arises from the integration of the simplified digital model with raster images that reproduce, at least in the material and geometric characteristics, the real object. The integration, which partially restores the perception of reality, increases the observer's awareness and communicates historical information with greater simplicity. The third model, Virtual model, simulates reality. It is immediately understandable and for this reason, it is functional in transmitting historical information which is sometimes canceled or transformed with particularly invasive restoration interventions.

It is desirable that these models can be an integral part of the restoration interventions like the photographic companions provided for the ministerial directives. Future developments concern the implementation of a model that, on the basis of continuous digital photograms, dynamically collects information on all stages of the restoration.

References

1. Munari, B. (2020). Design e comunicazione visiva, GLF Editori Laterza, Bari.
2. Paterna Baldizzi, L. (1913) Rocca Janula nell'arte e nella storia, in Memoria della R. Acca-demia di Archeologia, Lettere e Belle Arti (Vol. II). Napoli.
3. Della Noce, A. (1668). Chronica Sacri Monasterii Casinensis, Lutetiae Parisiorum.
4. Toubert, P. (1973). Les structures du Latium Médiéval-Le Latium méridional et la Sabine du IXe siècle à la fin du XIIe siècle, Bibliothèque des Ecoles Françaises d'Athènes et de Rome, Ecole Française de Rome.
5. Pistilli, E. (2000). La rocca Janula di Cassino, Edizioni Cassino s.r.l. Cassino.
6. Arnheim, R. (2002). Arte e percezione visiva. Nuova versione (Vol. 23). Feltrinelli Edito-re, Milano.
7. Pelliccio, A. (2020). The place of brownfields. GIS & HBIM for their enhancement, Edizioni Efesto, Roma.
8. Coulson, C. (2004). *Castles in medieval society: Fortresses in England, France, and Ireland in the central Middle Ages*. Oxford University Press on Demand, Oxford.

Digital Heritage: Before, During and After COVID-19: The Aurelian Walls as a Case Study

Marco Canciani

Abstract The aim of this research is the study of a complex monument, such as the Aurelian Walls, and of some portions of it (gates, towers and portions of mu-ra), using digital technologies of augmented and virtual reality.

Keywords Digital Heritage · Aurelian Walls · Virtual reconstruction · Augmented reality

1 Foreword: Research in the Field of Digital Heritage, Before, During and After COVID-19

The relationship between an asset of archaeological and architectural Heritage as a physical entity and its digital copy, that is *the virtual reconstruction of its original state*, is quite complex. Tomàs Maldonado had already shown how, in 1994, with the advent of the computer age, the question of the relationship between real space and represented space, whether real or virtual, still had no solution and that, indeed, still presented even significant problematic aspects. The relationship real/virtual in the last two years, due to the Covid-19 pandemic, has undergone even greater changes, assuming characteristics so diverse as to present three very distinguishable phases with respect to the pandemic: before, during, and after.

This contribution proposes an initial series of considerations based on research carried out in the last decade in the scope of virtual reconstructions in the Digital Heritage context, and, in particular, on studies in the last two years, regarding a highly representative monument, the Aurelian Walls. The pandemic has, on the one hand, produced a sudden halt to establishing new research and, on the other, a profound rethinking of ones already underway.

In the period from 1990 to 2019, the progressive evolution of digitization in the field of DH is responsible for the development of an ever-increasing number of vast

M. Canciani (✉)
Department of Architecture, University of Roma Tre, Rome, Italy
e-mail: marco.canciani@uniro.ma3.it

© The Author(s), under exclusive license to Springer Nature Switzerland AG 2023
I. Trizio et al. (eds.), *Digital Restoration and Virtual Reconstructions*, Digital Innovations in Architecture, Engineering and Construction,
https://doi.org/10.1007/978-3-031-15321-1_6

libraries of digitized data or already input in digital format, in many areas of research and, in particular, in architectural and archaeological heritage. This has led to an accumulation of digital archives that often did not communicate with each other nor were predisposed for external access [6].

Due to the multiplicity of information, it is necessary to establish a shared and scientifically rigorous definition of the term "Digital Heritage". These follow the principles established in 2003 in the Charter for the Conservation of Digital Heritage and the definition of the FAIR principles of 2016,[1] which include not only the digitization of analog data but also the whole of digital content, data and metadata, 3D models of virtual reconstruction, interconnected via an open system and a new digital data model, consisting of LOD (Link Open Data)[2] (Fig. 1a, b).

From late 2019, in this period characterized by confinement and the impossibility of accessing the artistic, architectural and environmental heritage physically, digital technologies, dedicated almost entirely to virtual and remote use via internet, have had an unprecedented acceleration and generated a sort of "*digital gold rush*" [10]. This has led to a reorganization of digital archives and to a uniform method of validation of the processes of virtual reconstruction, following established principles and the creation of newly created links, through a complex network of relationships, made possible thanks to the web, together with the multiplication of countless online applications, to permit virtual access to the available digital content [Dal [11]].

Partially underway, with the pandemic under complete control, the question that one asks is: what will the use of the immense patrimony constituted by digital archives be when the visitor of a cultural site can return to a normal and complete mode in

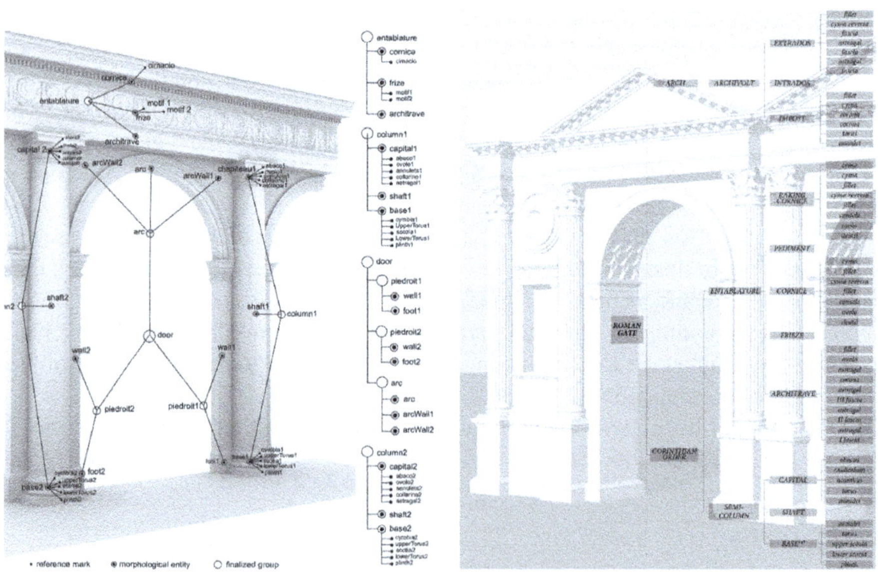

Fig. 1 **a** Semantic model of the Porta Aurea in Ravenna [12]. **b** Semantic model [8]

presen-ce? The aim of this study is an endeavor to contribute to the formulation of an answer that the sector ICAR17 can provide to the query, by reflecting on the importance of this **new digital revolution** that is leading to a sort of "disruptive innovation," capable of reformulating coexistence of the digital heritage with the physical heritage, experienced directly in situ, in which "*real reality merges with virtual reality*" [13].

2 Tree Phases in the Methodologies of Approach to Cultural Heritage

The methodology was developed through a line of research carried out since 2011, aimed at the study of innovative methods for the analysis, modeling, and virtual reconstruction of Cultural Heritage, put into practice and tested according to different levels of investigation and contexts, from the archaeological, to which the case study of the Aurelian Walls refers, to the urban and architectural.[3] This methodology includes three, intimately connected, procedures and implementation phases, which correspond to the phases mentioned above, both in terms of time sequence, procedures used, and type of data obtained.

2.1 First Phase—Acquisition of Digital Content, Elaboration, Virtual Reconstruction

In this first phase, the research was developed thanks to the creation of an interdisciplinary workgroup that included various scholars from the Department of Architecture of Roma Tre in the fields of Representation and Survey, History, Restoration, and Mathematics.[4] The data acquired according to the 3D survey procedures, now standardized,[5] were used to obtain three-dimensional digital models on various scales, from that of the urban fabric to that of the building and detail. These 3D models constitute the instrument by which it is possible to carry out the tasks of measurement, selection in distinct parts of the overall model, visualization according to established planes (horizontal and vertical sections), and orthographic views (elevations and axonometries), generation of additional specific geometric patterns and virtual reconstruction (Figs. 2 and 3).[6]

Multimedia data (historical documents, images, texts, other 3D models, etc.) are connected, by an Information System (IS) [9], to the 3D models, organized according to a database and a semantic structure, aimed at analyzing specific themes, such as the stratigraphic and material analysis or the virtual reconstruction relating to certain historical phases.

Fig. 2 Arch of Titus at the Circus Maximus: virtual reconstruction (Re-elaboration of M. Canciani)

Fig. 3 San Lorenzo in via Panisperna: virtual reconstruction at 1873 (Elaboration by M. D'Angelico)

2.2 Second Phase—Systematization and Development of Online Applications

In the second phase, researchers from the Department of Humanities of Roma Tre and ENEA[7] joined the first research group, with whom a study is developed aimed at systematizing data, following the LOD format, to validating the virtual reconstructions based on the semantic structure of the model and the definition of the level of uncertainty of the various elements which compose it [12, 15], and generating web applications, allowing data to be accessed remotely by different users, following required standards. This system, developed within the ECODIGIT project,[8] involves the use of dedicated applications, aimed at navigating virtual models enriched with semantic content, within which the user can navigate, explore and query in various way.

2.3 Third Phase—Development of Applications in AR

In this phase, the procedures are developed that allow the physical reality of the cultural asset to be connected with the virtual, consisting of multimedia digital contents (3D models, images, texts, etc.), in a kind of mixed reality—"a *partially immersive virtual environment preserving a certain quantity of reality*" [14, p. 5]—and certainly represents the most appropriate digital means of digital communication permitting great freedom of use by the observer as well as the interaction between physical and virtual space. The development of dedicated applications allows the user to experience, in presence, the heritage, constituted by the actual environment or by printed representation of the plan or section, augmented by digital content, *Mixed Reality* (MR), or *Augmented Reality* (AR).[9]

3 The Aurelian Walls Circuit: Studies and Research Before, During, and After Covid-19

The Aurelian Walls are one of the monuments most subject to studies of the Roman era, suffice it to cite [4, 5], and the proceedings "Le Mura Aureliane nella storia di Roma. 1. Da Aureliano ad Onorio" about 2017 [2]. They constitute the longest existing city walls, an imposing 19 km and originally included 381 rectangular towers and 14 gates (Fig. 4).

Constructed between 271 and 279 AD, under the emperors Aureliano and Probus, the walls reached a somewhat reduced height (8 m), with an uncovered defensive walkway, and crenellated towers at intervals of 30 m, in a rectangular crenellated and gates of three types. Starting from 401 AD, under Honorius, the walls underwent

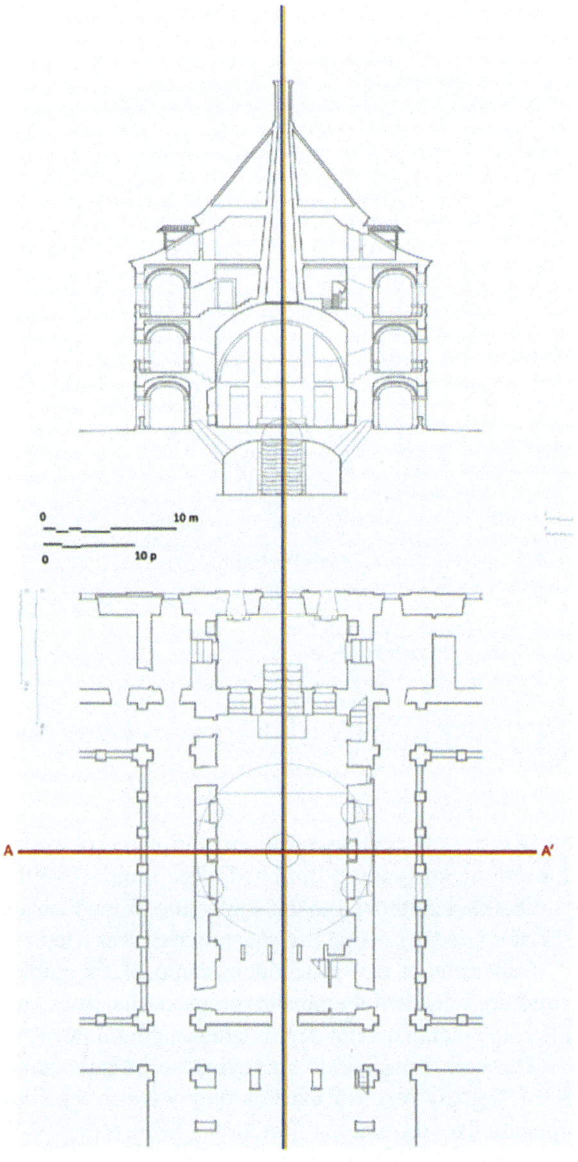

Fig. 12 Cross section AA' and main floor plan of the Convent's Kitchen area. (Chias, Abad and Fernández-Trapa 2021). The two symmetry axes are depicted in yellow and blue

and craftsmen in the Monastery. Skylights connecting cellars with cloister galleries are especially relevant because they were not studied and surveyed before. Likewise, skew openings under the kitchen sail vault open to the 15 ft level of the Infirmary cloister and the Guest cloister, facilitating kitchen ventilation (Figs. 6, 12 and 13).

Former uses and functions were successively replaced by new ones, and to meet the new requirements some refurbishments were undertaken.

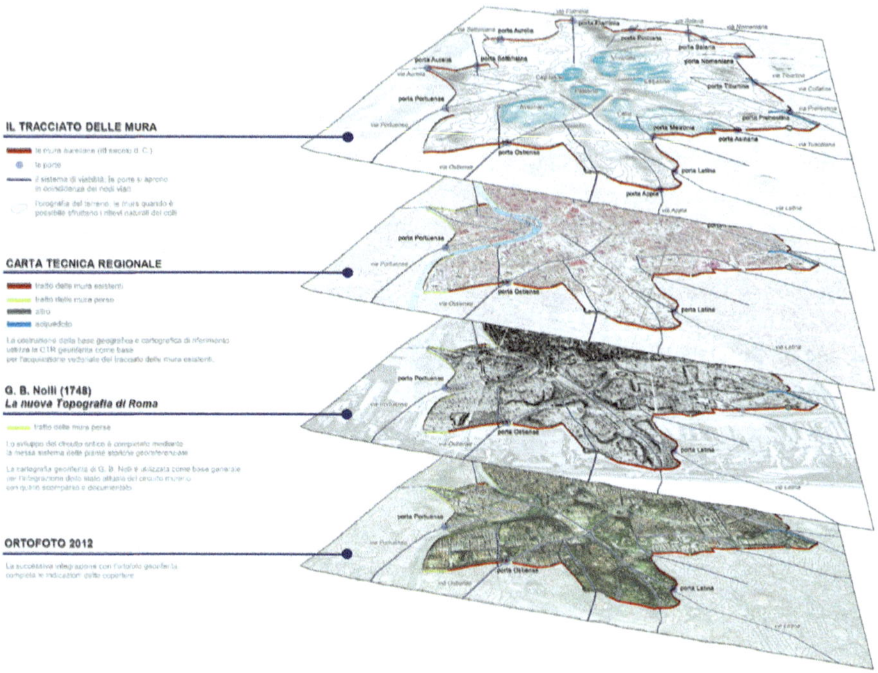

Fig. 4 Aurelian Walls: 3D maps with various information levels (Elaboration by M. Michelini)

profound modifications, reaching a total height of 18 m, the height of the towers was doubled, and the upper defensive gallery was vaulted and opened on the internal side.

In the centuries that followed, the maintenance and renovation of the walls and gates were carried out by papal administrations, with the addition of the ramparts in the sixteenth century and the architectural restructuring of many gates. The walls were kept as a border between urban Rome and the surrounding countryside until the reunification of Italy at the end of the nineteenth century, when the city began to expand further, and some gates were built according to significant architectural projects, while others were simply brutally truncated.

The elements studied in greater detail, consisting of two gates, two wall sections, and several towers, as detailed below, are integrated into the more general context of the walls (Fig. 5a–d).

(a) **Porta Latina**, one of the best preserved, despite the numerous modifications it has undergone, in particular under Honorius, is a portal of the second type of the Aurelian system (having a single archway with lateral semicircular towers and a still existing lateral covered rampart), presenting various characteristic elements (stone facing in travertine ashlars, the crenellated crowning at a higher elevated level double portal closing system, and an inner gate, towards the city, within which the staircase for access to the walkway rises.

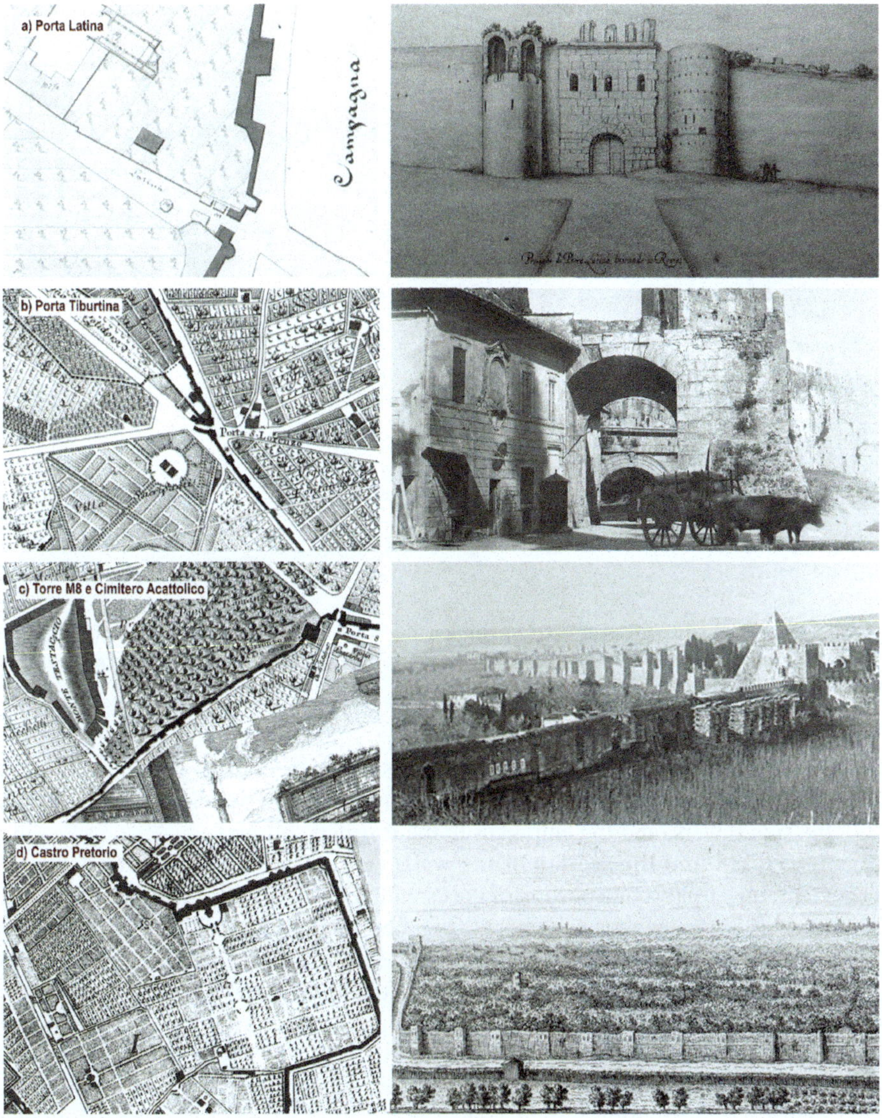

Fig. 5 Cartographies and Historical Photos: **a** Porta Latina; **b** Porta Tiburtina; **c** Tower M8 and Non-Catholic Cemetery; **d** Castro Pretorio (Re-elaboration by M. Canciani,)

(b) **Porta Tiburtina** is a unique monument, with a gate of the first type (*one fornix*), which incorporates the arch of Augustus, erected in 5 AD, with a wall structure allowing the aqueducts of the Marcia, Tepula, and Julia aqueducts to bypass over via Tiburtina. Under Aureliano and Honorius, the portal was endowed with side towers, a double door for customs duties, and a maneuvering room to operate

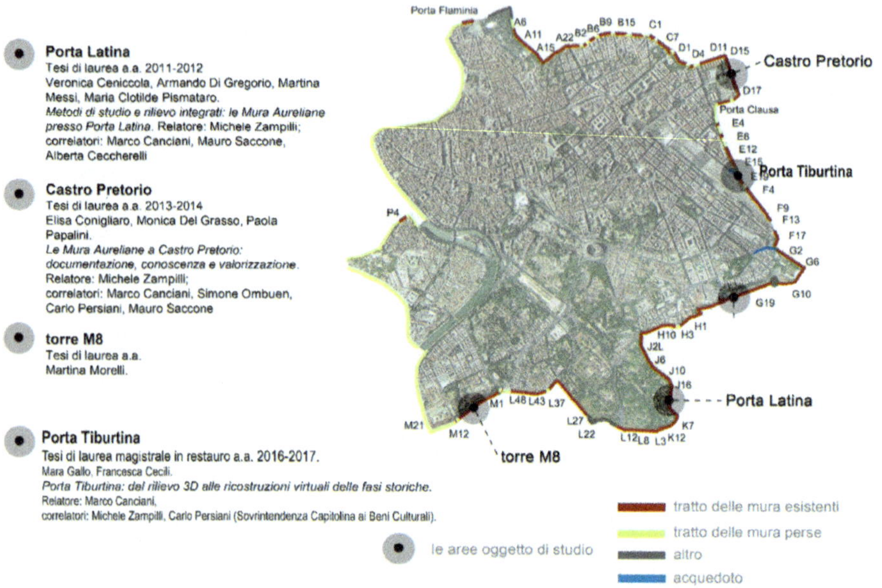

Fig. 6 Aurelian Walls with code references. Specific study areas: Porta Latina; Porta Tiburtina; Torre M8 and Non-Catholic Cemetery; Castro Pretorio (M. Michelini, Re-elaboration by M. Canciani)

the closing gate system. The major works took place under Nicholas V, in the fifteenth century, who improved the defensive system by reinforcing the walls, the side towers, and the entrances to the fortified complex.

(c) **Tower M8 and the section of the walls corresponding to the non-Catholic Cemetery** represent a portion of the "standard" walls, with construction characteristics homogeneous with respect to other sections. The wall structure, mainly in brick, is characterized by a series of additions and renovations visible on the wall face. The plan of the gate was that of the nearby Cestia pyramid which was preserved precisely because it became an integral part of the defensive system. The fortified enclosure, erected by Aureliano (271-279 AD) delimited the area of the river port; the towers, often truncated, instead, were subject of a fifteenth-century reconstruction commissioned by Nicholas V, and subsequently followed by restorations promoted by Julius III and Pius IX.

(d) **Castro Pretorio** represents a section of the wall that still maintains the original Roman planimetry, its design modeled after the castrum of the Roman legions, with the *cardo* and *decumanus* to structure the internal roadways and access ways through the walls. The construction, around the year 23, was composed of a system of buildings within a rectangular embattled enclosure with rounded corners, whose perimeter extended a total length of about 1.5 km. In the time of Aureliano and Honorius, there were significant adaptations to the wall structures, but the most radical changes took place in modern times, with the inclusion of the Macao Military Barracks and the National Library.

4 The Methodology Adopted for the Aurelian Walls

For the above reasons, the circuit of the Aurelian Walls represents an exceptional case study, given the complexity of the stratifications of the various historical phases, the imposing dimensions, and the variety of the elements of which it is composed, requiring the equally complex but, at the same time, flexible, methodology developed throughout the three phases described above.

4.1 First Phase (from 2011 to 2019)

In this first phase, a systematic study of the walls was undertaken which concerned, on a general level and by means of a GIS, the entire route of the section whether existing not, and, on a more specific level, the portions of Porta Latina and of the section between towers K7 and J16,[10] of Porta Tiburtina, of the section relating to Castro Pretorio and of tower M8 and the section corresponding to the Non-Catholic Cemetery. Some theses grew out of this as part of the five-year master's degrees, at the Department of Architecture, University of Roma Tre[11] as well as an interdisciplinary departmental research entitled "*The Aurelian Walls: knowledge, recognition, project*",[12] and some in-depth analyses developed in various publications. See references in the bibliography (Fig. 6).

The system involves the use of a platform based on a GIS (Geographic Information System) managed in collaboration with agencies for the protection of the territory, Superintendence of the Capitoline for Cultural Heritage, structured on two different information levels: one at the urban scale and the other the architectural scale. Both were developed using a shared data system that always allows connecting and reciprocal querying of the contents [1].

(a) The first level of information referred to the urban scale includes the entire extension of the walls and required the geo-referencing of the current and historical cartography in order to identify the sections of the walls documented but no longer existing. The system of relations is based on three information elements: the Portal, the Wall Section, and the Tower (Fig. 7).

(b) The second level, regarding the architectural aspects, common to the various specific objects of study, such as Porta Latina and Porta Tiburtina, has as its objective the construction of a three-dimensional model that permits, first of all, the creation of graphical drawings relating the various horizontal sections to each other and elevations showing the corresponding in situ sections (Fig. 8a–d).

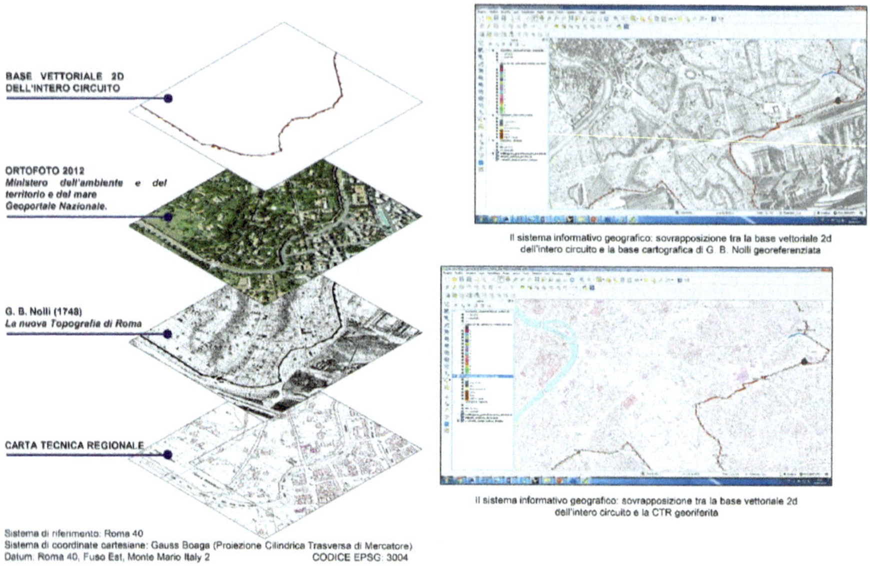

Fig. 7 1st level, GIS of the Aurelian Walls: Superimposition of the maps and screenshots of QGis (Elaboration by M. Michelini)

Fig. 8 2nd level, Survey 3D models with in situ sections: **a** Porta Latina; **b** Porta Tiburtina; **c** M8 Tower and Non-Catholic Cemetery; **d** Castro Pretorio. Elaborations by: **a** M. Michelini; **b** M. Gallo and F. Cecili, **c** M. Morelli; **d**.V. Cenniccola, A. Di Gregorio, M. Messi, M. C. Pismataro

In this phase, a procedure was developed to make the virtual reconstruction process rigorous, through the elaboration of three superimposed models: the triangulated mesh model of the 3D survey, the quadrangle mesh model, elaborated through a retopology process, and the geometric model of virtual reconstruction (Fig. 9). Subsequently, on these is developed an integrated reading of the data and metadata of the monument in its present state, thus obtaining, through an interpretation of the construction phases and a comparison with past hypotheses, a virtual reconstruction at specific historical phases.

In this phase, a method was tested that involves the use of an Information System, referring to various themes, such as chronological, material and state of preservation analysis, and allowing these analyses to be transposed from the two-dimensional processing to the corresponding superimposition on the 3D model (Fig. 10) [1, 2].

These studies were incorporated into the ECODIGIT project [7],[13] where prototypes of open GIS and 3D viewers were implemented, dedicated to remote and web access, capable of managing GIS data and 3D data, linked each other and semantic contents, such as images, documents, and others 3D models.

Fig. 9 Porta Latina: 3D model divided into three parts, on the left the mesh relief model with textures, in the middle the quadrangular mesh model, on the right the virtual reconstruction geometric model

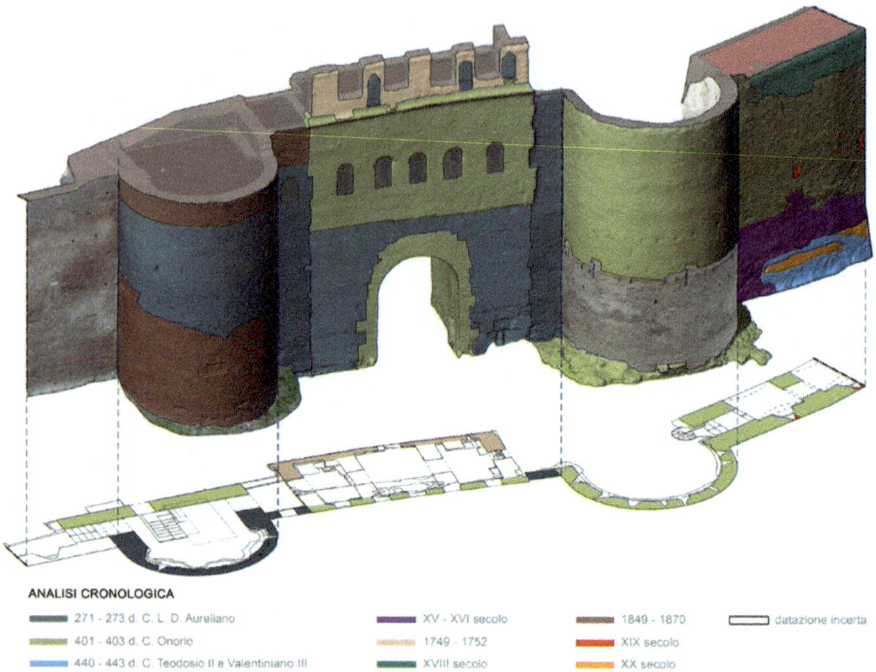

Fig. 10 2nd level, IS of Porta Latina: Superimposition of the chronological analysis on 3D model (Elaboration by M. Michelini)

4.2 Second Phase (from the Beginning of 2020 to 2021)

During the lockdown and confinement due to the pandemic, the research has made notable progress, analyzing the possibility of reprocessing the acquired data, which, not having been created with the aim of a uniform overall picture, either as a graphic instrument or as typologies, did not allow a transversal vision among the various parts composing the monument. The need to normalize the data according to the standards and principles recognized in the DH, meant that the data were reorganized according to the LOD format[14] and adapted to the needs of an open-source system, navigable on the net, accessible to more users, and updatable. This made it possible to use the same code for the various graphics, such as, in the case of the comparison between Porta Latina and Porta Tiburtina, the construction phases before and during the time of Honorius and, in the case of the sections of Castro Pretorio and the M8 Tower, the presence of the wall typology relating to the brickwork in alternate rows (Figs. 11 and 12).

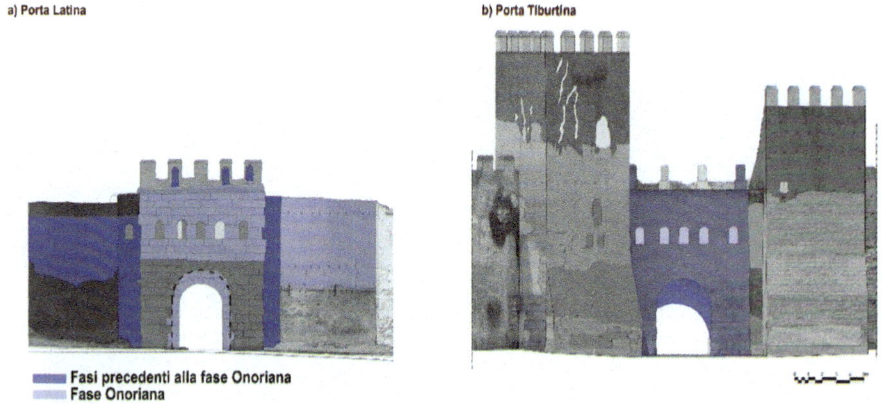

Fig. 11 Historical phases, Onorian and earlier: **a** Porta Latina; **b** Porta Tiburtina (Re-elaboration of Marco Canciani)

Fig. 12 Constructive phases (alternate bricks): **c** M8 Tower and Non-Catholic Cemetery; **d** Castro Pretorio. (Re-elaboration by Marco Canciani)

4.3 Third Phase (2021 to Today)

This phase of the research, carried out in the latest months of 2021 and still in an experimental stage, addressed the use of Augmented Reality applications to investigate the possibilities offered by AR systems for the enrichment of the Cultural Heritage with scientific contents and consisted of different levels of detail. The physical object of the Site is thus connected to additional digital information, consisting of three-dimensional content (3D survey models or virtual reconstructions), two-dimensional and multimedia content (images, texts, drawings), or content relating to the thematic

Fig. 13 Porta Latina: 3D virtual reconstruction superimposed on survey mode, using sections (Elaboration of Marco Canciani)

analysis produced. (constructional, material and historical development analyses, etc.) [15].

A specific study was developed *ex- novo*, by way of example, on the monument of Porta Latina, in two directions: on the one hand, to use 3D models in PDF format, extracting the vertical and horizontal in situ sections as overlapping elements of the survey model with the virtual reconstruction model (Fig. 13); on the other hand, based on the AR technique, to make the virtual reconstruction process rigorous and uniform, displaying the 3D survey model, the in situ sections and the virtual reconstruction model in a single navigable space at the time of Honorius with the survey model, overlaid on the horizontal section of the gate (Figs. 14 and 15).

5 Concluding Remarks

Many are the reflections elicited in the drafting of this text, also based upon the questions raised regarding the use of virtual models in the near future. First of all, it is important to highlight the necessity of systematization of the results of previous research and the formulation of future research to arrive at the constitution of digital archives that are compatible, integrable, and interchangeable using LOD data, following the principles, previously described and shared by now. Secondly, the study must not concern technologies end in themselves, but rather consider them as **containers**, within which **digital contents** are developed that constitute the true scientific value, which every researcher brings as disciplinary contribution and on which the ICAR sector17 can make significant contributions. From this point of view, the Aurelian Walls represent an exemplary case study.

Fig. 14 Porta Latina: 3D model of virtual reconstruction in Blippar (Elaboration of Marco Canciani)

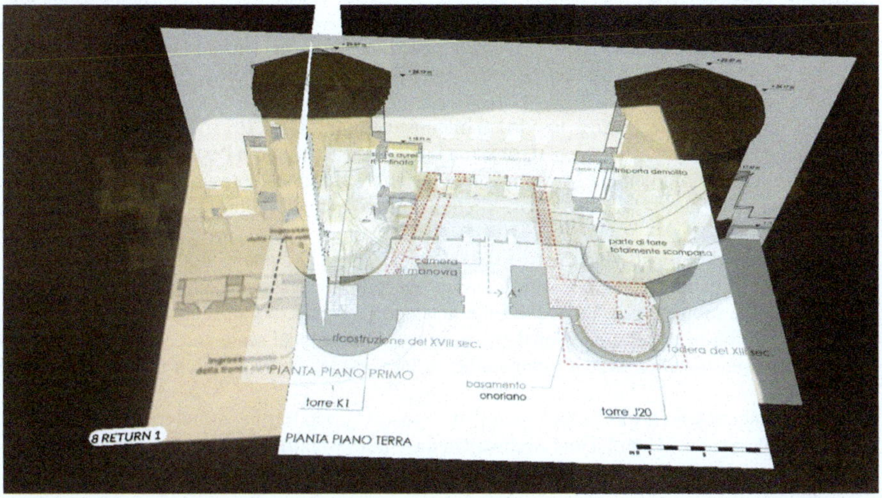

Fig. 15 Porta Latina: Augmented reality in Blippar of the virtual reconstruction, based on in situ sections and the print of the plan (Elaboration of M. Canciani)

Notes

1. The process of formalizing the criteria within the Digital Heritage is defined took a very long time, starting from 2003, with the Carta della Conservazione dei Beni Digitali (Charter of the Preservation of Digital Heritage), in which digital content was still defined as "information created digitally or converted into digital form from existing analogue resources, such as texts, databases, images, audio, graphics, software and web pages" (Carta della Conservazione dei Beni Digitali, 2003). In 2009, the London Charter makes it clear that the need was felt to establish a series of principles, given the increasingly intensive use of visualization methods, including: implementation, aims and methods, research sources, documentation, sustainability, and accessibility (London Charter, 2009). In 2011, these principles are referenced in the Seville Charter, dedicated to the use of technologies applied to archaeological sites, referring to Interdisciplinarity, Purpose, Complementarity, Authenticity, Historical Rigour, Efficiency, Scientific Transparency, Training and evaluation (Charter of Seville, 2011).The FAIR principles arise from a publication in the 2016 journal Scientific Data, by a consortium of scientists. From this publication, the GO FAIR Initiative was born, which is committed to the dissemination of these principles, now endorsed by the European Commission.
2. LOD (Linked Open Data) are meant as information having reciprocal links, organized according to a hierarchical structure (semantic), which organizes the objects, defined through ontologies and consisting of relevant entities, from the relations existing between them, from rules, axioms, and specific constraints of the domain [14].
3. These experiences are the result of research carried out in various fields, including: in archaeology: Arch of Titus at the Circus Maximus, Villa Adriana, Aurelian Walls; in urbanistics: Cave, Arquata del Tronto, Tivoli, Izalco (San Salvador); in architecture: San Carlino alle Quattro Fontane, Casamari Abbey, San Lorenzo in Panisperna. See references in the bibliography.
4. This workgroup, operating in virtual reconstructions since 2011, includes prof. Marco Canciani (coordinator) and prof. Giovanna Spadafora, arch. Maria Pastor Altaba, arch. Marco D'angelico, arch. Manuela Michelini, arch. Mauro Saccone (Representation and Modeling), prof. Saverio Sturm (History), prof. Michele Zampilli (Restoration), prof. Corrado Falcolini (Mathematics).
5. Using procedures involving the use of 3D laser scanners and photo-modeling methodologies and SFM (Structure from Motion) systems, not specified here (see the bibliography and specific texts on the subject), it is possible to obtain a cloud composed of millions of points, placed in their position in three-dimensional space, in real dimensional scale and equipped with color information, together with a triangulated surface model (mesh), which adapts to the point cloud.
6. A reference research was carried out by the research group for the virtual reconstruction in an archaeological context of Titus Archway in the Circus Maximus [12], and, in an architectonic context, the monastery at San Lorenzo in via Panisperna, absorbed by the structure of the Viminale [6].
7. In addition to the researchers of the Department of Architecture of Roma Tre, the research group involved: prof. Carla Masetti, Arturo Gallia (Department of Humanities), Maria Luisa Mongelli, Andrea Quintiliani, Marco Puccini (ENEA).
8. The original contribution of the research involves the reorganization of data for use on the Web. An example is the prototype developed of a GIS map viewer and the 3D models, connected to each other, that may be queried, and with multiple data banks (in particular, the Anagrafe Digital Library), created within the ECODIGIT project of the DTC of the Lazio Region between 2019 and 2020 (from ECODIGIT 11).

9. In this context, the study carried out for the 2020 research project D-Tech (*Digital-Twin Environment for Cultural Heritage*)#, aimed at developing services for sharing and viewing, on the web, digital copies of cultural assets and for the 2021 online REAACH conference [14] with a contribution focused on "*Augmented Reality as a research tool, for the knowledge and enhancement of cultural heritage*", constitute the theoretical bases for many experimental activities carried out in recent months, which also involved the study of the Aurelian Walls.
10. The numbering used is that adopted by the Superintendency, on that proposed by Cozza [4] and based on that of Richmond [3].
11. Porta Latina: Degree thesis a.y. 2011–2012. Veronica Ceniccola, Armando Di Gregorio, Martina Messi, Maria Clotilde Pismataro. Integrated study and survey methods: the Aurelian Walls at Porta Latina. Supervisor: Michele Zampilli; co-supervisors: Marco Canciani, Mauro Saccone, Alberta Ceccherelli.

    Castro Pretorio: Degree thesis a.y. 2013–2014. Elisa Conigliaro, Monica Del Grasso, Paola Papalini. The Aurelian Walls in Castro Pretorio: documentation, knowledge and enhancement.

    Supervisor: Michele Zampilli; co-supervisors: Marco Canciani, Simone Ombuen, Carlo Persiani, Mauro Saccone.

    Tower M8: Degree thesis a.y. 2015–2016. Martina Morelli. The M8 Tower of the section of the Aurelian Walls of the Non-Catholic Cemetery in Rome. Stratigraphic analysis of the walls.

    Supervisors: Marco Canciani, Paola Porretta, Francesca Romana Stabile.

    Porta Tiburtina: Master's degree thesis in restoration a.y. 2016–2017. Mara Gallo, Francesca Cecili. Porta Tiburtina: from 3D relief to virtual reconstructions of historical phases.

    Supervisor: Marco Canciani, co-supervisors: Michele Zampilli, Carlo Persiani (Capitoline Superintendence for Cultural Heritage).
12. Research funded by the Department of Architecture of the University Roma Tre for the years 2013 and 2014, carried out under the Convention between the same department (coordinator prof. Michele Zampilli) and the Superintendence of Cultural Heritage of Roma Capitale. To the research participate professors of the disciplines of Archaeology, Restoration, History, Surveying, Structures, Urbanism.
13. The EcoDigit project "Digital Ecosystem for the fruition and valorisation of Lazio's cultural heritage's and activities" is one of the initiatives of the Centre of Excellence of the Technological District for Cultural Heritage and Activities (DTC) of Lazio. The Work Pachage 4 "The system includes not only the connection between GIS and 3D data but also with open public data of Lazio, such as "Anagrafe" and Digital Library. Work Group for Work Package M. Canciani, coordinator, M. Saccone.
14. LOD, or Linked Open Data, provides information in multimedia format, including GIS and 3D data, organised in fields, which link from one piece of data to another.

References

Aurelian Walls

1. Canciani, M., Zampilli, M., Michelini, M., Saccone, M., & Scortecci, A. (2017). In Atti del Primo convegno Roma, 25 marzo 2015, *Le Mura Aureliane: dal rilievo 3D al GIS. Le Mura Aureliane nella storia di Roma. 1. Da Aureliano ad Onorio* (pp. 193–207). Roma: Edizioni Roma TrE-Press.

2. Canciani, M., Zampilli, M., Persiani, C., & Saccone M. (2017). Due casi di studio: Porta Latina e Castro Pretorio. In Atti del Primo convegno Roma, 25 marzo 2015, *Le Mura Aureliane nella storia di Roma. 1. Da Aureliano ad Onorio* (pp. 209–231). Roma: Edizioni Roma TrE-Press.
3. Richmond, I. A. (1930). *The city wall of imperial Rome*, Oxford.

Digital Heritage

4. Bianchini, C., Attenni, M., & Potestà, G. (2021). *Regenerative design tools for the existing city: HBIM potentials.* In Rethinking Sustainability Towards a Regenerative Economy (pp. 23–43).
5. Brusaporci, S., & Trizio I. (2014). *La Carta di Londra e il Patrimonio Architettonico: riflessioni circa una possibile implementazione.* In *SCIRES-IT scientific research and information technology-ricerca scientifica e tecnologie dell'informazione* (Vol. 3, Issue 2).
6. Canciani, M., et al. (2020). Modelli 3D e dati GIS: una loro integrazione per lo studio e la valorizzazione dei beni culturali. In *ARCHEOMATICA* (Vol. XII, no. 2, pp.18–23).
7. De Luca, L., Busayarat, C., Stefani, C., Véron, P., & Florenzano, M. (2011). A semantic-based platform for the digital analysis of architectural heritage. *Computers & Graphics, 35*(2), 227–241.
8. Maldonado T. (1994). *Reale e virtuale.* Milano: Feltrinelli editore.

Before, During and After COVID-19

9. Concas, A. (2020). L'arte post Coronavirus: Ripartire con il digitale: le strategie per i professionisti dell'arte. Segrate: ed. Piemme.
10. Dal Pozzolo, L. (2021). Il patrimonio culturale tra memoria, lockdown e futuro (ed. Italiana). Editrice Bibliografica.

Virtual Reconstruction

11. Apollonio, F. I., & Giovannini, E. C. (2015). *A paradata documentation methodology for the Uncertainty Visualization* in digital reconstruction of CH artifacts, «SCIRES-IT» (Vol. 5, pp. 1–24).
12. Canciani, M., Chiappetta, F., Falcolini, C., Michelini, M., Pastor Altaba, M., & Scortecci, A. (2019). A methodology for virtual reconstruction in different archeological heritage contexts. In *Der Modelle Tugend 2.0–Vom digitalen 3D-Datensatz zum wissenschaftlichen Informationsmodell* (pp. 479–496), Mainz: HochSchule.

Augmented Reality

13. Bonacini, E. (2014). La realtà aumentata e le app culturali in Italia: storie da un matrimonio in mobilità. In *Il Capitale Culturale. Studies on the value of Cultural Heritage* (No. 9, pp. 89–121).

14. Canciani, M., Spadafora, G., Camassa, A., & Saccone, M. (2021, in stampa). Augmented reality as a research tool, for the knowledge and enhancement of Cultural Heritage. In A. Giordano, M. Russo, & R. Spallone (Eds.), *Representation challenges: Augmented reality and artificial intelligence in Cultural Heritage and innovative design domain (REAACH), FRANCO ANGELI*, on line ed.
15. Pierdicca, R., et al. (2018). Un'indagine sulla realtà aumentata, virtuale e mista per i beni culturali. *Journal on Computing and Cultural Heritage (JOCCH), 11*(2), 1–36.

The “Amiternum Project” on Archaeological Site Valorisation

Stefano Brusaporci, Alfonso Forgione, Fabio Graziosi, Fabio Franchi, Silvia Mantini, Pamela Maiezza, Alessandra Tata, and Luca Vespasiano

Abstract In 2020 the University of L’Aquila bought the archaeological site of Amiternum, near the city of L’Aquila, where a series of excavations have been carried out from 2012. Through the so called “Amiternum Project”, the aim of the University is to create an international archaeological school-camp, and develop research activities for an adequate conservation, protection and valorisation of the site. The paper presents the first activities of an interdisciplinary research group of L’Aquila University—composed by archaeologists, historians, heritage scholars, ICT researchers—focused on the study of methodologies and tools for site interpretation, presentation, and narration, in particular through the use of digital approaches.

Keywords Archaeological heritage · Interpretation · Presentation · Public history · Digital survey · 3D modelling · Mixed reality · 5G

1 Introduction: The Amiternum Project

Since 2012, the University of L’Aquila has carried out a series of excavation campaigns at the archaeological site of the Roman city of Amiternum, located near the city of L’Aquila. These have unearthed the monumental remains of what was in all probability the first cathedral of the Diocese of Amiternum. In 2020 the University of L’Aquila bought the site where there are the ruins of the ancient cathedral of Amiternum, to create a permanent international archaeological school-camp, and develop research activities for an adequate conservation, protection and valorisation of the site (Figs. 1, and 2). In particular, the University has launched an interdisciplinary project aimed at developing national and international scientific and educational collaborations, both oriented to the excavation and discovery of new finds, and to the experimentation of advanced digital technologies, finally to the enhancement of the site itself, favouring its access, visit, use in presence and online. On the basis of

S. Brusaporci (✉) · A. Forgione · F. Graziosi · F. Franchi · S. Mantini · P. Maiezza · A. Tata · L. Vespasiano
L’Aquila University, L’Aquila, Italy
e-mail: stefano.brusaporci@univaq.it

© The Author(s), under exclusive license to Springer Nature Switzerland AG 2023
I. Trizio et al. (eds.), *Digital Restoration and Virtual Reconstructions*, Digital Innovations in Architecture, Engineering and Construction,
https://doi.org/10.1007/978-3-031-15321-1_7

the interdisciplinary study of the findings, and on the digital survey, critical diachronic 3D reconstructions of the basilica are proposed. Such models are displayed in virtual reality and mixed reality, also by experimenting 5G telecommunication applications.

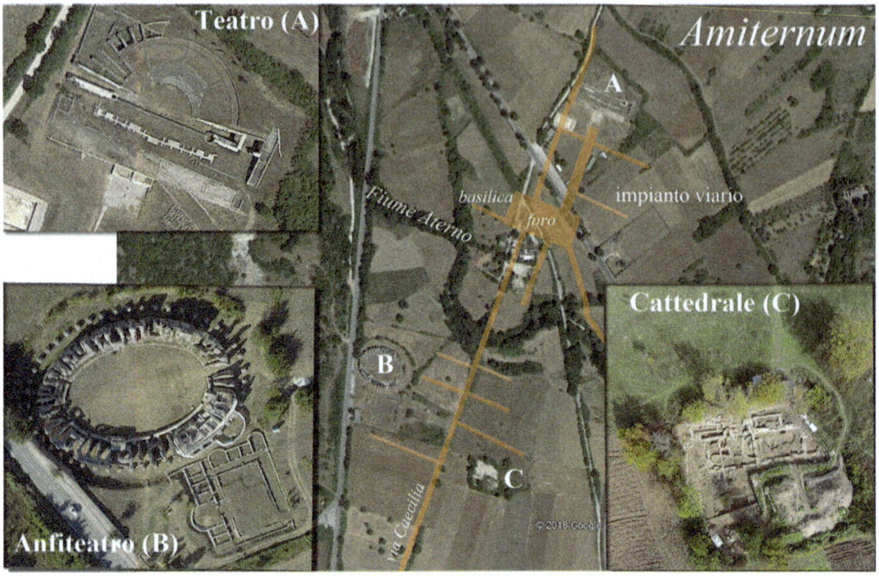

Fig. 1 The archaeological site of Amiternum: **a** the theatre; **b** the amphitheatre; **c** the cathedral

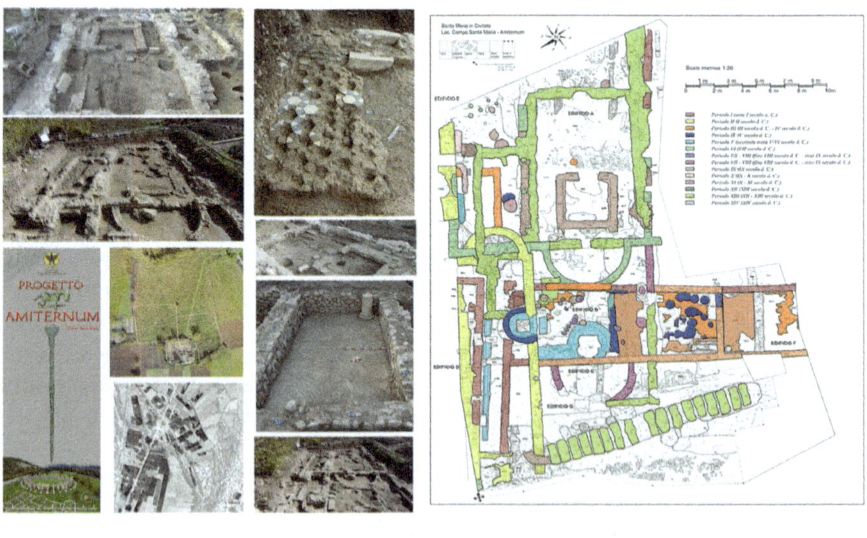

Fig. 2 Panel of the project with images of the archaeological excavations

As part of the general objectives, the paper presents the first activities of the research group interested in the study of methodologies and tools, primarily digital, for surveying, modelling, and visualization—both in Virtual Reality (VR) and Mixed Reality (MR)—for interpretation and presentation of the site. A reference to the theoretical-methodological lines set by the field of the so-called "Public History" is provided for storytelling processes, for the purpose of a useful use and enhancement of the site.

2 The Site

Eight excavation campaigns were conducted at the site of the ancient Amiternum. The analysis of the conspicuous excavated material has contributed substantially to identifying at least 14 periods of activity. The structures, belonging to at least 7 buildings, can be dated between the first century BC. and the late fourteenth century. The identified buildings, that are the result of successive additions and superimpositions, have occupied the excavation area on several occasions, incorporating or only partially reusing the previous structures, confirming the centripetal role of the cathedral within the city fabric, starting from the late fifth century. The oldest settlement found during the investigations consists of a domus from the Roman Republican period, which evolved and changed during the Imperial period. During the phases of the Christianization of the region, the site is of interest to the bishop who decides to establish the first religious settlement there. During the late antique period, the site came the main pole of the area, the real centre of the political and economic power of the territory. The baptismal building of that period has been currently identified. After the Lombard conquest of the territory, a large three-nave cathedral measuring 18 × 27 m was built, and it was subsequently transformed under Carolingian and Ottonian rule, when it lost its episcopal dignity in favour of the nearby city of Rieti, another nearby city. With the arrival of the Normans and during the Swabian domination, the ancient cathedral evolved further, transforming its structures once again. The history of the site is therefore closely connected to that of the entire territory and specifically to that of L'Aquila which inherited its diocese and its role as a pole of power (Fig. 3) [1, 2].

The digital survey campaign of the site was carried out using laser scanning technologies. Specifically, the instrument employed was the phase shift scanner Faro Focus S70, equipped with an integrated colour camera and characterized by a field of view (FOV) of 360° × 300°. The aim of the survey was to obtain a complete point cloud of the excavation with a homogeneous point density. To this end, 13 scan stations were selected, suitably positioned to minimize shadow areas. As scanning parameters, we chose to adopt a resolution corresponding to a point density of 7.7 mm at 10 m away from the instrument (resolution at 1/5) and a 3× scan quality. In addition, to improve the acquisition of the colour data of the points, it was decided to take advantage of the possibilities offered by HDR (High Dynamic Range) by setting 3 exposures for each scan. The processing of the acquired data was carried

Fig. 3 Aerial view of the excavation area of the ancient cathedral

out through the Scene software, within which the individual scans were combined through automatic registration procedures in order to obtain the total point cloud of the archaeological site of Amiternum (Fig. 4).

Fig. 4 Point cloud of the excavation site of the ancient cathedral

3 Between Interpretation, Presentation and Public History

"The ICOMOS Charter for the Interpretation and Presentation of Cultural Heritage Sites" (2008) highlights the importance of public communication as an essential part of the larger conservation process, according to the idea that "every act of heritage conservation is by its nature a communicative act" (p. 2). The Charter explains:

> Interpretation: "refers to the full range of potential activities intended to heighten public awareness and enhance understanding of cultural heritage site. These can include print and electronic publications, public lectures, on-site and directly related off-site installations, educational programmes, community activities, and ongoing research, training, and evaluation of the interpretation process itself." (p. 4).

> Presentation: "more specifically denotes the carefully planned communication of interpretive content through the arrangement of interpretive information, physical access, and interpretive infrastructure at a cultural heritage site. It can be conveyed through a variety of technical means, including, yet not requiring, such elements as informational panels, museum-type displays, formalized walking tours, lectures and guided tours, and multimedia applications and websites." (p. 4).

The charter roots on the concepts already presented by Tilden in 1957 in his essay "Interpreting Our Heritage" [3]. He starts from a dictionary definition: Interpretation is "An educational activity which aims to reveal meanings and relationship through the use of original objects, by first-hand experience, and by illustrative media, rather than simply to communicate factual information" (p.8). Tilden proposes six principles:

> I. Any interpretation that does not somehow relate what is being displayed or described to something within the personality or experience of the visitor will be sterile. II. Information, as such, is not Interpretation. Interpretation is revelation based upon information. But they are entirely different things. However, all interpretation includes information. III. Interpretation is an art, which combines many arts, whether the material presented are scientific, historical or architectural. Any art is in some degree teachable. VI. The chief aim of Interpretation is not instruction, but provocation. V. Interpretation should aim to present a whole rather than a part, and must address itself to the whole man rather than any phase. VI. Interpretation addressed to children (say, up to the age of twelve) should not be a dilution of the presentation to adults but should follow a fundamentally different approach. To be at its best it will require a separate program (p. 9).

The concepts of interpretation and presentation, rooting on public communication, also relate to the line of the so-called "Public History". This wording refers to a discipline dedicated to telling the history in public and with a close relationship with the public. In particular, the public history aims to share a "public sense" of history in order to foster a public awareness of one's own past. It follows a reflection on the methods of communication in museums, exhibitions, sites, etc. Clearly this is a fundamentally interdisciplinary activity, which has close contacts with the so-called Digital History [4, 5] and with Heritage Education [6].

A conference dedicated to Public History was held in Aquila, a city severely affected by an earthquake in 2009, where, in particular, the theme of the telling of the history referred to built heritage was reflected [7]. The research group of the

University of L'Aquila, now focused on the theme of Amiternum, has been working since a long time on the topic of the use of digital technologies for interconnected and participatory practices of interpretation and presentation of the history and memory of places and cultural heritage, especially through the use of virtual and augmented reality visualizations of past configurations of sites and monuments, in order to promote understanding and enhancement of the current state of the sites [8–10].

4 Visualizing Archaeology

The archaeology field has always dealt with the issues posed by the visualization of sites, buildings and finds, in their current state and of the interpretative hypotheses relating to their past configurations—and possible use-, i.e., reconstructions both related to the physical characteristics and to the cultural context.

With the development and diffusion of three-dimensional digital modelling technologies, archaeology was the first to experiment with the areas of virtuality for the visualization of finds and the reconstruction of configurations that no longer existing [10], and consequently developed reflections on the opportunities offered by this methodology [11, 12]. Subsequently, many scholars of various disciplines have dealt with related issues [13, 14].

"The London Charter" (2009) [15] and the "Principles of Seville" (2012) [16] were born from the field of archaeology. The first one, of general and interdisciplinary value, says "The London Charter seeks to establish principles for the use of computer-based visualisation methods and outcomes in the research and communication of cultural heritage in order to: Provide a benchmark having widespread recognition among stakeholders; Promote intellectual and technical rigour in digital heritage visualisation; Ensure that computer-based visualisation processes and outcomes can be properly understood and evaluated by users; Enable computer-based visualisation authoritatively to contribute to the study, interpretation and management of cultural heritage assets; Ensure access and sustainability strategies are determined and applied; Offer a robust foundation upon which communities of practice can build detailed London Charter Implementation Guidelines" (p. 4).

In accordance with the principle of "Implementation", the "Principles of Seville" proposes guidelines for archaeology. Of particular importance is the concept of "Transparency": "All computer-based visualization must be essentially transparent, i.e. testable by other researchers or professionals, since the validity, and therefore the scope, of the conclusions produced by such visualization will depend largely on the ability of others to confirm or refute the results obtained" [16, p. 8]. In this sense, the "paradata" are intended as a sort of "scholia", that is annotations and glosses that should accompany the modelling and visualization procedures to make explicit the scholar's critical choices [17]. In parallel, the issue of digital heritage has taken on increasingly important implications over the years [18, 19], also in relation to the participatory evolution of the concept [20–22]. Maurizio Forte, reflecting on the theme of digital reconstruction, specifies: "every simulation or virtual reconstruction will have its validation process in the coherence of the method and not in the truthfulness of the result [...] it does not matter that there is something objective because objectivity does not exist when one focuses on the simulations of the past. What matters is that my product, my reconstruction [...] is plausible according to the parameters that were used, that the premises, discussion and conclusions are sustainable and coherent" [23].

According to the excavation campaign and the consequent studies, as well as the surveying activities, a 3D model of the cathedral referable to the ninth century has been realized. From the analysis of the material remains and the archaeological analysed stratigraphy, therefore, it was possible to go back to a basilica with three naves with a semi-circular apse. The dividing arches of the three naves were set on two semi-pillars clamped to the rear wall of the building, then marked by a sequence consisting of two columns, a rectangular pillar 50 cm wide by 200 long and, very presumably awaiting the continuation of the investigations, from two other columns and a half-pillar clamped to the counter-façade. The scheme just described is extremely similar, among other things, to the Lombard phase of the cathedral of Salerno and other central-southern religious complexes.

This is the phase attributable to the Lombard cathedral modified during the Carolingian period with a presbytery enclosure aligned with the central nave and in front of the apse and the altar. It was decided to study this specific phase in more detail as it is the best preserved and revealed by the archaeological investigations still underway. The model, although elementary in its form, is useful for study purposes to better understand the original volumes of the surviving architecture and its contextualisation with the landscape and the ruins of a city definitively abandoned.

The model was developed from the geometric and dimensional data of the survey realized using laser scanner and based on the interpretative schemes of the site, processed following the excavation operations. The information available are essentially planimetric, therefore it is necessary to clarify that for the modelling in elevation it was necessary to proceed to a critical reconstruction, based on the interpretation of the only planimetric traces. However, the spatial layout with three naves, and the geometry of the apse, resulting from the excavation, identify a well-known ecclesiastical typology of a certain diffusion, and it was therefore considered to proceed by analogy.

To control the results, a VR headset was employed in navigation mode, directly within the model environment (Fig. 5).

For the next "presentation" step, the Unity rendering engine has been used (Fig. 6). The choice of this solution has been made in the frame of a workflow in progress of experimentation that it intends to systematize the process of elaboration from the phase of survey with digital technologies (carried out indifferently with laser scanner, photogrammetry, or SLAM) design of 3D models, up to the publication of virtual models explorable in MR. In this sense, the use of a rendering engine such as Unity allows an absolute versatility with respect to devices, technologies, and experiences: The experimental model has been implemented for Windows (.exe) and Android (.apk) environments and the Oculus headset compatible version is being developed

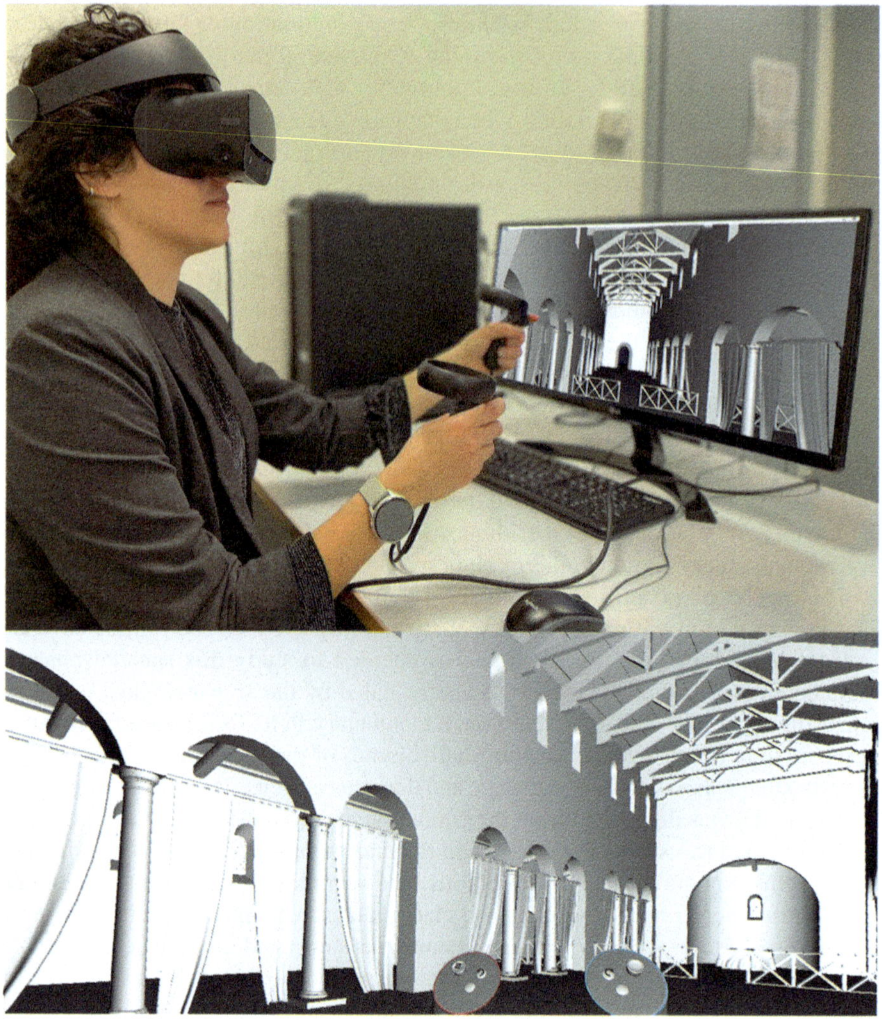

Fig. 5 VR interactive navigation of the 3D model of the ancient cathedral

(Fig. 7). This possibility offered by Unity to use the same model for export to different operating systems allows to optimize its use on different types of devices and consequently lends itself to experiences with different degrees of immersivity, from the pure VR experience to different degrees of AR. An alternative application, in the AR field, is the one experimented using the Sketchfab platform, and its AR service: the experience, limited to singular architectural elements, allows to visualize through smartphone, virtually modelled elements in the physical environment, with the possibility of moving around them. This application is designed for on-site experience, to allow to read in elevation traces currently detectable only planimetrically.

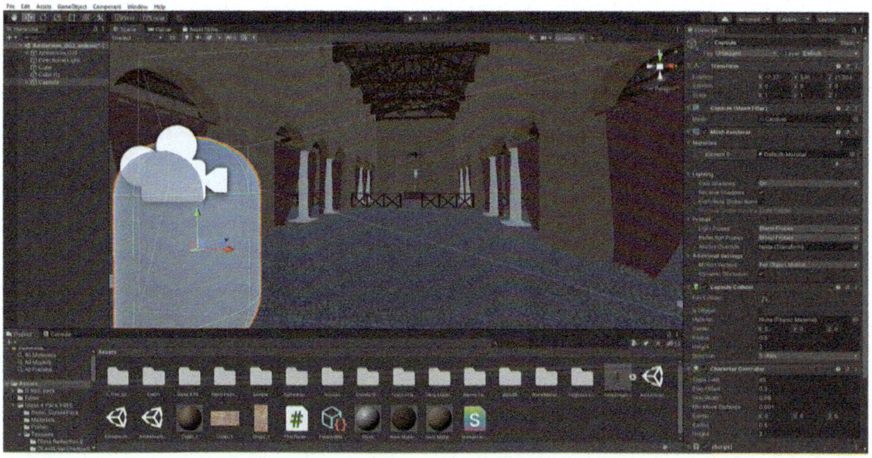

Fig. 6 Screenshots of the Unity interface of the model of the cathedral

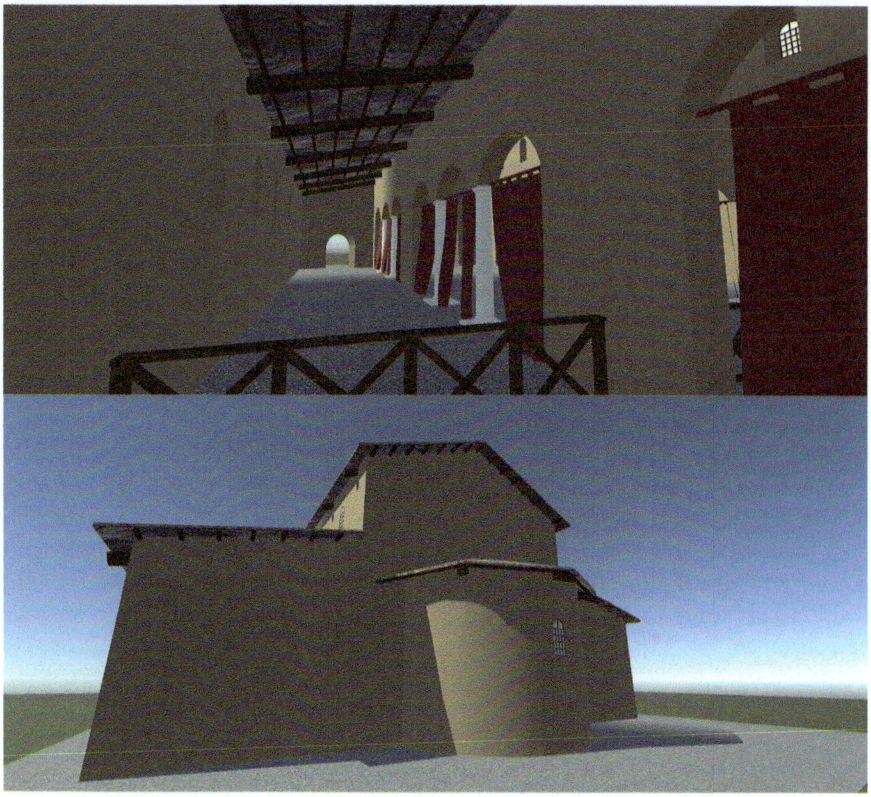

Fig. 7 Screenshot of the .apk application during a test on smartphone

5 Mobile AR for in Site Heritage Visualization

ICT solutions are very appealing for cultural heritage since they can contribute to the study, the disclosure, the promotion involving both the specific sector and the whole community.

The activities related to the area of "Campo Santa Maria" ad Amiternum close to the city of L'Aquila refer to the creation of services for the safeguard, preservation, promotion, and enjoyment of the regional cultural heritage (material and immaterial, historical, architectural, artistic, and naturalistic) and the promotion of innovative politics for tourism support. The initiative is concerned with the sites' heritage which is composed by cultural assets under preservation, restoration, promotion and safeguard and cultural activities contributing to the cultural industry.

The under deployment 5G network will be an enabler for experiential and sustainable tourism models for the enjoyment of cultural heritage with the possibility of experiencing increasingly engaging and personalized travel experiences directly from tourists' mobile devices (smartphones, tablets, headsets), without latency and availability issues, thanks to applications specifically developed in Augmented Reality (AR). Furthermore, the 5G network would represent a formidable tool in the hands of conservators, restorers and art historians allowing the use of diagnostic tools and their results for the proper conservation of cultural heritage in real time and in-situ.

The fifth generation of cellular technology (5G) will provide the computational and storage resources with very low latency: tourists will be able to "immerse themselves" totally in the reality discovering cultural and artistic events and information, as well as commercial and promotional information; the same will happen for conservators, who will be able to compare results of different diagnostic techniques and/or carried out at different times, as well as other useful data (e.g. historical reconstructions, documents, drawings and so on). The applications designed within the project allow the users to enrich the visit experience, simply targeting with the proper device, the observed asset. Thanks to the AR the viewport will be enriched by information and images (in addition to audio content) which will greatly increase the quality of the visit experience and provide elements otherwise not available "on site".

The 5G component will not be limited only to the data transfer but will be used in its whole levels using the storage and computational resources made available in the distributed nodes of the network. The idea is to use the different levels of location accuracy to identify the contents to be moved from the cloud, passing through MEC (Multiaccess Edge Computing) nodes up to the user devices from which they will be consumed [24, 25].

6 Conclusions

The paper presents the first results of the activities of a team of scholars of archaeology, built heritage, digital sensing, 3D modeling, and telecommunications. The research roots on the studies conducted within the INCIPICT project (http://incipict.univaq.it/) for the valorisation of cultural heritage through ICT and the research dedicated to the use of 5G technologies.

The interdisciplinary collaboration between scholars from different disciplines, through the joint development of a specific case study dedicated to the virtual reconstruction of the cathedral, fostered a theoretical-methodological reflection on the remediation of archaeological data, how to narrate the site, its history, characteristics and values, in particular through visual processes.

Traditionally, each discipline refers to its own methodologies, but interdisciplinary collaboration favours an applicative comparison, the identification of areas of overlap, the sharing of approaches and research tools. In particular, a trans-disciplinary reflection is encouraged by the common objective of developing multimedia storytelling paths, online and on site.

A specific objective of the present experiment is to test the potential offered by digital technologies in the context of the specific case study in order to prefigure a framework for the documentation, interpretation, presentation, enhancement and use of archaeological sites. To this end, based on the digital survey, 3D models relating to the diachronic phases of the basilica are created to build an adequate narrative of the finds and buildings and their values, understood both in the tangible and intangible dimension. The models are displayed in VR and MR overlapped to site excavations. According to this experience, future lines of research are the design of a valorisation program of the whole site, and the validation of the visual communication through multimedia interpretative models with of evaluation questionnaires dedicated to users of different types.

This activity combines with the general intent of the University to create in this site a virtuous example of cultural welfare. In this sense, the archaeology—conceived as a health genesis approach—would turn into a useful tool for preventing social and cognitive decay, through heritage education, inclusion and active involvement of the population.

References

1. Forgione, A., & Savini, F. (2019). Amiternum (AQ), Campo Santa Maria: dinamiche insediative e fasi sepolcrali di un nuovo polo di potere. Nuova sintesi delle ricerche in corso, Archeologia Medievale XLVI (pp. 197–232).
2. Forgione, A. (2021). Dalla terra alla storia: l'archeologia come motore di progresso sociale. In A. Hopkins (Ed.), *L'Aquila storia della città e del territorio* (pp. 69–82). Anicia, Roma.
3. Tilden, F. (1957). *Interpreting our heritage.* Chapel Hill: University of North Carolina Press.
4. Weller, T. (Ed.). (2013). *History in the digital age.* Abingdon: Routledge.
5. Mantini, S. (2019). Experiences of public history and ICT for the representation of cultural heritage. *DISEGNARECON*, *23*(12), 16.1–16.4.
6. Casonato, C., Vedoà, M., & Cossa, G. (2021) *Scoprire il paesaggio del quotidiano.* Lettera Ventidue, Siracusa.
7. Mantini, S. (Ed.). (2020). *Ricostruire storie.* Editoriale Scientifica, Napoli.
8. Brusaporci, S., Graziosi, F., Franchi, F., & Maiezza, P. (2019). Remediating the historical city. Ubiquitous augmented reality for cultural heritage enhancement. In: A. Luigini (Ed.), *Proceedings of the 1st International and Interdisciplinary Conference on Digital Environments for Education, Arts and Heritage. EARTH 2018* (pp. 305–313). Cham: Springer.
9. Brusaporci, S., Graziosi, F., Franchi, F., Maiezza, P., & Tata, A. (2021). Mixed reality experiences for the historical storytelling of cultural heritage. In: C. Bolognesi & D. Villa (Eds.), *From building information modelling to mixed reality* (pp 33–46). Cham: Springer.
10. Brusaporci, S., Centofanti, M., & Maiezza, P. (2017) MUS.AQ: A digital museum of L'aquila for the smart city INCIPICT project. In M. Ceccarelli, et al. (Eds.), *New activities for cultural heritage* (pp. 200–208). Cham: Springer.
11. Forte, M., & Siliotti, A. (Eds.) (1997). *Virtual archaeology. Re-creating ancient words.* New York: Harry N. Abrams.
12. Frischer, B. (2008). From digital illustration to digital heuristic. In B. Frischer (Ed.), *Beyond illustration: 2d and 3d digital technologies as tool for discovery in archaeology* (pp. v–xxii). British Archaeological Reports, Oxford.
13. Rodriguez-Navarro, P. (Ed.). (2017). Archaeological drawing. *DISEGNARECON*, *10*(19)
14. Giordano, A., & Huffman, K. (Eds.). (2018). Advanced technologies for historical cities visualization. *DISEGNARECON*, *21*(11).
15. The London Charter. Homepage. Retrieved May 10, 2022, from https://www.londoncharter.org/.
16. Principles of Seville, Homepage. Retrieved May 10, 2022, from http://sevilleprinciples.com/.
17. Bentkowska-Kafel, A., Denard, H., & Baker, D. (Eds.). (2012). *Paradata and transparency in virtual heritage.* Farnham: Ashgate Publishing.
18. Cameron, F., & Kenderdine, S. (Eds.). (2010). *Theorizing digital cultural heritage: A critical discourse.* Cambridge, MA: MIT Press.
19. Ch'ng, E., Gaffney, V., & Chapman, H. (Eds.). (2013). *Visual heritage in the digital age.* London: Springer.
20. Smith, L. (2006). *Uses of heritage.* Abingdon: Routledge.
21. Giaccardi, E. (Ed.). (2012). *Heritage and social media: Understanding heritage in a participatory culture.* Abingdon: Routledge.
22. Harrison, R. (2013). *Heritage critical approaches.* Abingdon: Routledge.
23. Ferdani, D. (2021). Disegnare con… Maurizio Forte. *DISEGNARECON*, *27*(14), DW.1–DW.10.
24. Lackey, S., & Shumaker, R. (Eds.). (2014). *Virtual, augmented and mixed reality: Applications of virtual and augmented reality.* Cham: Springer.
25. Coluccelli, G., Loffredo, V., Monti, L., Spada, M. R., Franchi, F., & Graziosi, F. (2018). 5G Italian MISE trial: Synergies among different actors to create a "5G Road". In *2018 IEEE 4th international forum on research and technology for society and industry (RTSI).* IEEE.

Castellaccio of Monreale: From the Survey to the Visualisation of Virtual Reconstructions

Vincenza Garofalo, Enrico Lepre, and Cristian Antonino Mancino

Abstract On Monte Caputo near Monreale stands a castle, presumably built in the twelfth century. The paper presents the first results of a still in progress survey activity, carried out on the entire monument using laser scanning and photogrammetric methods. This activity was carried out as part of a scientific agreement stipulated between the University of Palermo, Department of Architecture and the Club Alpino Siciliano, owner of the Castle. The graphic restitution of the survey provided new information and unpublished images of the castle. It was then possible to formulate hypotheses on the virtual reconstruction of lost configurations. This was achievable thanks to the survey analytical reading which is intended as a tool for the critical interpretation of the monument, together with the study of historical, archival and iconographic sources. A video was made by combining animation and visualisation techniques of 3D models with drone video shootings. It narrates the monument by proposing the virtual navigation of some of the castle areas, through the construction of three-dimensional digital models allowing to ideally move in space and time. The video lets to enjoy the castle even from a distance. The use of 3D visualisation techniques allows to transmit cultural content and historical information to a wide audience in a simple and immediate language.

Keywords Laser scanning survey · Virtual reconstruction · Graphical analysis · Norman architecture

1 The Castle

The Castle of San Benedetto or Castellaccio of Monreale stands in a strategic position in Monte Caputo, from where Palermo, Monreale and the surrounding territory were controlled (Fig. 1). There is no certain information on the time of its construction. However, many scholars date it back to the twelfth century, as per will of the Norman King William II. Maria Gabriella Montalbano ([16], p. 49 et seq.) provides an accurate

V. Garofalo (✉) · E. Lepre · C. A. Mancino
Department of Architecture, University of Palermo, Palermo, Italy
e-mail: vincenza.garofalo@unipa.it

© The Author(s), under exclusive license to Springer Nature Switzerland AG 2023
I. Trizio et al. (eds.), *Digital Restoration and Virtual Reconstructions*, Digital Innovations in Architecture, Engineering and Construction,
https://doi.org/10.1007/978-3-031-15321-1_8

Fig. 1 Castellaccio. Photo taken from a drone in flight

bibliography of scholars who trace the foundation back to the time of William II and others who date the monument to the time of the Islamic domination. Up until modern days, a certain dating has not been established.

The castle was a fortress and also a Benedictine monastery, donated by William II to the Abbey of Monreale. The castle overlooks the Cathedral of Monreale and the Benedictine Abbey of San Martino delle Scale. Therefore, it is not unlikely that it was built to defend both, as stated by Léon Dufourny in his Diario di un Giacobino a Palermo 1789–1793 [5] . Historical chronicles [4, 9, 10] report that the castle remained in use until 1370. Giovanni Chiaromonte, who had transformed it into a fortress, ordered its destruction in that year. This was so that the castle would not be occupied by his enemies. Pope Urban V ordered Chiaromonte to rebuild it, however Patricolo ([18], p. 7) reported that his will was disregarded since there is no evidence of reconstructions dating back from that period. As Mongitore (XVIII c.) reported [15], the Castle was a fortress for King Martin's army in 1393. It was then inhabited by a monk and a lay brother of the Abbey of San Martino until the end of the sixteenth century (1588 or 1589), when it was definitively abandoned. Del Giudice [4] reports on the state of ruin of the Castle, the collapse of the vaults of the "second order of the dwellings", as well as the impossibility of determining whether these were covered by vaults or flat floors. He provides information on the good state of conservation of the lower large cistern floors. The scholar describes them as divided into corridors, like the above structure. As early as 1702, the church no longer had columns. The Club Alpino Siciliano (Sicilian Alpine Club) bought it in a state of ruin in 1899 from the Monreale State Property Office, with the authorisation of the Ministry of Public Education. The club undertook the first urgent restoration works on the monument to use it as a mountain climate station. As of today, the club still owns the monument.

The castle architecture is austere and made of local stone. The external configuration resembles a fortress. There are noted four similar western quadrangular towers and three eastern ones which are different in size. Moreover, it is noted the presence of narrow slits on the lower level and a few ogival openings on the upper level. The internal distribution shows similarities with a monastery. There are found an adjoining quadrangular cloister, a church with three naves (once divided by columns which are lost today) and three apses inside the south-east tower. Most of the rooms, including the church, are now uncovered. All internal openings are ogival.

The plan has the shape of a parallelogram of about two thousand square meters, from where the towers protrude. Two parts can be distinguished in the plan. The first one is found in the northern area and it is accessed through two large ogival openings in the northern façade. This develops around an elongated courtyard and it was probably destinated to the monastery. Inter alia, the rooms included the refectory, the kitchen, the chapter house and a room with a niche faced to the east. According to Patricolo [18], the room may have been used as a private chapel. The east-facing niche was likely to be the apse. The upper floor was accessed from the courtyard via a staircase, of which traces are still visible.

The southern castle part houses the church and the quadrangular cloister. The church was probably open to the local religious people, given that there was an entrance in the southern front which was closed at the end of the nineteenth century. This is reported in the 1897 plan published by Patricolo [18], showing the survey by Francesco Valenti and Ettore Pietro (Fig. 2). The castle has four rainwater collection tanks covered by cross vaults and located under the cloister, church and refectory. The paper reports the first results of a research activity aimed at surveying the present condition of the Castellaccio. Furthermore, the research activity focused also on a virtual use to allow ideal reconstructions of historical configurations through multimedia systems. This activity was carried out as part of a scientific agreement for the survey, representation, documentation and dissemination of the Castle of San Benedetto (Castellaccio). The agreement was stipulated in 2017 and renewed in 2020, between the University of Palermo, Department of Architecture and the Club Alpino Siciliano, owner of the Castle—scientific director Vincenza Garofalo.

2 The Survey and Analysis for Virtual Reconstruction

The recent survey campaign is still in progress and it has been carried out on the entire monument using laser scanning and photogrammetric methods. This represents the first step towards the understanding of the monument. The survey campaign, coordinated by Vincenza Garofalo, was carried out in two different times: the laser scanner acquisitions were carried out with a Leica HDS7000 phase modulation scanner by Fabrizio Agnello and Vincenza Garofalo; the photogrammetric survey was carried out by Enrico Lepre and Cristian Antonino Mancino using a Yuneec Typhoon H 480 drone, which has a CGO3 + video camera with 4K resolution and 3-axis gimbal.

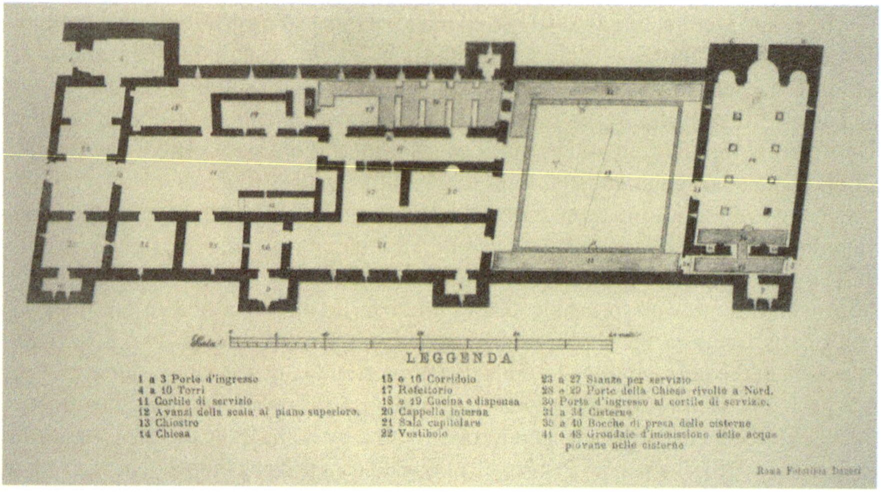

Fig. 2 Plan published by Patricolo [18]

It is available a small iconography in which the castle is always represented by the plan of the lower level, however no detailed surveys of the elevations are known.

Among the collections of the National Library of Paris, Giuseppe Pagnano ([17], p. 141) found a drawing by Domenico Marabitti from between 1789 and 1793. It is a sheet showing the plan and a section of the castle, including the cisterns under the cloister and the refectory. The drawing represents a very important testimony documenting the state of the castle in the eighteenth century [8].

The survey provided detailed information on the layouts, volumes and wall textures. These elements are particularly useful for a subsequent reading and analysis of the details and the entire architectural configuration (Figs. 3, 4, 5, 6, 7 and 8). The use of a drone was required due to the site orography, the castle position and the impossibility of reaching some of its parts which are at inaccessible heights. This allowed the integration of the data collected from the ground and the acquisition of some missing metric information. The laser scanning survey allowed to obtain detailed information on the measured surfaces. Added to this, a general model of the castle and its immediate orographic surroundings (Figs. 9, 10 and 11) were obtained through a photogrammetric survey. The use of both technologies made possible to acquire an almost complete information on the visible surfaces, although the presence of vegetation in some areas did not allow an optimal documentation of some of the castle surfaces. The surveys data were processed to obtain the 3D model of the castle current state (Fig. 12).

The drone photographs were processed with Agisoft Metashape software to get a point cloud, containing colorimetric information obtained from the pixels of the processed images, as well as a castle 1:1 scale textured polygonal mesh. From this, all the required graphic and numerical information can be extracted. The mesh was especially useful for the elaboration of the castle geometric model.

Fig. 3 East elevation, point cloud view (Vincenza Garofalo)

Fig. 4 North elevation, point cloud view (Vincenza Garofalo)

Fig. 5 West elevation, point cloud view (Vincenza Garofalo)

The analysis of the graphic data obtainable from laser scans and a photogrammetric survey allows to start a process of understanding of the monument, which is useful for processing additional information. The graphic restitution of the survey provided new information and unpublished images of the castle. It was then possible to formulate hypotheses on the virtual reconstruction of lost configurations concerning the staircase that served the upper level, the cloister portico and the church

Fig. 6 South elevation, point cloud view (Vincenza Garofalo)

Fig. 7 Axonometric view of the point cloud (Vincenza Garofalo)

aisles (Figs. 13 and 14). This was possible thanks to the survey analytical reading which is intended as a tool for the critical interpretation of the monument, together with the study of historical, archival and iconographic sources.

The presence of the staircase is testified by Luigi Lello [10] in 1588, as this is reported in the plan drawn by Marabitti and in the one published by Patricolo [18]. The stairwell with traces of steps is still visible today and the vaults supporting the flights are also preserved.

Excavations were carried out at the end of the nineteenth century by the Regional Office for the Conservation of Monuments of Sicily, under the direction of Giuseppe Patricolo. These brought to light some elements allowing the formulation of some hypotheses. The quadrangular cloister probably had a portico along all the four sides. This would be confirmed by two crucial elements. Firstly, the presence of a stylobate

Fig. 8 Axonometric view of the point cloud (Vincenza Garofalo)

Fig. 9 (Top left) 3D survey using a drone; processing of drone photos to obtain the point cloud. (Enrico Lepre and Cristian Antonino Mancino)

Fig. 10 Longitudinal section of the point cloud (Vincenza Garofalo)

Fig. 11 Longitudinal section of the church, point cloud view (Vincenza Garofalo)

which was found in its entirety during the excavations at two meters from each of the cloister sides. Secondly, traces of the rooflines which are visible on the walls. Along the cloister west wall, there are still traces of what appears to be battlements, which are also mentioned by Luigi Lello [10].

Excavations inside the church were also carried out under Patricolo's direction. These uncovered a brick base near the south-west corner where it was visible the imprint of one of the columns which originally separated the nave from the side aisles. The column presence in the sixteenth century is attested by Luigi Lello [10] and Tommaso Fazello [6]. Respectively, Lello reports having seen 'two orders of round columns made of bricks' in the church ([14], p. 26), while Fazello mentions that the church was supported by columns.

Based on such considerations, a conjectural reconstruction of some of the castle rooms was hypothesised.

3 The Visualisation and Navigation of Virtual Spaces

The results obtained in the phases discussed above allowed the creation of a video. This narrates the monument by proposing the virtual navigation of some of the

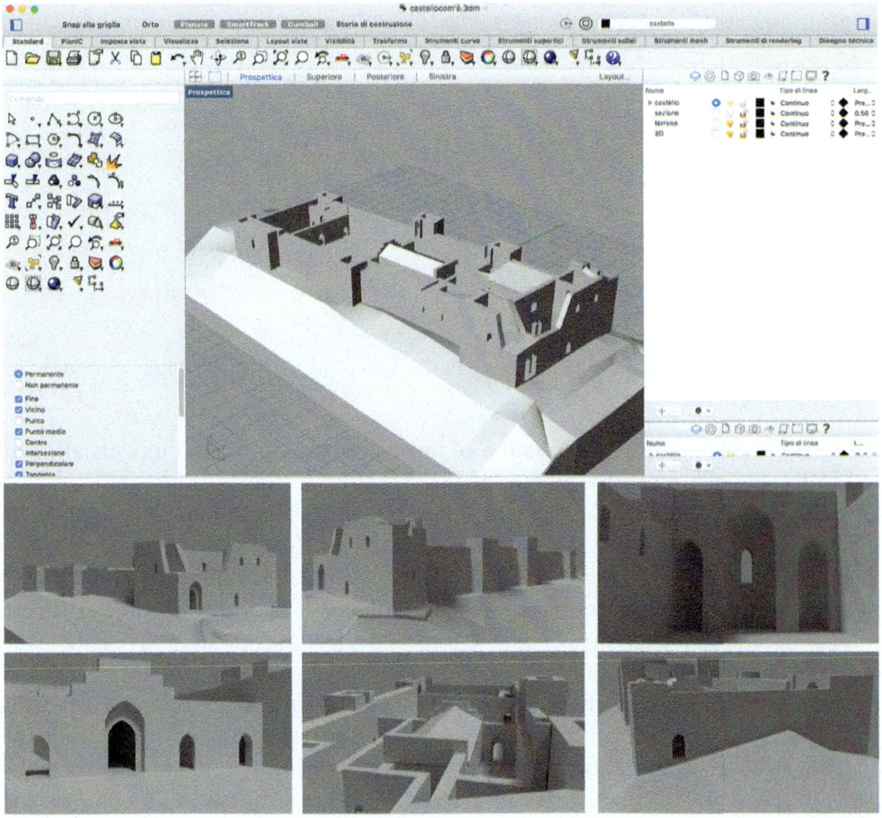

Fig. 12 (Top left) 3D model of the castle current state. (Enrico Lepre and Cristian Antonino Mancino)

Fig. 13 Hypothetical virtual reconstruction of lost configurations. (Enrico Lepre and Cristian Antonino Mancino)

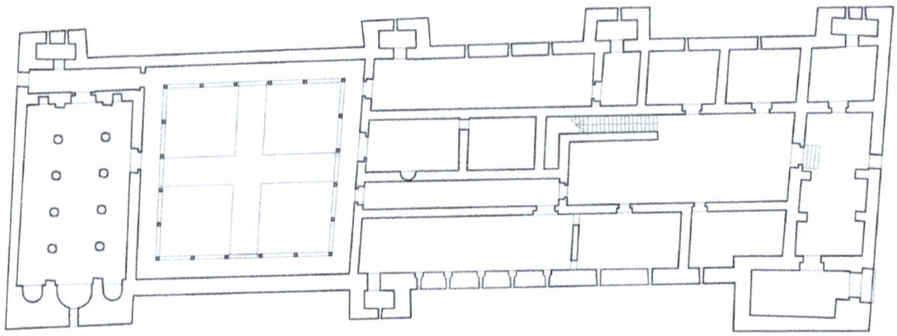

Fig. 14 3D Plan, reconstructive hypothesis. (Enrico Lepre and Cristian Antonino Mancino)

castle areas, through the construction of three-dimensional digital models allowing to ideally move in space and time [11, 13].

The video was made by combining animation and visualisation techniques of 3D models with drone video shootings. The video allows to enjoy the castle even from a distance (Fig. 15). The animations of the 3D models were obtained by simulating to place several virtual cameras inside them. The paths to follow inside the castle were traced once the characteristics of the focal aperture, lens and the shooting quality were defined. During the camera movements, a Boolean modifier was used to visualise the transition from the castle current state to its textured conjectural reconstruction. This allowed a real-time building effect from the ground to the top of the walls.

The virtual tour inside the 3D model proposes a reading of the monument taking place in space through three sequential steps. These are: the 3D model of the castle current state, the historical photos showing the castle as it looked at the end of the

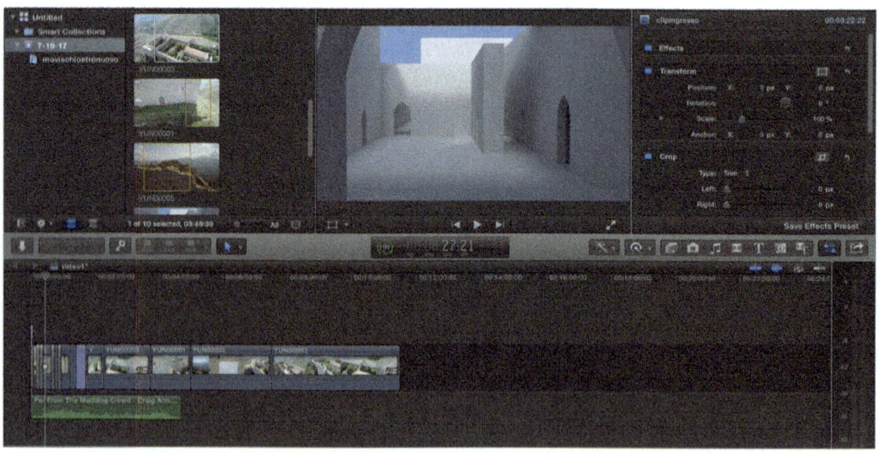

Fig. 15 3D Video editing process. (Enrico Lepre and Cristian Antonino Mancino)

nineteenth century and the ideal virtual reconstruction of the castle original configuration. The tour runs from the now disused northern access door to the first courtyard, where the staircase leading to the upper level is displayed. A passageway at the end of the courtyard leads to a long corridor, in which the cross vaults were reconstructed. Today, only a few fragments of these remain at the impost. The corridor leads to the cloister, where the portico was digitally rebuilt along all four sides. From here, it is reached the interior of the church, where the floor, roof and three naves are recreated through the virtual reconstruction of the columns and ogival arches which separated the main nave from the two aisles. The tour continues in the vestibule to the west where the cross vaults are ideally reconstructed. The tour ends outside passing through the southern passage, which was closed at the end of the nineteenth century and it is now ideally reopened. The video begins and ends with some drone footage of the castle and its surroundings. This is to contextualise the monument geographical position also in relation to Palermo, Monreale and the nearby towns.

A narrative voice is also integrated in the visual tour and it underlines the conjectural reconstructions of Fazello's and Lello's testimonies. These facilitate the understanding of the space and the correct reading of the elements.

The realisation of the final product was preceded by a detailed analysis of some case studies which used video technologies to narrate virtual reconstructions [1, 2, 7, 12].

The castle three-dimensional models can also be navigated through a virtual tour .This represents a visualisation tool which provides a totally immersive view of the virtual reconstruction using a visor and other mobile devices. Virtual tours are tools which have been commonly used for some years now for the remote enjoyment of the cultural heritage. These have been widely used by museums and cultural institutions during lockdown periods implemented by governments to deal with the Covid-19 pandemic [20].

The virtual model navigation allows the use of the castle in situ or remotely. This is possible in either its current state or in the three-dimensional reconstruction hypothesis. It is possible to see 360° images by wearing a VR visor and using a common smartphone equipped with a gyroscope. For this purpose, it is recreated a digital space where the user can freely move, allowing a 360-degree rotation of the user's gaze (Fig. 16). The virtual tour based on 3D models of the castle current state and three-dimensional reconstruction hypotheses can be integrated with the addition of textual, iconographic and multimedia contents. These can be activated during the remote navigation or used by mobile devices.

4 Conclusions and Future Aim

The experience presented here is an example of how surveys and data analyses allow the creation of apps for documentational purposes, scientific dissemination and the use of cultural heritage. This is possible through the immersive vision of virtual reconstructions and the use of commonly used devices.

Fig. 16 3D Elaboration of multimedia products: panoramic 360° views. (Enrico Lepre and Cristian Antonino Mancino)

The scans obtained from the laser scanning survey constitute digital copies capturing the reality objectively, archiving information which are available for potential further studies. Future project developments include the completion of the survey with the underground cisterns, a careful analysis of the metric data and the accurate reading of the masonry, which will hopefully provide answers to many unresolved questions.

The research path applied to the Castellaccio di Monreale is easily replicable in other similar contexts. The discussed study will be extended to other Sicilian castle sites, through studies and applied research in the field of survey, multimedia representation, design and visual communication. This will benefit the knowledge and enhancement of Monumental Heritage and the Territory. It is being planned a research activity defined by the Memorandum of Understanding between the University of Palermo, Department of Architecture and the Temporary Association for the "Promotion and cultural enhancement of the Castles of Sicily", a network of Sicilian municipalities with medieval and Renaissance castles in their territories. This is created to implement cultural projects aimed at enhancing the castle sites and the enjoyment of the cultural heritage and traditional products of its territory (scientific managers Vincenza Garofalo and Viviana Trapani). One of the objectives of the Memorandum of Understanding is to provide in-depth and coordinated documentation of the castles located in the region. The technical-scientific collaboration relationship has already had an outcome in an in-depth study of the castle of Calatubo in Alcamo, from virtual reconstructions of past configurations to VR/AR application [3].

Supporting the fruition of the cultural heritage, digital resources facilitate a process of democratisation. In fact, the main objective of the presented multimedia product is the understanding of the castle architectural features and the visualisation of its original context. This is through the integration of the real experience with the re-proposition of past configurations, which are obtained from the careful analysis of historical sources and photographs. The use of 3D visualisation techniques allows to also transmit cultural content and historical information to a wide audience in a simple and immediate language. The developed multimedia product will be further enhanced by the results of future investigations. This will be available on the Club Alpino Siciliano website and usable in a permanent installation at the Castellaccio. The 3D model, made usable through the video, allows to observe the monument and enjoy the architectural space without directly visiting it. Since the castle can only be reached by foot via an uphill path, the here presented product allows a virtual visit even to those who cannot physically access it.

Credits. Authors note: The paragraphs 1, 2, 4 are written by Vincenza Garofalo the paragraph 3 is written by Enrico Lepre and Cristian Antonino Mancino.

References

1. CNR Information Technologies Laboratory, http://itlab.ibam.cnr.it/index.php/teatro-di-catania/. Accessed July 2021.
2. CNR Information Technologies Laboratory, http://itlab.ibam.cnr.it/index.php/teatro-taormina-2017/. Accessed July 2021.
3. De Blasi, D. (2020). *Il Castello di Calatubo: dal rilievo alle applicazioni visive interattive*. Supervisor Prof. Vincenza Garofalo, co-supervisor Prof. Mirco Cannella, University of Palermo, Academic Year 2019–2020.
4. Del Giudice, M. (1702). *Descrizione del Real Tempio e Monasterio di Santa Maria Nuova di Morreale. Vite dè suoi arcivescovi, abbati e signori. Col sommario dei privilegi della detta Santa Chiesa di Gio. Luigi Lello ristampata d'ordine dell'Abbate Don Giovanni Ruano con le osservazioni sopra le fabriche e mosaici della Chiesa, la continuazione delle vite degli Arcivescovi, una tavola cronologica della medesima istoria, e la notizia dello stato presente dell'Arcivescovado*. Palermo: Regia Stamperia d'Agostino Epiro.
5. Dufourny, L. (1991). *Diario di un giacobino a Palermo 1789–1793*. Palermo: Fondazione Culturale Lauro Chiazzese della Sicilcassa.
6. Fazello, T. (1558). *De rebus Siculis decades duae, nunc primum in lucem editae*. Palermo: Ioannes Matthaeus Mayda et Franciscus Carrara.
7. Gabellone, F., Ferrari, I.,& Giuri, F. (2017). Un contributo alla ricostruzione del teatro di Taormina. In V. Greco (ed.), *Lifting Theatre. La straordinaria sfida al G7 di Taormina*, pp. 112–123. Milano: Mondadori-Electa.
8. Garofalo, V. (2021). Il Castellaccio di Monte Caputo a Monreale: tre piante a confronto. *LEXICON. Storie e architettura in Sicilia e nel Mediterraneo* (33), 79–83. Available at https://www.edizionicaracol.it/wordpress/wpcontent/uploads/2022/04/6_garofalo.pdf.
9. Inveges, A. (1650). *Parte seconda degli Annali della felice citta di Palermo prima sedia, corona del re, e capo del Regno di Sicilia la quale abbraccia quattro ere, ò dicciam periodi d'historia: parte della Romana, la Sacra, la Constantinopolitana ò Greca e Saracina...* Palermo: nella typographia di Pietro dell'Isola, impressor camerale.
10. Lello, G. L. (1588). *Descrizione del real tempio, e monasterio di Santa Maria Nuova di Morreale: vite de' suoi arcivescovi, abbati, e signori: col sommario de i privilegj della detta Santa Chiesa*. Reprint by Del Giudice, M. (1702). Palermo: Regia Stamperia d'Agostino Epiro.
11. Lepre, E. (2017). *Castellaccio di Monreale. Dal rilievo fotogrammetrico al racconto digitale*. Supervisor Prof. Vincenza Garofalo, University of Palermo, Academic Year 2016–2017.
12. Malfitana, D., Gabellone, F., Cacciaguerra, G., Ferrari, I., Giuri, F.,& Pantellaro, C. (2016). Critical reading of surviving structures starting from old studies for new reconstructive proposal of the Roman theatre of Catania. In J. L. Lerma, & M. Cabrelles (Eds.), *Proceedings of the 8th International Congress on Archaeology, Computer Graphics, Cultural Heritage and Innovation 'ARQUEOLOGICA 2.0' in Valencia (Spain)*, pp. 155–161. Valencia.

13. Mancino, C. A. (2017). *Castellaccio di Monreale. Il modello 3D per il racconto multimediale.* Supervisor Prof. Vincenza Garofalo, University of Palermo, Academic Year 2016–2017.
14. Millunzi, G. (1897). Il Monte Caputo e il Castellaccio di Monreale. *Sicula,* (II, 1), pp. 24–28.
15. Mongitore, A. (ms. XVIII c.). *Storia sagra di tutte le chiese, conventi, monasteri, ospedali ed altri luoghi pii della città di Palermo, Le chiese e le case dei regolari, parte I,* at the signs QqE5. Biblioteca Comunale di Palermo.
16. Montalbano, M. G. (1989). Il Castellaccio di Monreale. *Incontri e Iniziative: memorie del centro di cultura di Cefalù,* (n. IV. 2), pp. 49–72.
17. Pagnano, G. (2006). Da Dufourny a Hittorff. L'eredità dei disegni siciliani. In M. Giuffrè, P. Barbera, G. Cianciolo Cosentino (eds.), *The Time of Schinkel and the Age of Neoclassicism between Palermo and Berlin* (pp. 129–149). Cannitello: Biblioteca del Cenide.
18. Patricolo, G. (1897). Il castello di S. Benedetto: chiamato 'Castellaccio' sul Monte Caputo presso Monreale. *Giornale Scientifico di Palermo,* (IV, 8–9), pp. 1–18.
19. Schirò, G. (1990). *Castellaccio di Monreale.* Palermo: Accademia nazionale di scienze lettere e arti.
20. Valzano, V., Gaiani, M. (eds.) (2021). *SCIRES-IT—SCIentific RESearch and Information Technology* (11, 1). Available at http://www.sciresit.it/

Images of the Disappeared Puerta Real in Seville

Antonio Gámiz-Gordo and Pedro Barrero-Ortega

Abstract During the sixteenth century, Seville experienced a great flourishing due to the monopoly of trade with the Americas. It was consequently decided to renew the architectural image of the gates of the walled city so that their appearance would be in keeping with the new era. Their design followed a report issued in 1560 by the prestigious architect Hernán Ruiz II, author of the fine Renaissance finial of the Giralda tower of the Cathedral. The Royal Gate -(*Puerta Real*, also named *Puerta de Goles*) located near the river port- was completely renovated between 1560 and 1566. It had been known as the Royal Gate since King Philip II entered Seville through it in 1570. The gate subsequently appeared in important urban images by Bambrilla (1585), Janssonius (1617), Olavide (1771), Ford (1830–1833), and in a valuable photograph by Masson (c. 1858). In 1864, the Royal Gate was demolished and just one side of the city wall hosting a commemorative stone is preserved nowadays. By collecting its historical portrayals and analysing its urban environment, this research provides a virtual reconstruction of this unique Renaissance work. To this end, the dimensions of the preserved archaeological remains had been obtained with Masson's photograph, digitally processed, as foundation. Auxiliary geometric lines were used as references and both the Book of Architecture by Hernán Ruiz II and the treatise by Sebastiano Serlio as guidelines to define details. A photomontage of the gate in its current urban environment completes the reproduction, aiming to underline its great heritage importance.

Keywords Images · Disappeared heritage · Seville · Puerta Real · Architecture

1 Introduction: Historical and Urban Context

Medieval cities were surrounded by city walls which defended their inhabitants from external perils, enabled the confinement of the population in times of epidemics, and served also as a means of tax control. From its origin, the configuration of the city

A. Gámiz-Gordo (✉) · P. Barrero-Ortega
University of Seville, Seville, Spain
e-mail: antoniogg@us.es

© The Author(s), under exclusive license to Springer Nature Switzerland AG 2023
I. Trizio et al. (eds.), *Digital Restoration and Virtual Reconstructions*, Digital Innovations in Architecture, Engineering and Construction,
https://doi.org/10.1007/978-3-031-15321-1_9

wall of Seville differed as a result of urban growth and changes in the course of the Guadalquivir river. The city expanded during the Roman and Islamic empires, and in the Almohad Caliphate (1147–1248) its final and significant walled enclosure was consolidated [1]. This layout remained unaltered after the Reconquista of Seville by forces of Ferdinand III of Castile in 1248.

In the sixteenth century, Seville became one of the most important European capitals, due to the monopoly of trade with the Americas. The establishment of the *Casa de Contratación de Indias* in Seville in 1503 and the wedding of Emperor Charles V in 1526 marked the beginning of a period of great prosperity. Many significant monuments were built in that era, such as the *Lonja* or *Archivo de Indias*, the new Town Hall, the *Hospital de las Cinco Llagas*, the *Audiencia*, the *Casa de la Moneda*, and the finial of the *Giralda*.

The river port constituted the central image of Seville, next to the city wall with its many towers, gates, and wicket gates (Fig. 1). In approximately 1560, a renovation project was launched to endow the city with gates of a more spacious and functional nature. The aim would be to eliminate medieval bent entrances, no longer necessary for defensive reasons and very inconvenient in terms of accessibility. The plan was also intended to offer a renewed urban image that echoed the Roman triumphal arches. To this end, the Cabildo de Sevilla commissioned a report from Hernán Ruiz II, who introduced a new architectural language in accordance with the Renaissance taste [2].

The disappeared Puerta Real was situated near the port of Seville, although a bridge of boats prevented large ships from reaching its environs. In medieval times, the gate was enclosed by a large tower with a corner entrance. Its access path from

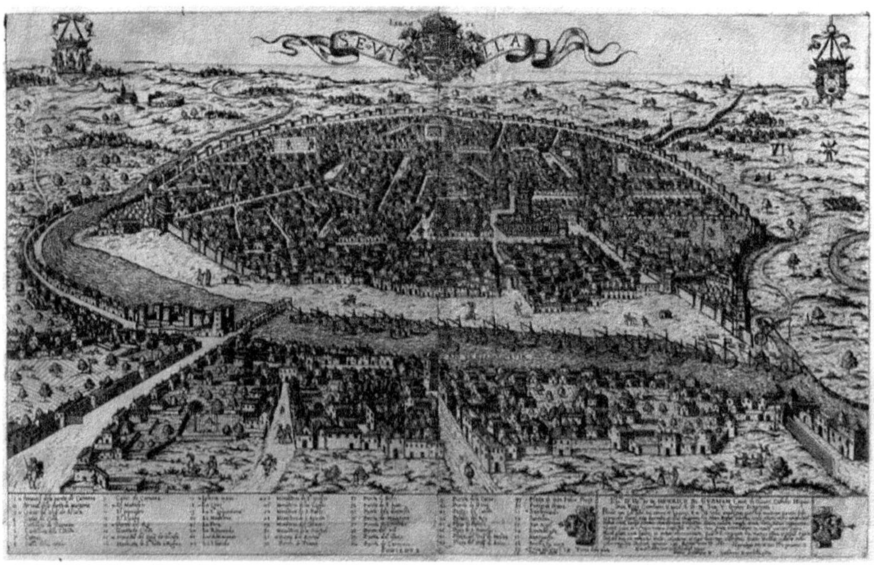

Fig. 1 View of Seville by Bambrilla (1585). National Library of Spain

the bridge of boats was attached to the city wall, leveraging the barbican rampart. Located in a terrain recess that diminished its defensive capacity, the medieval gate would be configured as a "tower-gate", a common order in Al-Andalus and Morocco in the eleventh-twelfth centuries. On various occasions, the medieval gate was rebuilt or reformed due to the damage caused by major flooding of the river. From outside the city, it was partially hidden by a mound on the banks of the Guadalquivir. The origin of said mound, which would serve as a defence against the flooding of the river, remains largely unknown, although it possibly emerged from the accumulation of rubble and refuse at the city gates.

In its surroundings, outside the city walls, Hernando Colón, son and heir of the admiral Christopher Columbus, built his palace on land granted to him in 1526 [3]. With a façade enriched with marble from Genoa, and housing his remarkable library, the vast gardens and groves transformed the topography, even having an ombu tree, brought from the Americas, planted in its upper terrace. The building was later used as the convent of San Laureano. Along the river there was also an area with orchards for washing wool and other secondary activities of the sailors living in the close neighbourhood of San Vicente. In the interior of the city, next to the Puerta Real, there was a small square where Alfonso XII (old Armas street), Gravina, and Goles streets converged with San Laureano gradient [4].

The place name "Goles" is mentioned in the *Libro de Repartimiento* following the Christian conquest of the city in 1248, and in other Castilian documents from the thirteenth and fifteenth centuries [5]. Several historians have attributed the origin of the toponym "Real [Royal]" to the entrance of King San Fernando in the conquest of the city. However, that name is only mentioned in documentary sources from 10 May, 1570, when King Philip II, the most powerful man in the world at that time, entered through this gate.

The Puerta Real was drawn by Hernán Ruiz II, the prestigious architect who devised the Renaissance finial of the Giralda of the Cathedral of Seville, as part of his powers as Master Builder of the City. Although his authorship is not officially documented, no scholar has questioned the attribution of this work to this architect, given the formal proximity between its architectural production and his design of other city gates in Seville: Puerta de Jerez, Arenal, Postigo del Carbón, and Puerta de la Carne, the latter having the greatest formal resemblance to the Puerta Real.

Hernán Ruiz II would also design the noteworthy gate of Triana, erected after his death [6]. When comparing these works with the drawings included in his Book of Architecture, an original manuscript preserved in the library of the Higher Technical School of Architecture of Madrid [7], similarities are detected with folios 131, 139 (Fig. 2) and others [8], plus with certain plates from the treatise of Sebastiano Serlio (Fig. 3) [9], which could serve as a reference to Hernán Ruiz II.

According to documentation preserved in the Municipal Archive of Seville, the intervention of the Puerta Real by Hernán Ruiz is dated between 1560 and 1566. It consisted of demolishing the existing gate, its vault and tower, for its reconstruction without the tower [3]. After this demolition in 1563, the work was concluded in 1564, according to a commemorative stone placed on the upper body of its interior façade,

Fig. 2 Drawing by Hernán Ruiz II, Book of Architecture, folio 139 (c. 1569). Library of the Higher Technical School of Architecture of Madrid (Raros, 39) [https://n9.cl/iwiu0]

which is addressed later in this paper. Furthermore, the adjacent urban space was conditioned and paved, according to another valuable document of the time [10].

Bearing this context in mind, this research has compiled and studied all the major graphic documentary sources between the sixteenth and nineteenth centuries in order to ascertain and publicise the Puerta Real devised by Hernán Ruiz II and demolished in 1864. Subsequent to an analysis of its surroundings and of its architectural composition, a schematic virtual reconstruction is provided for the dissemination and widespread appreciation of this heritage and its symbolic importance for the city of Seville.

Fig. 3 Plate by Sebastiano Serlio. Treaty of Architecture (1560). Library of the University of Seville, Fondo Antiguo [https://archive.org/details/HArteR06T02]

2 Urban Images from the 16th to the Nineteenth Centuries

From the second half of the sixteenth century, many images of Seville were published that recreated the city and its surroundings outside the city with varying degrees of precision, whereby special attention was paid to its river port as a symbol of prosperity [11]. One of the first views was drawn by Anton van den Wyngaerde (c. 1525–1571) around 1563–1567 [12]. The author labelled the name *Puerta de Goles* next to the house of Colón in one of his panoramic views, but its volume seems dubiously or carelessly outlined.

In 1585, Ambrosio Brambilla, architect, draftsman and engraver (active between 1579 and 1599), published an image of Seville with an elevated view, thereby showing the walled city in full for the first time [13]. The main gates were drawn with considerable accuracy and identified in the legend with the following numeration and names (Fig. 1): Arenal (31), Triana (32), Goles or Puerta Real (33), San Juan (34), Almenilla

Fig. 4 Detail of a view copied from Bambrilla's work in *Civitates Orbis Terrarum* (vol. IV, 1588). Private collection

or Barqueta (35), Macarena (36), Cordoba (37), Sol (38), Osario (39), Carmona (40), Carne (41), Jerez (42), and Postigo del Cárbon (43). The house of Hernando Colón (4) and its gardens (14) are also identified and drawn, although he had died in 1539. In this view, the Puerta Real designed by Hernán Ruiz II was represented for the first time, displaying its uniquely condensed upper piece and with great similarity in relation to other later images. The view was also published in volume IV of the *Civitates Orbis Terrarum* (1588) [14] with variations in the legend and labelled as "*guerta [huerta] de Colon*" (Columbus' orchard) on the drawing (Fig. 4).

A similar view, but much more schematic and idealised, was published in a reissue of the Cosmographia by Sebastian Münster (1626) [12]. Although, at first glance, it might seem like a simplified copy of the work by Bambrilla, surprisingly the Puerta Real was represented as a "tower-gate" with a detailed access path from Puente de Barcas (Fig. 5). As the first editions of this work appeared in approximately 1545–1550, this could be probably an imaginary vision of the medieval gate, prior to the new Renaissance construction.

The Puerta Real was referenced in another view by Joris Hoefnagel (1542–1600) taken from the opposite bank of the Guadalquivir and published in volume I of the *Civitates Orbis Terrarum* (1572) [14]. The label "Puerta de Goles" appears mistakenly displaced to the left of the sign "Casa de Colón", instead of to the right. However, it was never drawn, since it could be hidden by the above-mentioned mound, or perhaps because the construction was incomplete when the artist visited Seville around 1563–1567. In the foreground, small fishing boats were drawn, fishing being

Fig. 5 Detail of the view by Münster, in *Cosmographia Universalis* (1626). Private collection

a common activity in this stretch of the river, where large boats were unable to enter. It should be noted that the extramural area located next to the Puerta Real was called *Los Humeros* because it was the place were fish was smoked [15] (Fig. 6).

In 1617, Joannes Janssonius (1588–1664), Dutch cartographer, engraver, and merchant of prints, made a famous panoramic view of Seville seen from Triana, more than two metres in length, of which complete copies are only preserved in the British Museum, the Naval Museum of Madrid, and the National Library of Sweden [12]. In the foreground, Janssonius incorporated details of the port and splendid ships that illustrate the prosperity of the city, including a rhyming motto that would become quite popular: "*Qui non ha visto Sevillia non ha vista marravilla*" [He who has not

Fig. 6 Detail of the view by Hoefnagel, in *Civitates Orbis Terrarum* (vol. I, 1572). Private collection

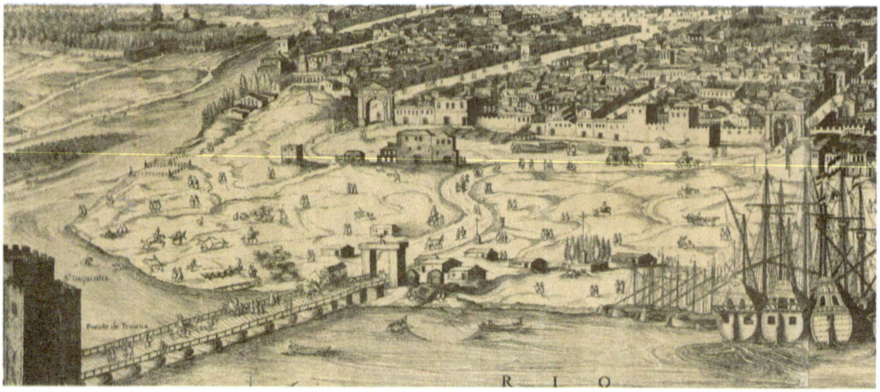

Fig. 7 Detail of the view of Seville by Janssonius (ed.) (1617). National Library of Sweden

seen Seville, has not seen wonder]. The towers and gates of the city wall are highlighted in a fanciful way. The house of Hernando Colón appears with its orchards and a peculiar mound; but the gate of Goles is represented in a remote position and rotated towards the south, unreally forming part of the main front of the city towards the port (Fig. 7). Its architecture seems idealised, as do other buildings that falsely showcase an accentuated architectural classicism, perhaps in an effort to evoke the image of a new Rome [2]. This view would serve as inspiration for another view by the Swiss engraver and publisher Mathäus Merian (1593–1650) in 1638, which was widely disseminated and became the subject of various simplified and farfetched plagiarisms.

The first city map of Seville, dated 1771 and known as the map of Pablo de Olavide (1725–1803), accurately details the layout of the city in the vicinity of the Puerta Real (Fig. 8). It shows three blocks built in front of the gate and close to the banks of the Guadalquivir, as well as the streets that made up the area of *Los Humeros* to the north. Furthermore, it represented the retaining wall parallel to the city wall, also drawn in subsequent maps of the city until its demolition in 1856. The staircase located next to the convent of San Laureano and the chapel Nuestra Señora del Rosario are also depicted.

The Hispanist and traveller Richard Ford accomplished around a hundred views during his stay in Seville between 1830 and 1833 [16], and many others throughout Spain. Two of these views detail the Puerta Real and its urban surroundings with considerable precision [17] (Fig. 9a, b). One view is depicted from outside the city walls in which the lower part of the gate is hidden and appears low in height due to the aforementioned topographic position. The other view was taken next to the arcades -that still remain in the square to this date- and details the small chapels that had been added to each side of the gate, which was consequently partially concealed.

Another beautiful yet little-known drawing of the Puerta Real, this time by the important Swedish painter Egron Lundgren [18], is stored in the National Museum in Stockholm. After having received a scholarship from the Stockholm Academy

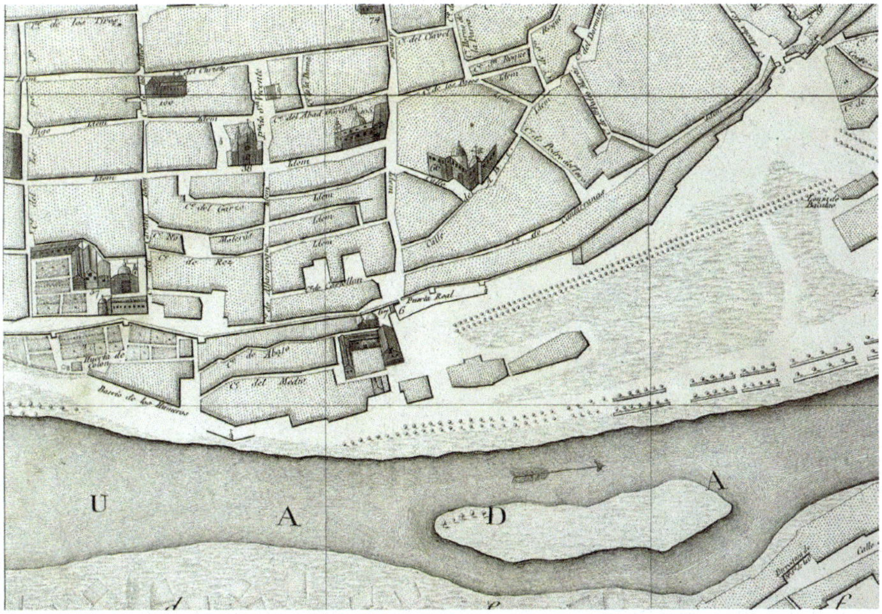

Fig. 8 Detail of the city map by Olavide (1771). Private collection

of Fine Arts to train in Paris and Rome, the artist decided to visit Spain. This trip brought him to Seville in August 1849, where he extended his stay until June 1853. He subsequently worked in England as a painter to Queen Victoria, who became interested in his Spanish drawings. He returned to Seville in 1857 and again after 1860. Lundgren produced various drawings and sketches of the artistic heritage and characters of the city. One has a similar point of view to the one by Ford in Plaza Real, and includes precise details by accentuating the light and shadows on grey paper (Fig. 10).

Of great documentary value is the photograph taken around 1855–58 by the photographer Luis Masson [19] (Fig. 11). It was included in an album dedicated to the Duke and Duchess of Montpensier in 1858, shortly before the demolition of the gate in 1864. This photograph provides invaluable information on architectural details consistent with the drawings by Ford and Lundgren. Moreover, visible through the gate, there are some buildings that would form part of a major transformation outside the city walls that involved the construction in 1863 of the Seville-Cordoba railway branch and nearby station Plaza de Armas, nowadays a mall. Said branch occupied the banks of the Guadalquivir, having a large section of the wall demolished from San Juan gate to Puerta Real. The resulting the new avenue of Torneo did not regained contact with the river upon the dismantling of the railway in this privileged spot on the occasion of the universal exhibition Seville Expo ‘92.

Finally, another drawing worth mentioning is that by Basilio Tovar, published in 1878 [20], fourteen years after the demolition of the Puerta Real. Perhaps, for this

(a)

(b)

Fig. 9 **a, b** Richard Ford's drawings of the exterior and interior of the walled city (1830 and 1832). Ford Family Collection

reason, the upper body was represented with an exaggerated width and unreliable proportions.

The transformations of the urban environment studied herein were also reflected in other later historical plans, among which the alignment plans of the façades of the historical centre from the late nineteenth century, preserved in the City Council of Seville, are of special interest. They were drawn on a scale of 1:300 and include dimensions and proportions. Two of these plans represented in red ink the alignment

Fig. 10 Drawing by Egron Lundgren (c. 1849–1860). National Museum, Stockholm [NMH A 501/1981]

modifications approved for the lane Calle de las Armas in July 1874, the square of Puerta Real in April 1880, and San Laureano street in March 1888. In both plans, the exact position occupied by the former Puerta Real is specified (Fig. 12a, b).

3 Archaeological Remains

The loss of the defensive value of the walls in the nineteenth century led to their abandonment, considered at the time an obstacle to development and progress. After the demolition of the Puerta Real between September and October 1864 (according to the municipal agreement of September 14, 1864), only one side of the nearby city wall has been preserved. Its remains were kept for a period in the San Fernando cemetery, which at that time had been open for only eight years, possibly for its later reconstruction therein. Unfortunately, no such rebuilding was ever carried out and thus the remains of the gate would be scattered and lost forever [21].

Fig. 11 Photograph by Louis Masson, included in an album dedicated to the Duke and Duchess of Montpensier (1858). Private collection

In 1995, the City Council conducted an archaeological excavation at the confluence of San Laureano street with the square of Puerta Real to study its original structure. A wall section, rampart, and different remains from the Middle Ages were then discovered [22]. A pavement in a different shade was laid in the place previously occupied by the missing gate, thereby providing a crucial reference for its virtual reconstruction in this research (Fig. 13a, b).

In 1995, the memorial stone previously located in the entablature of the gate was placed on the upper part of the wall remains (Fig. 14). The stone had been trusted to the Provincial Archaeological Museum on 12 March 1880 by the Monuments Commission (R.E. 269) [3]. In order to calculate the dimensions of the gate through the preserved Masson photograph, the stone has been measured, resulting in a piece of 160 × 70 cm. Its inscription reads:

Reigning in Castilla the very high and powerful and catholic king don Philip the second, the very illustrious lords of Seville commanded this work, accomplished in the presence of the very illustrious lord don Francisco Chacón, lord of the villas of Casarrubios and Arroyomolinos and mayor of the alcazars and dome of Avila. It was finished in the month of May 1564.

In addition, the original Goles gate would have a Latin diptych in honour of King Ferdinand III The Saint, placed by Ferdinand Columbus in 1535. Said diptych, perhaps renovated by Hernán Ruiz, was placed on the frontispiece of the new gate and then disappeared, although it was depicted in Tovar's drawing in 1878. Regarding its composition, thanks to historiography, to the drawings by Ford and Tovar, and the photograph by Masson, it is known that there were two coats of arms on the

(a)

(b)

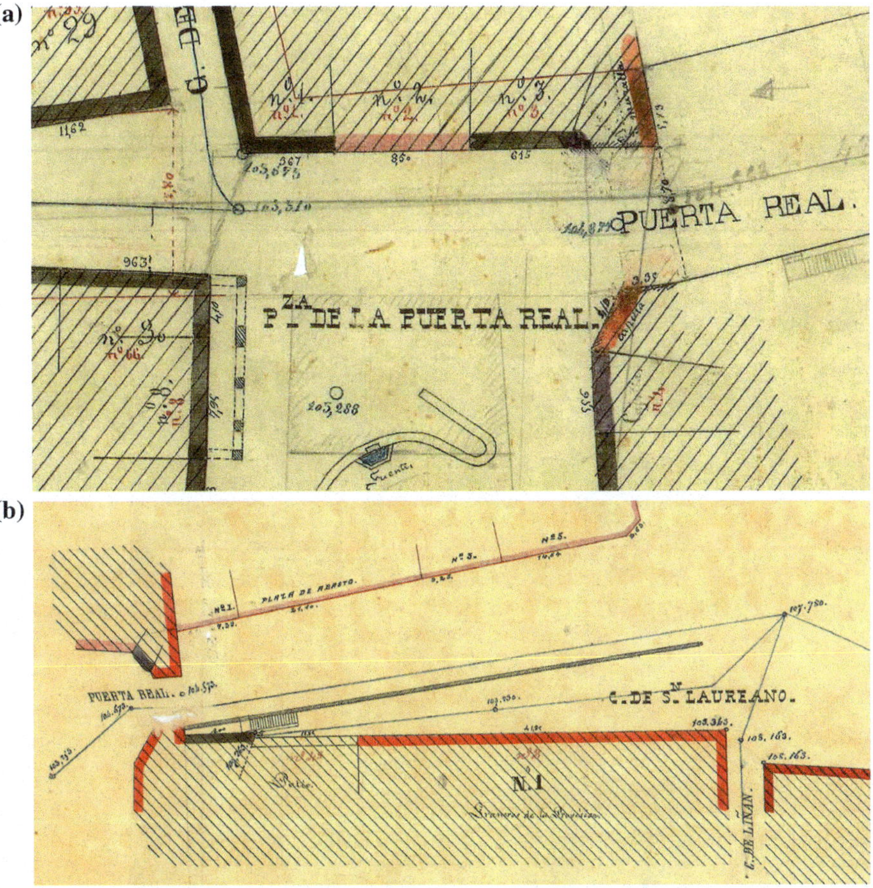

Fig. 12 **a**, **b** Details of façade alignment plans near the Puerta Real (1874–1880 and 1888). Urban Planning Management, City Council of Seville [113/2019; 207/2019]

frontispiece of the second body: one on the interior with the city's coat of arms, and another on the exterior side with the royal coat of arms. Probably one of the three stone coats of arms of the city currently preserved in the Provincial Archaeological Museum of Seville was the one adorning this gate [5].

4 Formal Composition and Layout

The new Puerta Real was undoubtedly one of the most beautiful gates of the walled enclosure of Seville. It was reminiscent of the Roman triumphal arches honoring

(a)

(b)

Fig. 13 **a**, **b** Current environments of the former Puerta Real and abutting city wall section. Authors' own photographs

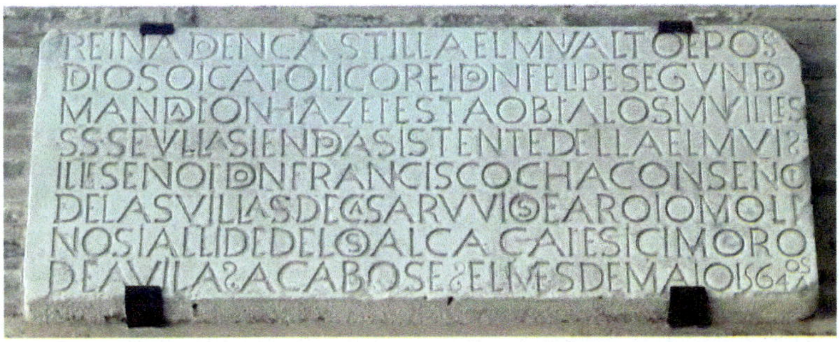

Fig. 14 Commemorative stone originally located on the interior façade of the Puerta Real (today placed on the adjacent section of the city wall). Authors' own photography

prominent leaders and executed with great rigour and precision in their measurements. It had a marked Mannerism style, typical of both its author and of the late Renaissance period.

The gate was composed by two clearly differentiated bodies. In the lower body, the central arch or vault was semi-circular, with a lightweight archivolt and a keystone decorated with a corbel. This arch had a proportion of one and a half times its height to the keystone, which was very Hispanic and sensible for an urban gate. It was framed by two very rough Tuscan pilasters, and on a backward plane with respect to the massifs supporting the arch, in accordance with a common formula in architectural mannerism. The pilasters carried individual fragments of architrave and frieze, which extended their dimension in an epigraphic band on the arch, divided into three sections. The central memorial stone, which is still preserved today, includes the aforementioned text (Fig. 14). The Romanised municipal initials, S. P. (Senatus Populus) and Q. H. (Que Hispalensis), appear on the two lateral pieces. In accordance with the taste of Hernán Ruiz II, the scarce decoration was concentrated on the entablature and on the second body; the composition was finished off with balls and pyramids which were commonly found in the work of this architect.

The peculiar upper body of the gate, otherwise known in less academic terms as the *attic*, is a cubic volume of reduced size, crowned by a triangular pediment. Its square façades include a circular coat of arms of the city, with the Holy King seated between Archbishops Isidore and Leander. This detail is very similar to one still placed on the wicket gate Postigo del Aceite, also attributed to Hernán Ruiz II, which may serve as a reference for its reconstruction. Four semi-acroteria are attached laterally, similar to those atop the pilasters, and grant the whole piece a certain military appearance. All the stonework would be made with stone brought from El Puerto de Santa María (Cadiz) and Espera (Seville), as it was customary at that time in Seville.

In this research, the photograph by Masson, processed with Asrix software, has been employed as a geometric and dimensional basis of the virtual reconstruction undertaken, to obtain a restored view similar to that of its elevation. In order to ascertain the photograph dimensions, measurements have been obtained from the preserved archaeological remains: the adjacent wall, the pavement that since 1995 has marked the original location of the gate, and the previously mentioned memorial stone. The measurements that appear in the alignment plans of the late nineteenth century have also been reassessed, as well as the classical architecture principle of relation among parts and proportions related to each other and to a whole.

According to the manuscript by Hernán Ruiz II and the treatise by Serlio, geometric lines formed the genesis of all architectural design [8]. The architecture of that time employed basic shapes, such as the triangle or the square, which sometimes generated lines of a more complex nature, always with precise dimensions, to enhance the beauty of a whole.

Therefore, simple lines have been used to frame the overall composition. The auxiliary layout proposed has included several significant points (Fig. 15): the point that crowns the central pinnacle of the pediment (1); the upper points of the lateral pinnacles (20, 21, 22, and 23); the line of the impost (11–12), which passes through point 13 defining the centre of the arch; segment 1–4, the axis of symmetry of the composition; segment 24–25, which delineates the pediment from the upper body; and line 9–10, which crowns the cornice between the lower section and the highest one.

The Book of Architecture by Hernán Ruiz II [7] and the treatise by Sebastiano Serlio [9] have been consulted, in addition to the available historical images, in order to specify the shape and dimensions of the mouldings and details. For example, the arch of the first body is related to the drawing on folio 131 by Hernán Ruiz II, while the arrangement of the pediment, spheres and finishing pyramids is related with folio 139 (Fig. 2).

Likewise, the reduced attic with a triangular pediment framing a circle between its recesses and decorated by three acroteria, or the side setback pilasters, also appear in Serlio (Fig. 3). Finally, the drawing of the Puerta Real has been completed with

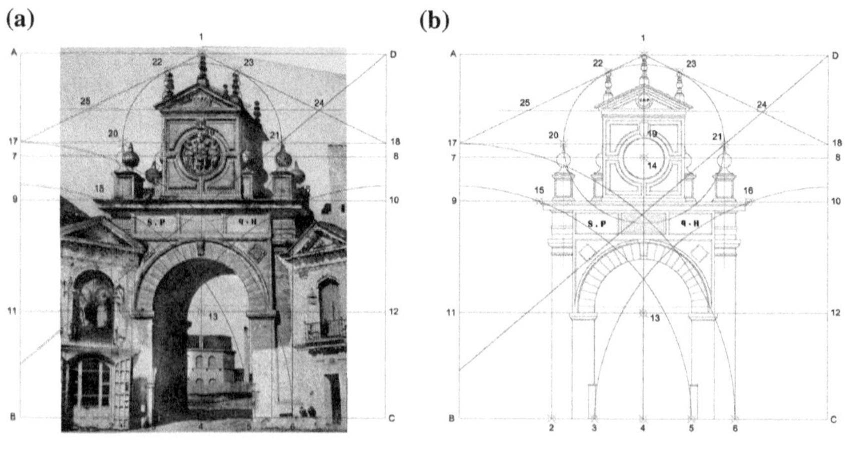

Fig. 15 **a, b** Restored photograph by Masson (c. 1855–60) and hypothesis of auxiliary tracing. Authors' own work

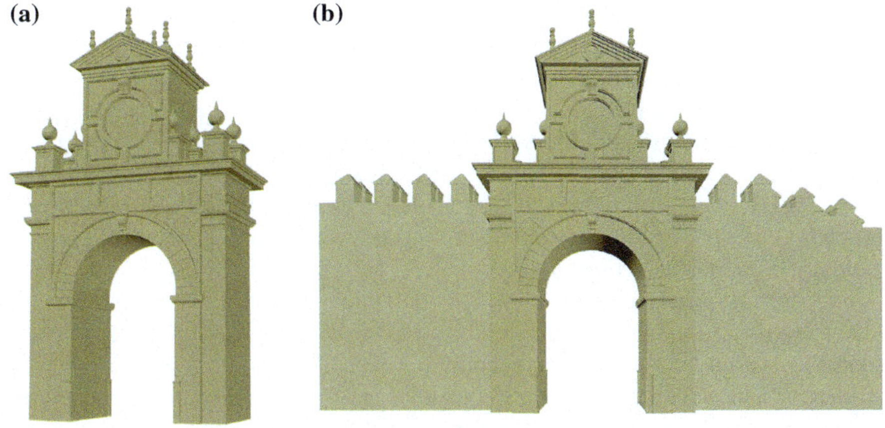

Fig. 16 **a**, **b** Volumetries of the Puerta Real and its adjacent walls. Authors' own work

a CAD program, thereby obtaining a three-dimensional geometric model that also includes the adjacent wall (Fig. 16a, b).

5 Final Considerations

The iconic interpretation of both imaginary and disappeared buildings has a long tradition. Thus, the Seven Wonders of the World, monuments now disappeared almost completely, could be only portrayed when the new graphic techniques during the Renaissance allowed their reproduction in multiple engraving collections. Mythical edifices such as the Babel Tower or Solomon's Temple were depicted in purportedly scientific studies. It is not uncommon to find representational sculptures or ancient building models in museums and art galleries. Each period used its own technical means to reproduce the architectural reality, either fictional or lost, and preserve its great value.

Around 1560, architect Hernán Ruiz II created a major renovation plan for the old medieval gates of the city of Seville, so they became an emblem of sixteenth century modernity. In their place, new and functional gates were erected, of a more monumental and symbolic character. The Puerta Real, located near the port and the Guadalquivir river, was one of the most beautiful gates in Seville. It appeared in images or engravings of that time that were widely disseminated throughout Europe, together with the famous phrase: "He who has not seen Seville, has not seen wonder". It was also drawn in the nineteenth century by prominent travellers, such as Richard Ford and Egron Lungren, and, prior to its demolition, the Puerta Real attracted the attention of Luis Masson, one of the first photographers in Seville, shortly after the invention of this technique.

However, the loss of the defensive value of the city walls led to their general abandonment, once considered a constrain to progress and the demolition of most of their parts and gates. The Puerta Real disappeared in 1864. Today, the place where it was once located offers no charm to both the city inhabitants and the numerous tourists interested in Seville's iconic corners. For this reason, there have been proposals for the reconstruction of these historic gates in the twenty-first century, such as the notable freehand drawings, created circa 2006 by Professor architect Rafael Manzano Martos, winner of the 2010 Driehaus award, featuring some of prominent disappeared gates, such as the Puerta Real (Fig. 17).

Graphic advances available today make possible to recover disappeared architectural and urban heritage in the form of precise virtual images, which may generate greater awareness in both society and public powers of their importance as a symbol of identity and a source of progress. In future research, it is intended to apply the methodology followed here to other demolished gates in Seville. A central aspect

Fig. 17 Drawing of the Puerta Real reconstruction as seen from the interior. Rafael Manzano-Martos (2006)

Fig. 18 Photographic montage of the reconstructed Puerta Real, seen from the outside. Authors' own work

to this research involves having access to extensive and precise graphic documentation on their architecture and environment. Masson's excellent photography and the preserved archaeological remains have made it possible for the first time to undertake the graphic reconstruction with great geometric precision of the Puerta Real, a heritage of major importance for the city.

Finally, the graphic recreation accomplished could also be transformed from a drawing into constructed reality. For this reason, an elementary graphic montage has been provided within its current urban environment (Fig. 18) in order to awaken public interest and debate, beyond the academic scope. Seville, a city with an exceptional historical architecture legacy, now holds the possibility of reconstructing, with scientific rigour, architectural icons tragically disappeared, such as the Puerta Real devised by architect Hernán Ruiz II.

References

1. Valor-Piechotta, M. (1996). El último siglo de la Sevilla islámica 1147–1248. Universidad de Sevilla & Gerencia Municipal de Urbanismo, Sevilla.
2. Lleó-Cañal, V. (1979). Nueva Roma: Mitología y humanismo en el renacimiento sevillano. Diputación Provincial de Sevilla, Sevilla.

3. Albardonedo-Freire, A. J. (2000). Documentación sobre la reforma de la Puerta de Goles (Real) entre 1560 y 1566. *Laboratorio de Arte, 13*, 11–37.
4. Collantes-de-Terán, A. [dir.], & Cortes-José, J. [coord.] (1993). Diccionario histórico de las calles de Sevilla. Ayuntamiento de Sevilla, Consejería de Obras Públicas, Sevilla.
5. Jiménez-Maqueda, D. (1999). Las Puertas de Sevilla. Una aproximación arqueológica. Fundación Aparejadores & Ediciones Guadalquivir, Sevilla.
6. Gámiz-Gordo, A., & Barrero-Ortega, P. (2019). Imágenes de un patrimonio desaparecido: La Puerta de Triana en Sevilla. *EGA Expresión Gráfica Arquitectónica, 24*(36), 80–91. https://doi.org/10.4995/ega.2019.10905
7. Ruiz, H. Libro de Arquitectura (c. 1569). ETSAM Homepage: https://n9.cl/iwiu0. Accessed 28 November 2021.
8. Navascués-Palacio, P. (1974). El libro de Arquitectura de Hernán Ruiz el Joven. E.T.S.A.M., Madrid.
9. Serlio, S., Extraordinario Libro di Architettura di Sebastiano Serlio, Architetto. Giouambattista & Marchio Selfa Fratelli, Venetia (1560). Homepage: https://archive.org/details/HArteR06T02. Accessed 28 November 2021.
10. Mal Lara, J., Recibimiento que hizo la muy noble y muy leal ciudad de Sevilla a la C.R.M. del Rey don Felipe N.S. (1570). BVMC Homepage: https://n9.cl/7xrpq. Accessed 28 November 2021.
11. Díaz-Zamudio, T., & Gámiz-Gordo, A. (2018). Views of Sevilla environs until 1800. In: Graphic Imprints The Influence of Representation and Ideation Tools in Architecture. pp. 1177–1188. Springer, Cham. https://doi.org/10.1007/978-3-319-93749-6_97
12. Cabra Loredo, M. D., & Santiago Páez, M. E. (1988). *Iconografía de Sevilla 1490–1650*. Fundación Focus & Ediciones El Viso.
13. Gámiz-Gordo, A., & Díaz-Zamudio, T. (2019). Sevilla extramuros en el siglo XVI: Tres vistas del Civitates Orbis Terrarum. Boletín de la Asociación de Geógrafos Españoles, 80, 2592. https://doi.org/10.21138/bage.2592.
14. Braun, G., & Hogenberg, F. (eds.), Civitates Orbis Terrarum, vol. I y IV. Colonia and Amberes (1572 y 1588)
15. Pozo-Barajas, A. (1996). *Arrabales de Sevilla, morfogénesis y transformación: El arrabal de los Humeros*. Universidad de Sevilla.
16. Rodríguez-Barberán, F. J. (ed.). (2007). La Sevilla de Richard Ford. 1830–1833. Fundación El Monte, Sevilla.
17. Rodríguez-Barberán, F. J., Gámiz-Gordo, A., & Robertson, I. (2014). Richard Ford: Viajes por España (1830–1833). Fundación Mapfre & Real Academia de Bellas Artes de San Fernando, Madrid. RABASF Homepage: https://n9.cl/tc495. Accessed 28 November 2021.
18. Plaza-Orellana, R. (2012). Egron Lundgren: un pintor sueco en Sevilla. Diputación de Sevilla, Sevilla.
19. Fernández Rivero, J. A., & García Ballesteros, M. T. (2018). Descubriendo a Masson, fotógrafo en la España del XIX. Ediciones del Genal, Málaga.
20. Tovar, B. (1978 [1878]). Las puertas de Sevilla en dibujos de B. Tovar. Colegio Oficial de Aparejadores, Sevilla
21. Raya-Rasero, R. (2006). Historia secreta de los derribos de conventos y puertas de Sevilla durante la revolución de 1868. Asademes ediciones, Sevilla.
22. Ramírez-Reina, F. O., & Vargas-Jiménez, J. M. (1996). Las murallas de Sevilla: Intervenciones arqueológicas municipales. In: El último siglo de la Sevilla islámica: (1147–1248), pp. 83–95. Universidad de Sevilla, Gerencia Municipal de Urbanismo, Sevilla.

Digital Reconstruction for the Analysis of Conservation State: The Transmission of Historical Memory of St. George and the Dragon Tile in San Michele Basilica Facade

Elisabetta Doria, Hangjun Fu, and Francesca Picchio

Abstract The case study presented describes the experimentation of 3D digital reconstruction carried out on a portion of the facade of San Michele Basilica in Pavia. The facade is characterized by numerous sandstone bas-reliefs, that have been deteriorating throughout the years, composing narrative cycles of historical kings and myths of the Lombard Kingdom. Starting from acquisitions carried out on the field and through the study of the historical conservation of the decorative tiles, researchers worked to create a digital model representing the actual state of conservation of the *St. George and the Dragon tile.* Such model has been compared to the one based on historical images and surveys carried out during the past century. The goal is to obtain an information system for conservation management protocols and for dissemination of this disappearing heritage. The method applied to the pilot case consists of geometric and material survey, documentation of the state of conservation and analyses with non-invasive techniques, leading to the three-dimensional reconstruction of the tile. The result allows to perceive shapes of the tile that are now illegible and understand the volumes as they should have been. The different conservation status of the tile can be appreciated on the physical object or in virtual mode, through VR system and FDM 3D printing.

Keywords Architectural survey · Cultural Heritage digitisation · Digital reconstruction · 3D printing · St. George and the Dragon · San Michele Basilica

E. Doria (✉) · H. Fu · F. Picchio
University of Pavia, 27100 Pavia, Italy
e-mail: elisabetta.doria@unipv.it

H. Fu
e-mail: hangjun.fu@unipv.it

F. Picchio
e-mail: francesca.picchio@unipv.it

© The Author(s), under exclusive license to Springer Nature Switzerland AG 2023

I. Trizio et al. (eds.), *Digital Restoration and Virtual Reconstructions*, Digital Innovations in Architecture, Engineering and Construction,
https://doi.org/10.1007/978-3-031-15321-1_10

1 Introduction

1.1 State of Art and Case Study

The current research concerns the virtual representations of architectural sites for the transmission of memory for knowledge and conservation of the cultural heritage [13]. The documentation process is developed as a pilot project on a portion of the facade of San Michele Maggiore Basilica in Pavia: the tile of *St. George and the Dragon*, which has an extension of 1.3×0.5 m.[1] This tile has been chosen as the pilot case because, starting from the beginning of the twentieth century, it has been the subject of numerous analyses and comparisons, that allow understanding how much the geometry and shape of this bas-relief have changed over time [18]. As analyzed in numerous studies "geometric memory is essential for the knowledge, protection, and conservation of architectural and historical heritage" [3]. Survey and representation via drawings and models introduce heritage conservation actions as tools for protecting the object and its memory [22]. Nowadays, the use of 3D replicas for the preservation practices and analysis of cultural heritage is well established. Physical replicas are perceived as "immediate, surface, temporary, modern, popular, and democratic" [12]. Add to this are immersive fruition and the ways of interacting with the digital data, in the interoperability, and in the possibility of adapting the information to the different communication strategies that the object reconstruction takes shape [24]. The development of three-dimensional digital models allows virtual and physical experiences via 3D printing. In cultural heritage, such technology represents a "challenging domain of application, in relation to the print quality and the low production costs" [28] and the 3D printing of cultural assets allows an immersive tactile experience for users [11]. 3D prints offer opportunities to change experiences for users: the prints can be easily reproduced in multiple replicas and their use is increasingly widespread nowadays [17]. The 3D printing of elements of the Cultural Heritage is nowadays considered one of the possible revolutions in the field of cultural heritage, thanks to its possibility of activating innovative uses in the conservation and communication field [2].

1.2 Historical Background

The Basilica of San Michele is a characteristic example of a Lombard Romanesque gabled basilica and can be dated back to the twelfth century thanks to a vast archive

[1] The research project involves the laboratories DAda- LAB and PLAY of University of Pavia. Project managers are Prof. Marco Morandotti and Prof. Sandro Parrinello. Research group is composed of Dott. Francesca Picchio, Ph.D. Stud. Elisabetta Doria, research fellow and Ph.D. Stud. Hangjun Fu and Ph.D. Stud Silvia La Placa (postproduction vectorial drawings) of University of Pavia.

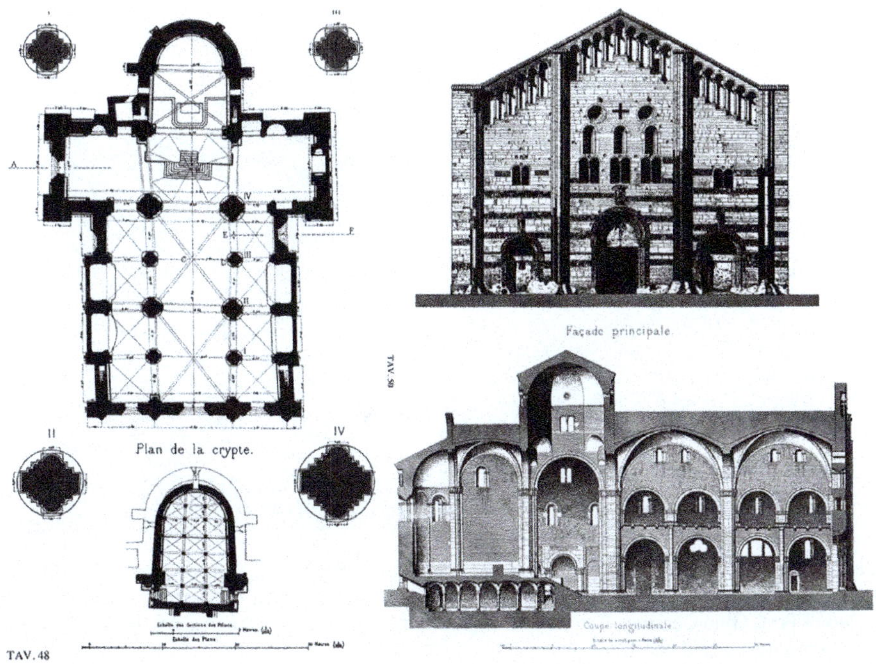

Fig. 1 Historical drawings of the architecture of the Basilica made by F. De Dartein: planimetry, main facade where the tile of the pilot case is present, and longitudinal section [10]

documentation [27]. Archival sources testify the existence of a special throne dedicated to the coronation of sovereigns of the Lombard kingdom during 642–1155 A.D. Within a portion of this timeframe, Pavia remained the administrative capital of the finished kingdom of the Lombard domination [27]. The Basilica has a Latin cross planimetry, with the central body consisting of three naves. Walls are characterized by ashlars of sandstone from the hilly quarries of the Province of Pavia [32] (Fig. 1).

Over the centuries, the Basilica has undergone important partial reconstruction and restoration interventions.[2] Starting from the mid-1900s, Gino Chierici, director of the Superintendence of Medieval and Modern Art in Milan, started a documentation methodology process involving the entire Basilica, via photography and the casting of decorative elements.[3] While the photographic acquisition procedure was carried

[2] An important structural intervention was made in 1489: 4 cross vaults were added, and the side aisles were enlarged. The major restoration works were carried out between 1860 and 1875; Those restorations have a historical-philological aspect and were aimed at restoring the original forms, involving the replacement of mouldings and decorative elements, both internal and external with some copies. Supervision of works has been in succession: Arch. Giovanni Battista Vergani and Eng. Sirio dell'Acqua.

[3] For more details see the SIRBeC PDF file available at: Lombardia Beni Culturali, Pavia. Chiesa di San Michele https://www.lombardiabeniculturali.it/stampe/schede/H0110-12345/. Compiler: Seccareccia, S. (1996). Upgrade: Schiavi, A., Seccareccia, S. [33].

Fig. 2 Comparison between the current status of conservation of a sandstone decoration (on the right) and its state of conservation in 1942 (on the left) thanks to the photographs of the working group of Prof. Gino Cassinis. The historical photograph (on the right) represents the status prior to the restoration carried out in 1963

out in detail on multiple tiles, casts were never modelled to safeguard the friable decorative stone. Photographic analysis was key for the development of a comparative analysis between the current status and the one surveyed in the mid-twentieth century [1, 9]. Attention to the state of conservation of the facade culminates with a series of intervention proposals starting in 1958. The realization of these proposals took place in 1963 with an assignment entrusted to Piero Sanpaolesi. Following the intervention of Sanpaolesi, in the 70 s campaigns began to investigate the state of conservation of the stones of the Basilica, which had not undergone interventions after those of 1963 [19] (Fig. 2).

The treatment carried out by Sanpaolesi[4] slowed the degradation process but with an unsatisfactory final effect at medium distance, as highlighted by studies conducted in the 1970s [18]. The products chosen for the intervention as well as the lack of routine maintenance led to the detachment of flakes of various sizes.[5]

[4] Piero Sanpaolesi (1904–1980) was an engineer, architect, restorer, architectural historian, and Italian academic. Protagonist of restoration culture of the second half of twentieth century, he was among the first to experiment with methods for the consolidation of stone materials used in architecture. See in references: Piero Sanpaolesi digital Archives University of Florence.

[5] For more information and technical references related to the conservation intervention and its criticisms see: Lombardini N. in *I restauri della facciata della Basilica di San Michele Maggiore a Pavia* (2018), pp. 68–72.

Fig. 3 Position of the *St. Geroge and the Dragon tile* with respect to the Basilica facade

Due to the aforementioned degradations, surveys and monitoring of the state of conservation activities were carried out. Gino Cassinis, full professor of topography at the Politecnico di Milano, made the photogrammetric survey of the most deteriorated bas-relief of the facade, identified with St. George and the Dragon [18]. The state of high deterioration of the stone surface after the 1963 intervention is evident from the comparison with historical drawings of the late nineteenth century by Fernand De Dartein, and photographic images were taken before and after the restoration, from the early twentieth century and in the 1970s [10]. Surveyed data has been used to produce comparative models aimed at the reconstructive analyses, allowing for the digital modeling and virtual reconstruction of *St. George and the Dragon tile* in its original state of conservation before the 1963 restoration interventions. The facade of San Michele Basilica is characterized by several bas-reliefs creating a decorative story and preserving the glorious past of the city and its territory. *St. George and the Dragon tile,* positioned in the facade next to the central portal at a height of 2.5 m, is one of the most degraded and compromised [25, 26, 34]. For this pilot case, three-dimensional modeling and historical reconstruction analysis were conducted to build a proposal for the valorisation of the cultural heritage and the development of a management system for conservation and an immersive virtual reality system (Fig. 3).[6]

[6] The geometric and material survey of the external walls of the Basilica is part of two consulting contracts as a research activity aimed at digital documentation and advanced vector representation: South external wall and roof survey stipulated between the Department DICAr of the University of Pavia and the company REA – Restauro e Arte S.r.l., Facade survey stipulated between the DICAr Department of the University of Pavia e Superintendence of Archeology, Fine Arts and Landscape for the provinces of Como, Lecco, Monza Brianza, Pavia, Sondrio, Varese (Functional area of architectural heritage).

2 Project Development Methods

2.1 Survey and Acquisitions of Data

The tile of *St. George and the Dragon* has been analysed since the 1940s[7] with state-of-the-art photogrammetry techniques.[8] Thanks to historical drawings, photographs, and studies conducted throughout the 1900s, this tile is the ideal case for experimenting with historical modeling and detailed documentation. The oldest reference to the state of conservation of this tile is due to Fernand De Dartein, who represented in detail numerous bas-reliefs of the Basilica in his work "*Etude sur l'architecture lombarde*" [10]. Starting from 2020, the DAda-LAB and PLAY laboratories of the University of Pavia started a documentation campaign on the external perimeter and on the facade of the Basilica, focusing on *St. George and the Dragon tile.*

Different digital tools have been used and integrated with the acquisition campaign: TLS laser instrumentation (Faro CAM2 S150), SfM photogrammetric techniques with UAVs and reflex cameras (DJI Mavic Mini, DJI RTK, Nikon D850/24–70) [5, 21], and Artec Eva 3D scanner for metric acquisitions of the tile.[9] The Artec Eva 3D scanner uses a close scanning technology (range 50–70 cm) with structured light pulses provides, in the scope of this project, an optimal mesh detail to acquisition time ratio. Acquisitions have to be performed at a close distance, using ladders for the raised position of the analysed object. The survey phase, involving the structured light laser Artec Eva, was conducted on the field by acquiring 4 separate scans, subsequently merged in post-production using Artec Studio software by Artec3D.[10] To facilitate the integration between laser and photogrammetric acquisition, six black, and white targets were positioned around the tile, to be captured by all the instruments involved (TLS, MLS, Artec Eva and cameras) enabling reliability checks through the registration different point clouds and mesh models (Figs. 4 and 5).

---

[7] Convention for the Safeguarding of the UNESCO Intangible Cultural Heritage, 2006: adds the enhancement of intangible values to the Franceschini Convention for the Safeguarding of Cultural Heritage of 1967.

[8] The photogrammetric acquisition saw the detection of 500 points in a time of one hour. See: Lombardi, 2018, pp. 39.

[9] The research was enforced in the collaboration "Agreement for the development of research activities about the digital documentation of cultural heritage and landscape using drones" between DICAr Department of Civil Engineering and Architecture of University of Pavia and iFlight Technology Company Limited, signed in February 2020, lasting three years.

[10] The Artec Eva is a handheld, colour scanner released in 2012, that can capture and process up to two million points per second. The scanner was designed for the capture of medium to large objects, as the tile case study. The device has a scan area of 214 × 148 mm at its closest range and 536 × 371 mm at its furthest, a 3D resolution of up to 0.5 mm, and a 3D point accuracy of 0.1 mm. Eva can operate at distances between 0.4 m and 1 m from the object, capturing up to 16 frames per second. Data can be exported as a mesh model in a.obj file. For more information see: Evers, P.: Eva 3D Scanner delivers accurate colour and structure data without markers, In: 3ders website, Archived from *the original* on 5 May 2016.

Fig. 4 *St. George and the Dragon tile.* Above: the historical drawing by F. De Dartein [10]. Centre: Images of the studies conducted by G. Cassinis: topographic points (center left) and isolines (center right). Below: comparison between the mid-1900s photo (left) and the current (right) highlights the severely altered state of conservation

Fig. 5 Comparison between the archive sources used to develop the historical reconstruction model. From left to right: drawing by F. De Dartein, [10], topographic points and isolines curves developed by prof. Gino Cassinis. For more details about the work conducted by Cassinis, see: Cassinis, G.: *Riproduzione di un bassorilievo con procedimenti fotogrammetrici.* In: Palladio, a. 6, n. 5–6. C. Colombo, Roma, 1942

Fig. 6 Survey phase conducted with the 3D Artec EVA via structured light on the pilot case of *St. George and the Dragon tile*. On the right: data post-production phase. Above are visible the 4 different scans acquired, highlighted in different colours. On the lower part, the final model obtained via scans merging

Following the registration of the four Artec Eva scans, a global mesh model was developed with the application of the texture. The final digital object is a textured, open mesh 3D model in which the level of metrical and geometrical accuracy has been validated with the laser scanner point cloud (Fig. 6).

2.2 *Virtual and Physical 3D Reconstructions*

The three-dimensional model obtained with the Artec Eva scanner represents the digital copy of the tile in the current state of conservation [31]. The project is aimed at preserving an architectural artifact of such artistic relevance, at the activation of museumization and protection processes as well as the enhancement of cultural heritage [6]. The digital reconstruction of the tile in a past state, based on historical archive sources, was used to feed immersive virtual reality systems [14]. Starting from the current state of conservation, via additive techniques, the backward process from the eroded object to the original shape was rebuilt via ZBrush, a digital sculpting software.

The default ZBrush sculpting tool, without modifiers or custom settings, displaces outwards the vertices over which it passes, simulating the addition of clay to a sculpture. Different effects and additions can be achieved with various brush modifiers, such as Strokes, Alphas, and edit curves [35]. The sculpting process was carried out by adding material from the model of the current status, using the drawing by De Dartein as a reference. The drawing was sized on the digital survey thanks to notable points that are still visible nowadays and used as a basis for 2D geometries (Fig. 7).

The third dimension was interpreted thanks to the oldest available isolines survey and topographic points, representing an intermediate state of conservation between the De Dartein drawings and the current image of the tile. Multiple aspects of the original documents were considered during the reconstruction phase. For example the intensity of the shadows visible in the historical images, useful to understand the high of the bas-relief geometries; the trend of isolines curves and of acquired points by Cassinis and Bezoari, useful to understand the differences in-depth, the comparison with the historical drawing of De Dartein, acting as a reference for the sculpturing process of the reconstructive model [18]. The digital model representing the actual image of the tile was transformed into a solid closed mesh model and then exported to proceed with the digital reconstruction [29]. The next phase of both

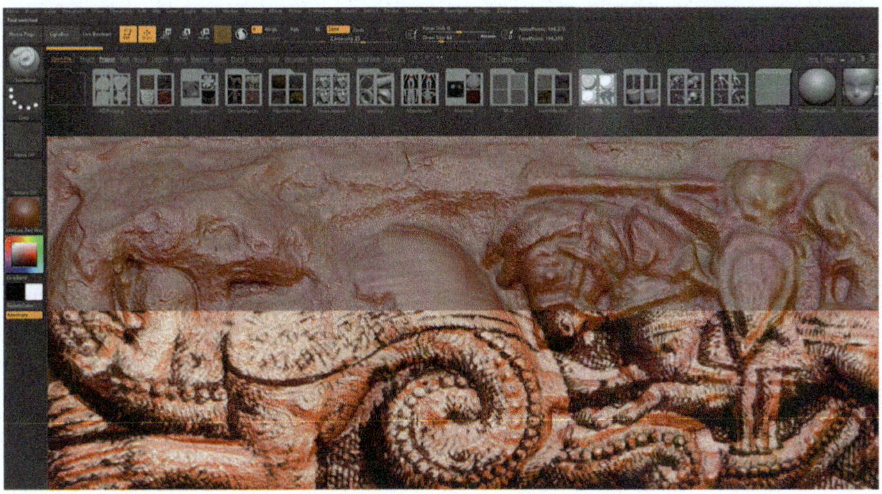

Fig. 7 Image representing the sculpting phase using ZBrush. Above: material addition phase. Below: overlap with the drawing by F. De Dartein

the enhancement of the current state and the historical reconstruction consists of the physical transposition of the digital models. The physical construction makes the tile an object that can be experienced from a tactile point of view, making the knowledge of its shape accessible to different categories of users and with different informative purposes [23] (Fig. 8).

The 3D printing of the two models, the current state of conservation and historical reconstruction, was made using FDM fused deposition printing with Ender 5-plus printer. The details of the models allow a 1:1 scale print reproducing exactly the real dimensions of the decoration [16]. The tile printed as a test for the calibration of the instruments was made on a 1:5 scale. Further experimentations were performed to test the feel of the printed material to the touch and to reproduce the material effect of the original bas-relief. The negative cast of the contemporary *St. George and the Dragon tile* was printed to create a plaster and a cement model to simulate the graininess and friability of the real stone (Figs. 9 and 10).

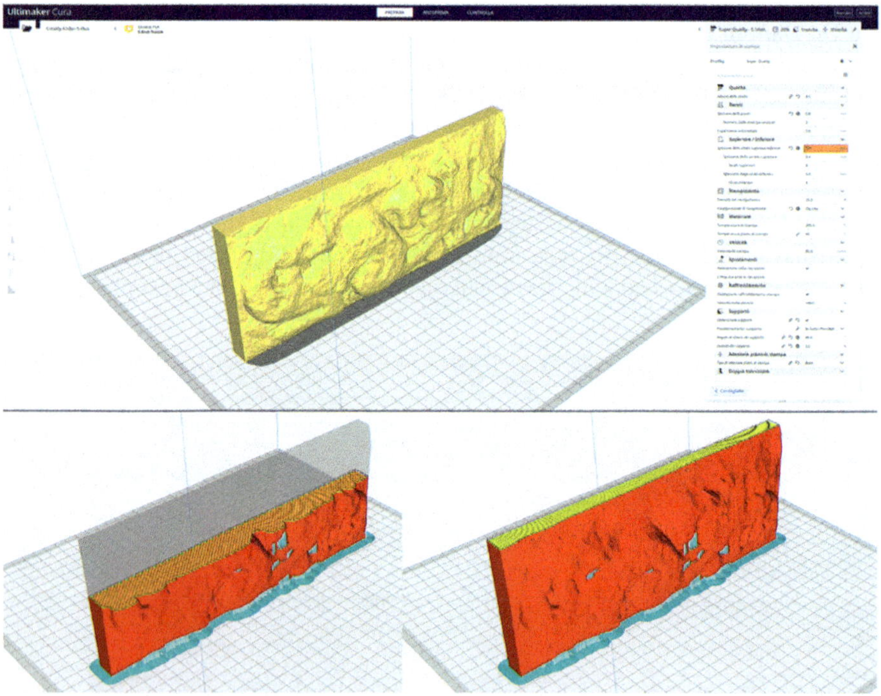

Fig. 8 Management of the 3D printing phase of the tile. The model was printed along the z axis, leveraging the properties of the printer, since such axis grants highest possible detail with a 0.4 mm extruder. In the images below: progress of the printing phase by levels with the parts added to adhere to the base and the supports for the decorations, highlighted in blue

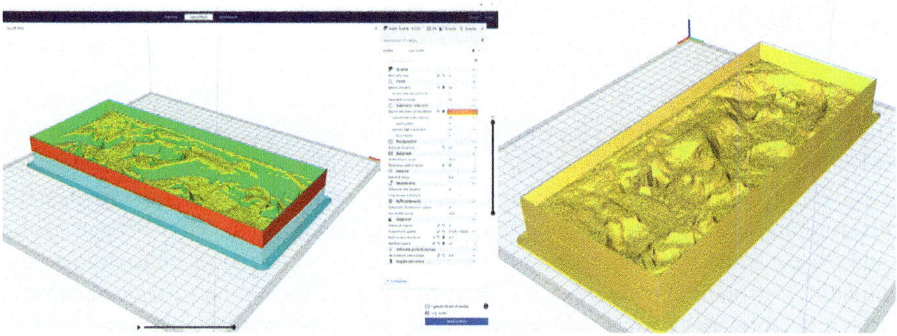

Fig. 9 Design of the 3D printing of the tile to produce a concrete and plaster cast

Fig. 10 On the left, from top to bottom: Tile reconstructed in PLA filament, tile in the current state of conservation in PLA filament, concrete tile in its current state of conservation. For the PLA prints, two filaments with different colours were used to test the visual rendering of the material. The concrete tile was produced with a 3D-printed cast. Concrete was tested with inert of different sizes to find the solution that was closest to the real material effect. Although the current tile is clearly visible to anyone who goes to the Basilica, the position in which it is located does not allow direct observation. This is possible with a three-dimensional physical reproduction which can be printed on a 1: 1 scale

2.3 *Usability and Potential of the Digital Three-Dimensional Reconstruction*

The sculpting process allows representing different states of conservation. The historical phases enable a visual timeline of the historical evolution, accelerations, and stabilizations of visible surface decay and pathologies, linking such alterations to the restoration and interventions as well as to the passage of time. The heritage dataset can be used to structure an information system for documentation, conservation, and enhancement planning. Digitization becomes the interpretation or transposition of an architectural or decorative heritage value over time, acting as a representative form of physical memory [13] and, in its perceived identity form, a symbol of collective memory. The transmission of memory, despite being intangible, is one of the purposes of digitization of built heritage, enhancing the aspects of both the physical conservation of the asset and its strength as an identifying element for the community (Fig. 11).

The enhancement of non-material aspects of the asset allows the development of conscious management of the object, to define the asset itself as "valued" [22]. Valorisation of built cultural heritage is connected to the policies and practices of use and preservation. Despite being a possible goal for the process of conservation, the valorisation phase can also represent a set of actions organized and coordinated

Fig. 11 Rendering of digital models that can be used as a basis for the dissemination and usability of the cultural product. Models can be easily inserted into immersive or orbiting reality platforms that can be connected to websites and tourist web portals [31]

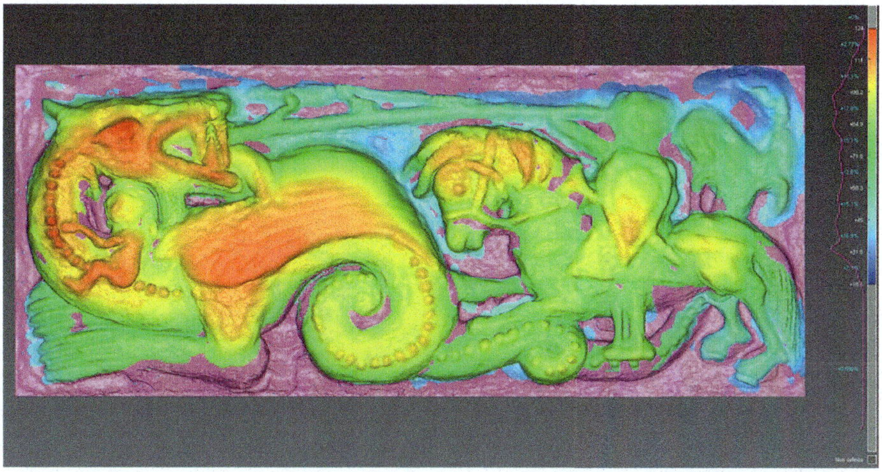

Fig. 12 Comparison between the two states of conservation of the superimposed tiles. The current state of conservation is coloured in pink and is visible only where no further disintegration is present. The reconstructed tile thickness (from blue to red) was mapped with progressive colours. Cooler colours indicate where it has been least rebuilt, while warmer colours indicate protruding points with more rebuilding. Thanks to this process are possible to evaluate the amount of eroded material and the relationship between time and material loss

in time (intervention time schedule), aimed at increasing the quality and identity of the individual heritage [15] (Fig. 12).

3D scanning tools and advanced modeling programs have encouraged studies on modeling as instruments for visualizing, designing, and enhancing new methods of analysis within the field of architectural heritage [8]. Documentation and valorisation can be performed through digital 3D models enriched with metadata, broadening to scope both for informative system level and specific technical level uses. Digital technologies applied to cultural heritage are the platforms in which models and metadata can be linked in interoperable and interactive formats [7]. In the field of Cultural Heritage, such technologies include 3D models and GIS data visualizers [4]. For this specific case, experimentation began to integrate into an ArcSCENE platform a mesh model of the facade of the Basilica with a simplified representation of the decorations and overlapping the selectable surface of the tile.

All mesh models are exported from the modeling software in their respective format (.rsh,.obj) are imported in ArcSCENE as shapefiles. Textual data related to the tile can be entered through the upload of excel files, as images or pdf through hyperlinks (see Fig. 13). The information system is still in early development, with CIDOC-CRM compatibility being targeted among future developments. Nowadays the information system contains the tile of *St. George and the Dragon* with associated photographs, links to the models, images of comparison between the models, information related to the dates of restorations and interventions carried out; the

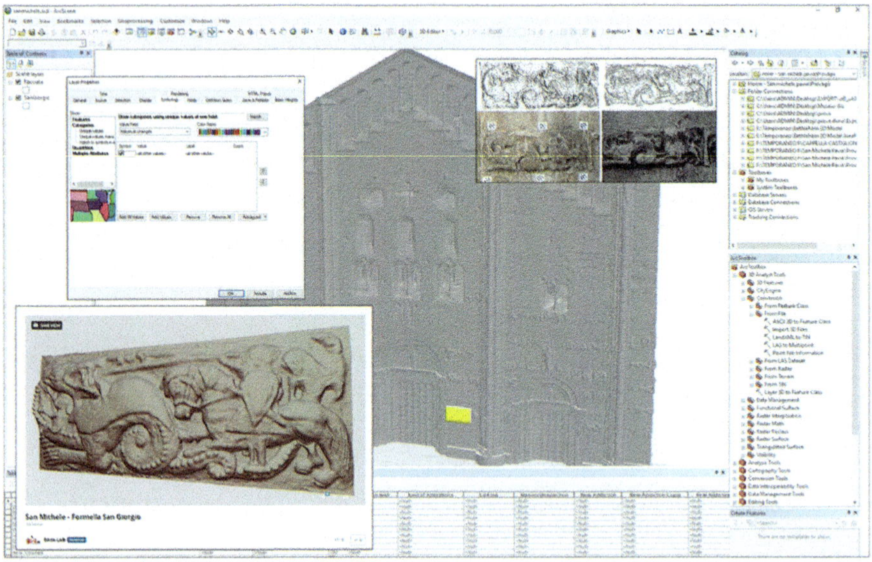

Fig. 13 The inclusion of the models in the georeferenced information system is one of the possible approaches for the planning of conservation operations. Once the tile model (in yellow) has been inserted into the reference system, it is possible to link metadata to the model. In this case study metadata refers to historical, technological, and material information acquired. Additional data, such as historical and current photographs, pdfs, and links are connected as pop-up hyperlinks

reconstructed model link is available online.[11] With the increase of a large number in the tile dataset, it will be possible to create thematic maps via queries insisting on the dataset (Fig. 14).

3 Conclusion

The process conducted on the pilot case follows the documentation of different states of conservation and their three-dimensional reconstruction via non-invasive techniques, suitable for friable and non-cohesive materials. The dataset acquired and post-produced during the experimentation phase represents a case study to structure the documentation and enhancement protocols of the decorative surfaces and elements of the Basilica external fronts. The goal is to be able to proceed with the historical reconstruction for several other tiles and decorations to have a digital and three-dimensional archive of the friable decorations. An information system for data

[11] It is possible to view the model of the tile reconstructed on historical data at the following link: Dada LAB: 3D digital model, the reconstruction of the ancient tile on the facade of Sab Michele in Pavia on Skecthfab: https://skfb.ly/6XPEB [30].

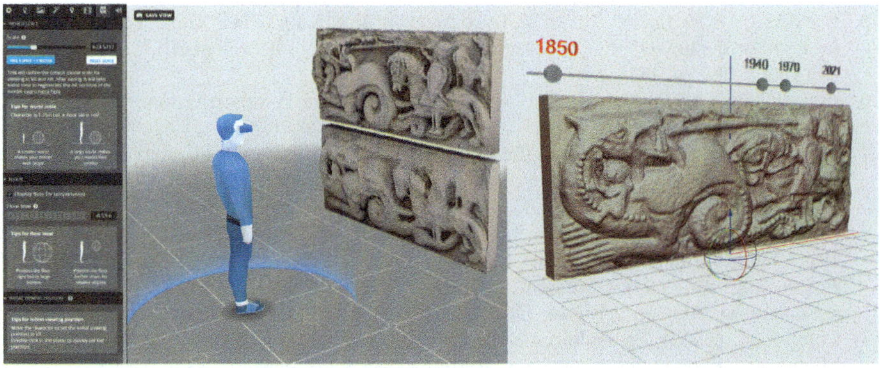

Fig. 14 Positioning of the three-dimensional model of the tile in a virtual environment for immersive reality aimed at the usability of the decorative object and at knowledge sharing. It is possible to view the digital model online [31]

integration was designed and built within the aim of this pilot case, as a single information platform for combined data analysis across multiple sources. Information systems, allowing for the integration, entry, and update of data, allow the development of analyses to monitor the state of conservation over time and guarantee an updatable archive of the asset and management protocols. The historical progress of the conditions of the tiles can also be used to share knowledge of the heritage, including models in usability platforms to improve the virtual musealization of the Basilica and its possible application in the information-tourism field. The 3D models shown in this contribution were created to develop the perception of cultural heritage and decorations no longer existing and no longer completely visible and appreciable nowadays, both via virtual representation in information systems and with physical artifacts.

References

1. Albertini Ottolenghi, M. G. (1986). La pietra del San Michele: i restauri ottocenteschi. In C. Re-possi, & M. Dragoni (Eds.), *La pietra del San Michele: restauro e conservazione*. Pavia: Società per la conservazione dei monumenti dell'arte cristiana in Pavia.
2. Amico, N., Ronzino, P., Vassallo, V., Miltiadous, N., Hermon, S., & Niccolucci, F. (2018). Theorizing authenticity—practising reality: the 3D replica of the Kazaphani boat. In P. Di Giuseppantonio Di Franco, F. Galeazzi, & V. Vassallo (Eds.), *Authenticity and cultural heritage in the age of 3D digital reproductions*. McDonald Ints. for Archaeological Research, Cambridge.
3. Balzani, M., & Maietti, F. (2017). Architectural space in a protocol for an integrated 3D survey aimed at the documentation, Representation and conservation of cultural heritage. In Diségno. 1, Jul. 2017, pp. 113–122.
4. Batarin, L., Bertozzi, S., & Moretti, E. (2014). Tecnologia Gis per la manutenzione programmata dei beni culturali. In *Proceedings of the International Conference, Preventive and Planned Conservation*, Monza, Mantova 5–9 May 2014.

5. Bertocci, S., & Parrinello, S. (2017). The drawing of Hadrian's Villa in Tivoli. Extensive survey for heritage documentation. In Disegnarecon, Vol. 10/ n.19 - december 2017. Archaeological Drawing ISSN, pp. 1828–5961.
6. Bonacini, E. (2011). Nuove tecnologie per la fruizione e la valorizzazione del patrimonio culturale. Aracne Editrice, Roma.
7. Brusaporci, S., & Centofanti, M. (Eds.) (2010). Sistemi Informativi Integrati per la tutela, la conservazione e la valorizzazione del Patrimonio Architettonico Urbano. Gangemi Editore, Roma.
8. Brusaporci, S., Graziosi, F., Franchi, F., & Maiezza P. (2019). Remediating the historical city. Ubiquitous augmented reality for cultural heritage enhancement. In A. Luigini (Eds.), *Proceedings of the 1st International and Interdisciplinary Conference on Digital Environments for Education, Arts and Heritage EARTH 2018*, pp. 305–313. Cham, CH: Springer.
9. Chierici, G. (1942). Le sculture della Basilica di San Michele a Pavia. Edizioni de L'arte, Milano, re-edition by the association "Il Bel San Michele" onlus.
10. De Dartein, F. (1995). Architettura Romanica a Pavia, con disegni e rilievi, 1865–1882. Alberto Arecchi (eds). Associazione Culturale Liutprand. Pavia.
11. Di Giuseppantonio Di Franco, P., Camporesi, C., Galeazzi, F., & Kallmann, M. (2015). 3d printing and immersive visualization for improved perception of ancient Artifacts. *Teleoperators and Virtual Environments, 24*(3), 243–264.
12. Di Giuseppantonio Di Franco, P., Galeazzi, F., & Vassallo, V. (2018). *Authenticity and cultural heritage in the age of 3D digital reproductions*. McDonald Institute for Archaeological Research, Cambridge.
13. D'Urso, S. (2018). Il riuso delle memorie dei luoghi. In G. Minutoli (Eds.), *VI Convegno Internazionale sulla documentazione, conservazione recupero del patrimonio architettonico e sulla tutela paesaggistica*. 11–14 ottobre 2018. Gangemi Editore, Roma.
14. Galasso, F., Parrinello, S., & Picchio, F. (2021). From excavation to drawing and from drawing to the model. The digital reconstruction of twenty-year-long excavations in the archaeological site of Bedriacum. *Journal of Archaeological Science: Reports, 35*, 2021.
15. Gasparoli, P. (2012). La manutenzione preventiva e programmata del patrimonio storico tutelato come prima forma di valorizzazione. In: TECHNE, Vol. 02/2012. Firenze University Press, Firenze.
16. Inzerillo, L., Di Paola, F. (2017). From SfM to 3D print: automated workflow addressed to practitioner aimed at the conservation and restauration. *International Architecture Photogramm. Remote Sensing Spatial Inf. Science, XLII-2/*W5, pp. 375–382.
17. Joy, J., & Elliot, M. (2018). Cast aside or cast in a new light? The Maudslay replica maya casts at the museum of archaeology and anthropology, Cambridge. In: Di Giuseppantonio Di Franco, P., Galeazzi, F., & Vassallo, V. (Eds.), *Authenticity and cultural heritage in the age of 3D digital reproductions*. McDonald Institute for Archaeological Research, Cambridge.
18. Lombardini, N. (2018). I restauri della facciata della Basilica di San Michele Maggiore a Pavia. Tipografia PI_ME editrice S.r.l., Pavia.
19. Lombardini, N. (2021). Il restauro della facciata della basilica di San Michele Maggiore a Pavia e l'opera di Piero Sanpaolesi. Aracne Editrice, Roma.
20. Lo Turco, M., Piumatti, P., Calvano, M., Giovannini, E. C., Mafrici, N., Tomalini, A., & Fanini, B. (2019). Interactive digital environments for cultural heritage and museums. building a digital ecosystem to display hidden collections. In: DISEGNARECON, vol. 12/n.23.
21. Miatton, C., & Parrinello, S. (2017)/ Descriptive models for the management of the maintenance work on the facade of the Basilica of San Michele in Pavia. In: 3DMODELING &BIM. Dei tipografia del Genio Civile, Roma.
22. Morandotti, M., Besana, D., Zamperini, E., & Cinieri, V. (2014). La gestione sostenibile del patrimonio immobiliare tra riuso e valorizzazione. In S. della Torre (Eds.), *La strategia della Conservazione programmata, dalla progettazione delle attività alla valutazione degli impatti*. Proceedings if the International Conference Preventive and Planned Conservation, Monza-Mantova 5–9 maggio 2014 (pp.131–140). Nardini Editore, Milano.

23. Neumüller, M., Reichinger, A., Rist, F., Kern, C. (2014). 3D printing for cultural heritage: Preservation, accessibility, research and education. In M. Ioannides, & E. Quak (Eds)., *3D Research Challenges in Cultural Heritage*. Lecture Notes in Computer Science, vol 8355. Springer, Berlin, Heidelberg.
24. Parrinello, S., & Dell'Amico, A. (2021). From survey to parametric models: HBIM systems for enrichment of cultural heritage management. In *From Building Information Modelling to Mixed Reality*. Springer Tracts in Civil Engineering, Switzerland.
25. Peroni, A. (1967). San Michele di Pavia. Cassa di Risparmio delle Prov. Lombarde, Milano.
26. Peroni, A. (2017). San Michele di Pavia. PI_ME editrice, Pavia.
27. Soriga, R. (2012). Pavia e la Certosa. Guida storico-artistica alla città di Pavia e alla sua Certosa. Edizioni Selecta, Pavia.
28. Scopigno, R., Cignoni, P., Pietroni, N., Callieri, M., & Dellepiane, M. (2015). Digital fabrication techniques for cultural heritage: A survey. In *Computer Graphics Forum*, number 1, Vol. 36, pp. 6–21.
29. Tucci, G., & Bonora, V. (2011). From Real to ... "Real". A review of geomatic and rapid prototyping techniques for solid modelling in cultural heritage field. In: International Archives of the Photogrammetry, Remote Sensing and Spatial Information Sciences, Volume XXXVIII-5/W16, Trento, Italy.
30. Artec 3D Homepage, https://www.artec3d.cn/cases/chinese-temple-archway-3d-scanning. Accessed 15 November 2021.
31. Dada LAB: 3D digital model, the reconstruction of the ancient tile on Skecthfab, https://skfb.ly/6XPEB. Accessed 3 January 2022.
32. Lombardia Beni Culturali. Basilica di S. Michele Maggiore, https://www.lombardiabeniculturali.it/architetture/schede/PV240-00004/. Accessed 23 September 2021.
33. Lombardia Beni Culturali, Pavia. Chiesa di San Michele, https://www.lombardiabeniculturali.it/stampe/schede/H0110-12345/. Accessed 4 January 2022.
34. Piero Sanpaolesi Archives University of Florence, https://archivi.unifi.it/entita/83d538e3-4ead-4595-8c29-36e2fe1dd47d/Sanpaolesi,%20Piero%20%28Rimini%201904%20-%20Firenze%201980%29/informazioni. Accessed 24 November 2021.
35. Pixologic Homepage, http://docs.pixologic.com/reference-guide. Accessed 20 November 2021.

Architecture and Archeology. Virtual Reconstruction of Ipi's Tomb TT315 in Deir-el-Bahari, Theban, Egypt

Ernesto Echeverria Valiente, Flavio Celis D'Amico, and Fernando da Casa Martín

Abstract This article describes data acquisition and processing operations with a 3D laser scanner for the tomb TT315 in the archaeological area site in Deir-el-Bahari, Theban (Egypt). We collected data manually, with drawings and photographs of the most relevant elements, together with a TLS (Terrestrial Laser Scanner) campaign, that has allowed us to create a high-resolution point cloud with the morphology of the outer area and galleries, rooms, and interior wells. All data was processed to obtain a 3D virtual object. This paper will be used as a template to develop the specific images required for each of the survey purposes: knowledge transfer, dissemination and the creation of the cartographic base for the whole researchers' team. All the material allows for knowledge transfer to society through built-in 3D models. The research conditions at the archaeological site of Deir-el-Bahari forced us to create our own methodology. These conditions are linked to the site conditions: climate, limited access, bureaucracy permits; the distance between the university and the site and the site's morphology, with galleries excavated into stone caves with structural irregularities. The object of this study is important on its own, because we know very little about funerary complexes of the Middle Kingdom. On the other hand, the process to carry out the research is allowing us to assign new uses to virtual representation and reconstruction techniques.

Keywords Archeological survey · Laser scanner · Egyptology · 3D digital fabrication · Heritage preservation

E. E. Valiente (✉) · F. C. D'Amico · F. da Casa Martín
Department of Architecture, University of Alcalá, Madrid, Spain
e-mail: ernesto.echeverria@uah.es

F. C. D'Amico
e-mail: flavio.celis@uah.es

F. da Casa Martín
e-mail: fernando.casa@uah.es

© The Author(s), under exclusive license to Springer Nature Switzerland AG 2023
I. Trizio et al. (eds.), *Digital Restoration and Virtual Reconstructions*, Digital Innovations in Architecture, Engineering and Construction,
https://doi.org/10.1007/978-3-031-15321-1_11

1 Introduction

This article describes the process of documenting graphically the tombs of Deir-el-Bahari archaeological complex, in Luxor (ancient Thebes), Egypt. The case study to exemplify this process is carried out on vizier Ipi's tomb (TT315), located on the northeast slope of the site. We documented how we collected several types of data, from approaching the study area and its surroundings to the description of the structure of the tomb and its various dependencies; the collection of geometric data; the virtual reconstruction of the space, and the reconstruction to scale in 3D of parts of it.

A rigorous digital collection of geometric data has been carried out and once it was processed and interpreted, it was converted into a point cloud. Subsequently, and from this cloud, we produced the working documents that were shared with all researchers working on this project. The whole of the excavation is documented in a multipurpose database in 2D and 3D. This documentation also supports the progress of the rest of the team by providing them with a multimedia support in which they can reference their findings and work, considering that the team includes subjects such as Archaeology, Egyptology, Restoration, Epigraphy, Architecture, Forensic Anthropology, Geology, Photography or Papyrology, among others.

All these files were also used for outreaching purposes, to disseminate findings and to advance the typological study of these tombs. With that in mind, we also created a VR of the space to preserve the real one and offer the possibility of printing it by using CAD/CAM techniques (Computer-Aided Design/Computer-Aided Manufacturing) in 3D at scale [1].

On the other hand, it should be noted that the work as specialists in architecture on this type of archaeological excavation has transcended the merely disciplinary field of architectural drawing. It extends to the recognition of the structural safety of the complex, studying the degree of conservation and stability of the site, and when detecting any danger, giving reinforcement proposals. The graphical documentation of the site supports the correct decision-making in this regard [2]. The documentation work that is presented here, still in the research phase, together with other investigations currently underway [3], will generate an innovative database on the funerary architecture of ancient Egypt [4].

The study is divided into several stages which correspond to the travel periods of the expeditions authorized by the MSA (Ministry of State of Antiquities of Egypt). Field work usually lasts between 30 and 45 days a year. The team includes around 45 international researchers plus almost 100 local workers. The excavations and all the findings are documented in situ while each expedition is on. During the rest of the year, the researchers use the recovered material to continue developing their investigations in their institutions [5].

The field work carried out, both in the architectural field and in other specialties, must meet a series of conditions:

- *Legal*: The local administration issues work permits with severe limitations. It is absolutely forbidden the removal of any material from the excavation areas. Therefore, all findings should be documented in situ regardless of their size or the possibility of being removed.
- *Temporary*: It involves working in a site located thousands of kilometres away from the University and in which accessibility is limited by government permits issued long time in advance and for a specific time frame.
- *Climatic*: Field work must be carried out in extreme hot weather, bringing people and machines to the limit.
- *Accessibility*: The challenging accessibility to the work site forces us to have all processes carefully planned to reduce the time and cost of transporting the equipment. Also, the difficult orography (in general of the area, and particularly of the site) makes it difficult to move around and use the equipment throughout the day (Fig. 1).
- *Transversality*: Field work should serve several purposes. The results must be obtained in different formats to serve all the teams involved. Each department must be able to use it both for their field work, and for the academic research that must be carried out during the rest of the year. It should also serve to develop scientific dissemination material based on 3D models, to develop material to share with society, for public exhibition and to produce scale models using 3D digital manufacturing [6].

Fig. 1 Scanner leveling maneuvers inside the tomb

2 Bakground: Historical Analysis of the Monument and Construction Stages

Although the Egyptian civilization is well documented, the sites that are the object of our study, belonging to the nobility, were heavily plundered and there is not much documentation about them beyond some hieroglyphs found in Ipi's sarcophagus. These tombs have been looted and emptied of their contents at different times in history [7], both at the beginning of their construction and in more recent times during the twentieth century. They were also reused as minor tombs for years.

At the end of the 11th dynasty (2005, BC, approx) the victory of King Mantuhotep II (Nebhepetra Mentuhotep) established a new and powerful dynasty at Thebas [8]. The construction of the funeral complex of Mentuhotep II in Deir el Bahari and the construction of the private tombs of the most powerful members of the government and the army, which contributed to the reunification of the empire, signifies the beginning of the Middle Kingdom of Ancient Egypt (Fig. 2).

The peak of el-Qurn was the key element that made them choose this area to set up the funeral temples of Mentuhotep II and Hatshepsut at its base [9], as it happened with the tombs that lie under the great pyramids of the Egyptian empire [10] (Fig. 3).

In order to achieve the maximum number of discoveries in a short period of time, Winlock allocated an average of one month per grave [11], until in 1923, another MMA team discovered the main ramp of Hatshepsut's mortuary temple, and therefore, Winlock had to leave the Deir el-Bahari necropolis and focus all his attention into the royal monument [12].

Ipi's tomb is the first studied by the MKTP expedition, following Winlock's footsteps [13]. It is a monument excavated in the bedrock with 22 m. length corridor leading to a square chamber of around 5×5 m. At mid distance, we found a shaft in the middle of the corridor (model chamber) with another chamber to the left (north-western section) of the shaft. In the left section of the chamber there are two side

Fig. 2 Aerial view of the work area. Funerary complex with Hatshepsut and Mentuhotep temples

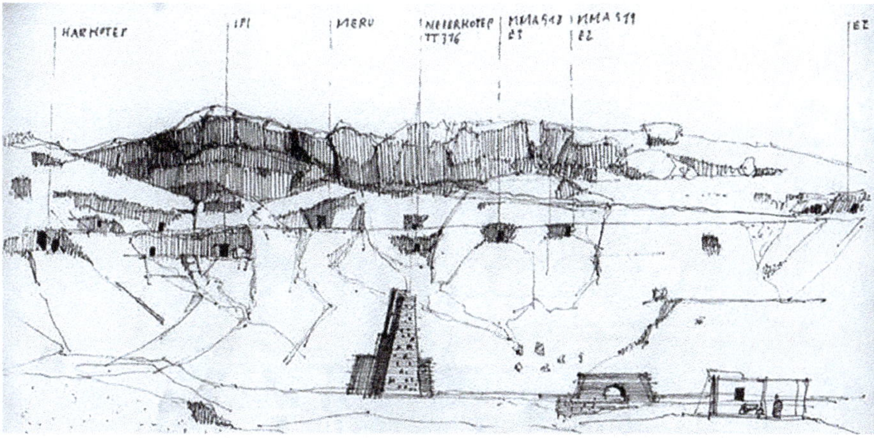

Fig. 3 Sketch of the tombs in the work area in Deir-El-Bahari

chambers, mainly one in the left side with a dimension of 4 × 4 m. and another one in the shape of a niche (probably for a single coffin) in the frontal wall, measuring 3 × 1 m. In the middle of the square chamber's floor, we find the beginning of a ramp toward the funerary chamber, with a descending slope of 30° that leads to the sarcophagus chamber of 4.7 × 4.5 m. In this chamber, there are two horizontal levels, one is a virtual level at the height of the ramp, and another one in around 1 m down in depth. In the lower section, there is a limestone sarcophagus, while in the upper section we would find –if it had not been taken away– the pavement of the chamber floor. The geometry of this chamber is complex, mainly because multiple fragments of stone make it difficult to understand the original plan of the room (Fig. 4).

3 Methodology

The research work corresponds to the study of the integrated lifting operations of the archaeological remains, including the study of the outdoor spaces and the indoor spaces of Deir el-Bahari complex's tombs (galleries, tomb shafts and annexed dependencies), beginning with the study of Ipi's (Fig. 5). The work belongs to the processes of research on the digitization of data collection based on previous photogrammetry techniques, used for decades [14]. In addition to the metric knowledge of the site [15], the re-elaboration of the 3D documentation can ser to improve the typological and constructive knowledge of it [16]. At a later research stage, we plan to produce a virtual reconstruction of the original state, where the linings of the tomb, currently looted, could be visualized [17]. There are a few preserved fragments of such linings that could serve as a template for this reconstruction.

The method used adjusts itself to the previously exposed conditions (temporary, climatic, accessibility and transversality) and to the different technologies available

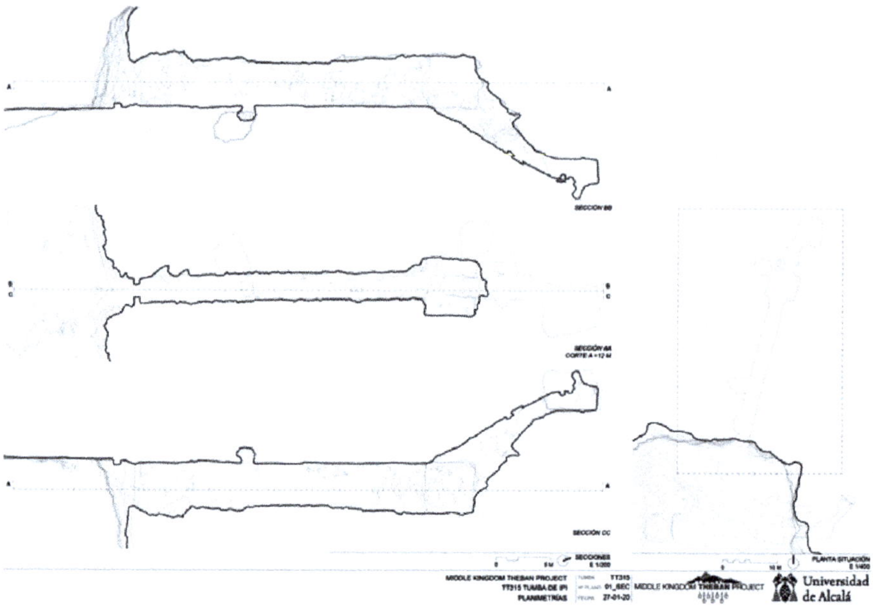

Fig. 4 General plan and sections of Ipi's tomb (TT 315). Drawing after data processing

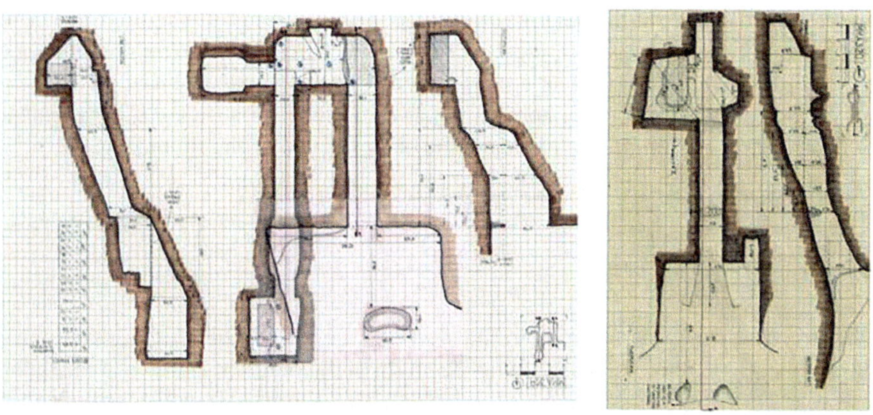

Fig. 5 Some plan sketches of Egyptian tombs in Deir-El-Bahari work Area. Drawings by the authors made with tape measure and laser range finder

at any given time. Depending on the object of study and its context, the methods of study and analysis keep adapting [18].

The data collection is carried out during a week of field work, avoiding interference with other researchers or with the rest of workers. It starts once the tomb is cleared of debris and search remains. In a first stage, plan and section sketches of the spaces are made to establish a work plan to optimize the actions and adapt them to the short time available (Fig. 5). These sketches are also used by the rest of the researchers as temporary support for their work, since detailed digital planimetry can only be carried out once all the information has been processed.

After this first manual or "analogue" approach, it can start the process of scanning. The key ideas to consider are:

- Creating an indoors point cloud.
- Creating an outdoors point cloud.
- Connecting outdoor space and indoor space.
- Connecting with the rest of the sites in the same area.

The tombs are not architectural constructions with regular layouts. They are cavities in living rock with multiple inflections and arbitrary relief, conditioned by the geology of the place. A balance must be struck between the degree of detail and the number of precise scan positions of the scanner. On top of that, each time the scanner is repositioned, a stable and levelled set up must be achieved.

In a regular and orthogonal construction, a gallery could be scanned with 2 or 3 scanning positions, so that there are no "shadows," and the entire space is covered. In our case, as it is a "cave", it is necessary to increase the positions of the scanner if we want to avoid dead areas that might affect the image and the point cloud model.

In the outdoor areas, the determining factors are the great luminosity, the intense heat, and the roughness terrain. This requires working in the early hours of the morning, even before sunrise. When the heat increases, the scanner automatically shuts down because is set to do it when it exceeds its upper operating temperature range. This forces teams to stop working and keep the scanner in a shady area, inside one of the galleries, while waiting to reach a temperature apt to work again.

The other challenge is the irregularity of the terrain that delays and hinders access to the marked positions to set the scanner up and lengthens the placement time by having to apply continuous levelling adjustments to the tripod. In the outdoor area, the work is done with less precision and fewer scan points (Fig. 6).

Fig. 6 Placing the scanner in outdoor areas

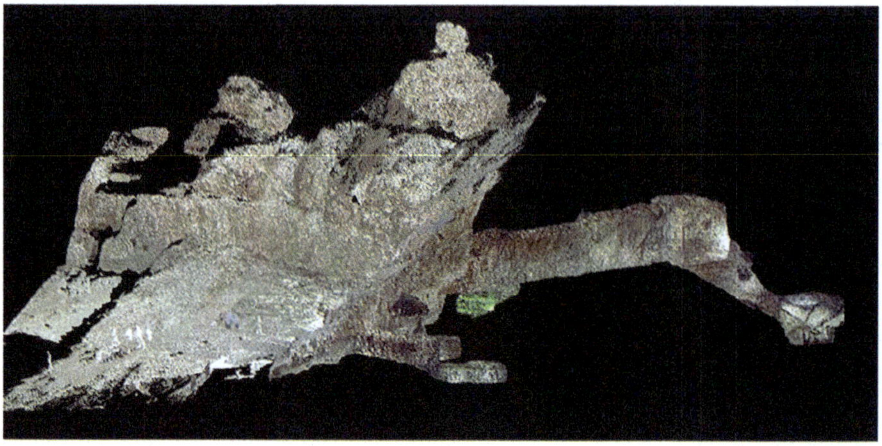

Fig. 7 External and internal scanning of Ipi's TT 315 tomb

The goal is to get a general image of the terrain in which the tombs are located (Fig. 7). To guarantee that the connection between the outdoor and indoor spaces is as precise as possible, the common points must be reinforced to allow a correct georeferencing and connection between both areas.

The georeferencing of the site is carried out by total station and GPS obtained outside, and the reference points are physically marked within the scanned area to incorporate them into the survey. The Egyptian team carries out a photogrammetric survey through pictures taken with a Canon 6000 professional camera, treated later on Agisoft Photoscan software. This associated photogrammetry, although it is carried out independently of the laser scanner survey, serves as a double check and allows comparison between both techniques, which improves the collection of data.

4 Results

The works carried out consisted of manual data collection, with drawings and photographs of the most relevant elements, together with a TLS (Terrestrial Laser Scanner) campaign. They allowed for the creation of a high-resolution point cloud of the morphology of the outer area, galleries, rooms, and interior wells. All this data was processed to obtain a virtual 3D object, which allows for greater knowledge and dissemination and provides the cartographic template for the research.

In the case study at hand, the tomb of Ipi, a FARO scanner focus 3D multi-sensor 120 laser scanner was used for the data collection. In the study of the inner area, 73 scanning positions were carried out, with an average duration 6.5 min, 1/5 resolution, a range of less than 5 m (typically 1.5 m), 6.3 megapixels per shot, and 3200 × 2040 pixels. To achieve the maximum possible scanning and overlap surface, we used the

Fig. 8 Noting the position of the scan points

scanner's maximum sweep both vertically (90º to −60º) and horizontally (−180º to 180º). Where this was not possible, we sought the maximum possible range.

In the exterior scan, being a mountainside, the sweep was limited to the essentials. In this way, it was possible to reduce the weight of the files and the working times, which is critical due to the aforementioned climatic conditions. Generally, the scanner sweep was vertically from 30º to −60º and horizontally from −60º to 60º, avoiding shots of open sky or parts of the slope that were more than 20/25 m away (Fig. 8).

The default resolution and quality were made leveraging accuracy and storage capacity, previously tested in test scans (Fig. 9). It should be noted that the complex morphological structure of the site asks for a significant number of scans, and that assembling them in a single file implies memory management that both the program and the computers must be able to manage.

Each scanning generates a group of folders and files. On one hand, a spherical photograph (*.jpg) is saved, based in a flat cylindrical projection of the object in each

Fig. 9 Laser scanner parameters

LASER SCAN PARAMETER	FARO FOCUS 3D
CHOSEN OUTLINE	Outside from 40m Inside from 15 m
RESOLUTION AND QUALITY	Resolution: 1/6 Quality: 4x
SCAN RANGE	Vertical range: from -60º to 90º (variable) Horizontal range: from 0º to 360º (variable)
SENSORS	Use clinometer: activated Use compass: activated Use altimeter: activated
COLOR PARAMETERS	Weighted measuring to the center
ADVANCED CONFIGURATION	Clear contour: activated Clear sky: activated

Fig. 10 Spherical photograph attached to scan. Chamber of the sarcophagus of Ipi TT 315

position (Fig. 10). On the other hand, the scanner generates a database file (*.xml), which includes all the positioning data of the scan, as well as different types of files related to the point cloud data of each position. Once all the scans are performed, a total point cloud of the object is created using FARO's Scene Software. It is the result of the sum or fusion of all the scans, overlapping the common points.

Based on this total point cloud, it is possible to save images of it, or to conduct visual tours through the model, which regardless the practical use, gives a realistic sensation of the object. In order to obtain graphic documents and plans to work on, it is necessary to isolate a point cloud area using the "clipping box" tool to create a 3D mesh. From this model, the key points are chosen and exported in *.pts, *.ply or *.e57 formats to be accessed from the Autodesk ReCap software [19].

The use of ReCap as the main software allows for the recovery of additional existing information in the scan files such as name, location of scan points, or information about each point in relation to the cloud position or its colour. The information is debugged by removing uninteresting or duplicated elements. In shady areas of the scans that remain open, it is possible to close the meshes. Ultimately, the information is uploaded to Autodesk servers and the mesh is generated, obtaining a *.obj file. This service is currently provided free of charge to authorized Autodesk users and helps obtain results without a high-performance computer.

When the continuous 3D mesh is generated (Fig. 11), it is possible to use editing software to process it, such as "ZBrush", Maya, 3dsMax, or Meshmixer. In this study, Meshmixer program was selected because of its free license and intuitive software. If errors and glitches remain on the object's faces or shadow areas, we must select mesh faces using "clean mesh" tool of Rhinoceros.

It is difficult to represent these tombs with typical architectural plans (plan, elevation and section), because they have a very irregular geometry (Fig. 12), so we must make a series of cuts, parallel to a specific direction (horizontal or vertical).

The use of meshes supports accessing to information for use in 3D models [20], overcoming the mere global views, which are classic documentary reconstructions of Architectural Heritage [21]. From the mesh you can obtain both scale printed

Fig. 11 From the access to the sarcophagus, final scan Ipi's TT 315

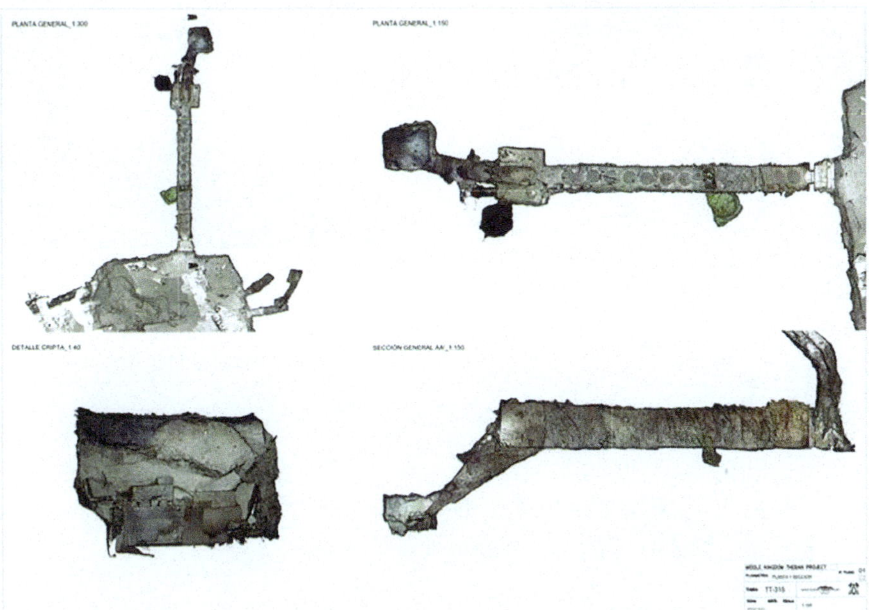

Fig. 12 Plan and general section of the tomb of Ipi from a 3D model

models, as well as any section, at the desired distance The thickening tasks allow for lots of possibilities of digital manufacturing of prototypes and models from meshes, and they are carried out thanks to the "Workflow" proposal.

With SLS and SLA technologies, we can obtain lightweight files with high-resolution models. When creating the mesh, it is important to thicken it evenly and regularly, as this will avoid gaps that would store resin in SLA technology, or nylon powder in SLS technology (Fig. 13). In manufacturing by CNC milling, this method accelerates the Gcode generation and avoids problems such as the access of the software to internal faces or the flipping of the mesh (Fig. 14).

These scale models are very useful for dissemination in seminars and conferences to the public. Some outstanding initiatives could be found in the British Museum [22] or in the exhibition "The other Nefertiti" [23]. There are also initiatives carried out by public agencies to digitize the management of Heritage elements, as in the National Institute of Archeology and History (INAH) in Mexico D.F. [24], or at the UAH [25]. Another field of work in which 3D Heritage virtual reality is used is in video games [26], such as the one that has helped to rebuild Notre Dame after the fire of April 2019 in Paris [27].

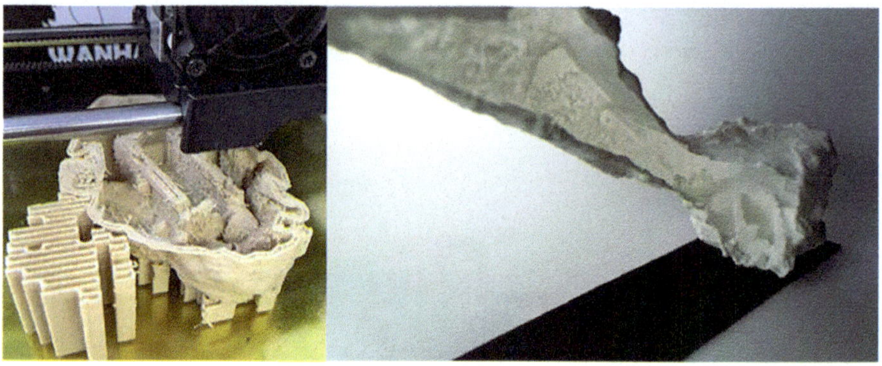

Fig. 13 SLS Nylon 3D printed model

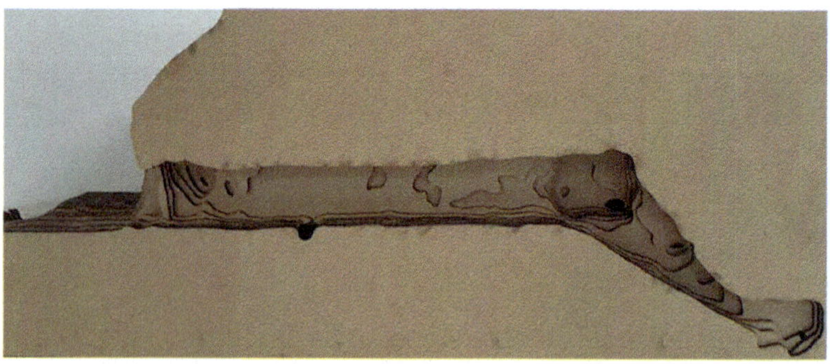

Fig. 14 CNC printed model using layers of MDF wood boards

5 Conclusions

The research work on the archaeological site of Deir-el-Baharí, Thebes, Egypt was developed in a context quite unusual if compared with the common case in the study of the heritage. For this reason, our team had to work with a methodology that is refined and improved in each expedition for the study of such a unique site.

These specific conditions are linked to the characteristics of the place: climate, accessibility, administrative management, remoteness and limited working time. They are also linked to the morphology of the object of study: galleries excavated in living rock and structural irregularity.

All these conditioning factors imply a different work planning from the usual work carried out under normal circumstances in Europe, and its most outstanding characteristic is redundancy: to the normal work campaign, sequenced by working days, a series of days are added to cover possible contingencies, adding a 30% increase to the expected time. This increase in time also affects the cost of the survey, which is increased by similar amounts, having an impact on the logistics and budget of the expedition.

After data collection, the information is processed in the UAH digital graphics laboratory, obtaining three-dimensional objects to work on. The success of the research is to obtain reliable data from these tombs, and to systematize their treatment, for better knowledge and dissemination of this particular heritage. With these studies, we achieve the proposed goals, as well as ensure the preservation for the future, avoiding the loss of heritage in the event of an accident or attack. In recent years, some organizations have launched platforms in which they collect all the data and documentation of the Heritage for its conservation and dissemination. Some examples are the CIPA Congresses (International commitée of Architectural Photogrammetry), Factum Foundation [28] or Cyark [29].

6 Innovation and Original Aspects

The research carried out has two sides: regarding the object of study, it is possible to apply new approaches to typological, constructive, and historical understanding of the funeral complexes of the Middle Kingdom of Egypt. Regarding the process itself, we tried new methodologies supported by virtual representation and reconstruction techniques that improved the scope of those processes. In addition, it was possible to try the data collection system in a hostile environment. Finally, it was also possible to use the virtual digital three-dimensional models and 3D prints for the dissemination of archaeological and architectural heritage.

References

1. Daneshmand, M., Helmi, A., Avots, E., Noroozi, F., Alisinanoglu, F., & Sait Arslan, H. (2018). *3D Scanning: A comprehensive survey*. arXiv.org:1801.08863 [cs.CV].
2. Adembri, B., Cipriani, L., Fantini, F., & Bertacchi, S. (2015). Reverse designing: An integrates method for interpreting ancient architecture. *Scientific Research and Information Technology, 5*(2), 15–32.
3. Mozas, A. T., Pérez, J. L., Gómez, J. M., Martinez, J. L., & Jiménez, A. (2020). 3D models of the qh31, qh32 and qh33 tombs in Qubbet el Hawa (Aswan, Egypt). *The International Archives of the Photogrammetry, Remote Sensing and Spatial Information Sciences*, Vol. XLIII-B2–2020, XXIV ISPRS Congress, pp. 1427–1434.
4. Terlikowski, W., Gregoriou-Szczepaniak, M., Sobczyńska, E., & Wasilewski, K. (2021). Advantages of using 3d scanning in the survey of architectural monuments on example of archeological sites in Egypt and Russia. *Archives of Civil Engineering, LXVII*, 189–201.
5. Morales, A., Abd El-Hady, R., Accetta, K., Arranz, M., Bardají, T., Carrillo, M. F., Celis, F., Díaz, C., Dorado, E., Echeverría, E., Falk, S., Gracia, C., Ikram, S., Illana, S., Kruck, E., Luciañez, M., Martínez, O., Meza, D., Mora, P., Ortiz, J., Osman, M., Sánchez, R., Serova, D., Shared, H., Spinelli, D., Tarek, A. & Yamamoto, K. (2018). The Middle Kingdom Theban Project: Preliminary report on the University of Alcalá Expedition to Deir el-Bahari, Fourth Season 2018. *Studien Zur Altagyptischen Kultur, 47*, 183–222.
6. Neumüller, M., Reichinger, A., Rist, F., & Kern, C. (2014). 3D Printing for cultural heritage: Preservation, accessibility, research and education. In M. Ioannides, & E. Quak (Eds.), *3D Research Challenges in Cultural Heritage*. Lecture Notes in Computer Science, vol. 8355. Springer, Berlin-Heidelberg, pp.119–135.
7. Carter, H. (1912). *Five years' explorations at Thebes: A record of work done 1907–1911*. Frowde.
8. Bull, L. S. (1924). A new vizier of the Eleventh Dynasty. *Journal of Egyptian Archaeology, 10*, 1, 15.
9. Szafrański, Z. E. (2014). The exceptional creativity of hatshepsut. In *Creativity and Innovation in the Reign of Hatshepsut, Studies in Ancient Oriental Civilization*, 69, 125–137. Chicago, University of Chicago.
10. Soliman, R. (2009). *Old and middle kingdom tombs*. Golden House Publications.
11. Winlock, H. E. (1914). Excavations at Thebes in 1912–13 by the Museum's Egyptian expedition. *Bulletin of the Metropolitan Museum of Art, 9*(1), 18–19.
12. Winlock, H. E. (1947). *The rise and fall of the Middle Kingdom in Thebes*. Macmillan.
13. Morales, A., Abd El-Hady, R., Accetta, K., Alarcón, S., Bardají, T., Celis, F., Echevarría, E., Falk, S., Hussein, M., Ikram, S., Ortiz, J., Osman, M., Sáez, A., Sánchez, R., Serova, D., Shared, H., Yamamoto, K., & Zidan, E. (2017). The middle kingdom Theban project: Preliminary report on the University of Alcalá Expedition to Deir el-Bahari, Third Season. *Studien Zur Altagyptischen Kultur, 26*, 143–168.
14. Navarro, P. L. (2012). fotogrametría digital automatizada frente a los sistemas basados en sensores 3d activos. *EGA Revista de Expresión Gráfica Arquitectónica, 20*, 100–111.
15. Echeverría, E., Celis, F., & Casa, F. (2015). Drawing as a research tool: Reconstruction of the trip time of the urban image of Alcala de Henares. *EGA, Revista Expresión Gráfica Arquitectónica, 25*, 180–191.
16. Raposo, J. F. (2010). Identification of architectural drawing and design as methodological processes of scientific reseach in architecture. *EGA Revista de Expresión gráfica Arquitectónica, 15*, 102–111.
17. Goitia, A. (2010). Restituir, Redibujar, Aventurar. Estrategias para documentar tres puertas monumentales de Madrid. *EGA Revista Expresión Gráfica Arquitectónica, 15*, 74–83.
18. Mesa, A., Regot, J., Núñez, Mª. A., & Buill, F. (2007). Estrategias de modelado en la Sagrada Familia. *EGA Revista Expresión Gráfica Arquitectónica, 12*, 92–101.

19. Moreno, K., & Echeverría, E. (2019). The use of digital tools for the preservation of architectural, artistic and cultural heritage, through three-dimensional scanning and digital manufacturing. *The International Archives of the Photogrammetry, Remote Sensing and Spatial Information Sciences, XLII-2/W9*, 501–506.
20. Chías, P., Abad, T., De Miguel, M., & Llorente, M. P. (2020). The Transparente in the Basilica of the monastery of El Escorial. *Nexus Network Journal, 22*, 1133–1154.
21. De Miguel, González, M., Fernández-Cabo, M. C. (2020). Mannerist polyhedral wood ceiling in Castroverde de Campos (Zamora, Spain). *Informes de la Construcción, 72/557*, e326.
22. Bryce, E. (2018). Reprinting history (The British Museum uses 3D scanning to bring artifacts to life). from http://www.wired.co.uk/article/reprinting-history British Museum Repository, and https://sketchfab.com/britishmuseum. Retrieved December 2018.
23. Nelles, N., The Other Nefertiti, Venice Architectural Biennale 2016. from https://www.allove rsky.com/The-Other-Nefertiti. Retrieved December 2018.
24. Navarro, P, Herráez, J., Mora, Á., Barros, H., & Denia, J. L. (2011). Aplicaciones de la Tecnología de digitalización tridimensional por la coordinación de monumentos históricos del instituto nacional de arqueología e historia (INAH) en México DF. (2009 y 2010). *EGA Revista Expresión Gráfica Arquitectónica, 17*, 42–53.
25. Olmo, L., Castro, M., & And López-Macia, M. (2012). La utilización del Scanner en el registro arqueológico: La experiencia de la Universidad de Alcalá. *Visual Archaeology Review, 3*(5), 93–97.
26. Porcuna, D., Córdoba, R., Sanz Cabrera, R., & Montes Tubio, F. (2016). Metodología para la reconstrucción virtual interactiva en modo videojuego del Patrimonio cultural. Aplicación al castillo medieval de Torreparedones (Baena). *EGA Revista Expresión Gráfica Arquitectónica, 28*, 278–287.
27. Tallon, A. (2012). La technologie 3D au service de Notre-Dame. In D. Sandron, J. P. Cartier, & G. Pelletier (Eds.), *La grâce d'une cathédrale—Notre-Dame de Paris*. París: La Nueé Bleue, 171–181.
28. Lowe, A., Digital Recording in a time of Iconoclasm, Mass Tourism, Celebrity Culture and Anti-Ageing. British Council. from https://www.factum-arte.com/ind/569/articles. Retrieved December 2017.
29. Chías, P., & Abad, T. (2015). Spatial data infrastructures and Spanish cultural heritage: The INSPIRE framework applied to the monastery of El escorial. *Journal of Map & Geography Libraries, 11*(2), 245–265.

Diachronic 3D Reconstruction of a Roman Bridge: A Multidisciplinary Approach

Germano Germanó

Abstract The study analyses multidisciplinary approaches in the reconstruction of the different construction phases of a Roman masonry bridge, implemented through the application of archaeological stratigraphic method and 3D reconstruction. The case study, a bridge in Canosa di Puglia over the Ofanto river (Italy), is today a masonry structure with five arches of similar dimensions. However, historical sources report a three-arched monument, built in the second century CE as part of the vast imperial construction project of the Via Traiana, of which only the piers, the abutments and the foundations are left. Despite numerous restorations, the large central arch collapsed in 1751 and the bridge was rebuilt in a very different shape, losing all evidence of its Roman monumentality. By cross-referencing data collected from historical documents, on-site surveys, photogrammetric processing and metrological analysis, the research aimed to reconstruct three-dimensionally the diachronic evolution of the monument.

Keywords Masonry bridge · 3D reconstruction · Archaeology of architecture · Cultural heritage · Roman archaeology

1 Introduction

The evolution in recent decades of 3D technologies and graphic representation has significantly contributed to the accuracy of historical and scientific analyses and to the development of increasingly plausible reconstructive hypotheses of archaeological sites and architectural artifacts [1–3].

A specific research field is represented by masonry bridges, particularly of Roman and Medieval age, found in a large part of the Mediterranean and European areas, abandoned or still in use, hidden or visible. The complexity of approaches to the study of ancient bridges is due to the physical conditions and degree of deterioration in which the monuments have survived to the present day, to the different architectural

G. Germanó (✉)
Scuola Superiore Meridionale, University of Naples Federico II, Naples, Italy
e-mail: germano.germano@live.com

© The Author(s), under exclusive license to Springer Nature Switzerland AG 2023
I. Trizio et al. (eds.), *Digital Restoration and Virtual Reconstructions*, Digital Innovations in Architecture, Engineering and Construction,
https://doi.org/10.1007/978-3-031-15321-1_12

stratifications that have often changed the original shape, to the frequent lack of historical and graphic sources, and to their location on rivers or in other inaccessible sites that hinder the survey and the understanding of all their parts.

Starting from the comprehensive studies on ancient bridges [4–6], the recent technical literature is very rich and varied in the topics addressed.

A recurrent aspect is the assessment of safety and seismic risk [7–9] often related to research, restoration, and conservation projects [10–14].

Several studies analyse how to integrate digital photogrammetry procedures with non-destructive investigation techniques, structural analysis and virtual reality [15–18].

In the experience of recent years, combining survey and stratigraphic analysis has proven to be an efficient method to develop a three-dimensional reconstruction of single phases or diachronic evolution of monuments [19–23], but also to propose a visualisation of reconstruction of different contexts, from small spaces to urban scale [24–26].

This brief summary highlights the importance of a multidisciplinary approach to the analysis of ancient masonry bridges and the necessity of connecting archaeological documentation and virtual reconstruction in the earliest stages of the survey [27].

A case study is represented by the masonry bridge of Canosa di Puglia, in southern Italy, which historical sources report as dating back to Roman times and consisting of three large arches, instead of the current five. Although an archaeological campaign was conducted in 1985, directed by Professor Raffaella Cassano from Bari University [28], to study the *platea* (a flat foundation in the riverbed made of *opus coementicium* and paved with stone slabs), methodologically accurate construction hypotheses have not been made, especially for the Roman phase.

2 The Case Study

The bridge (Fig. 1) dates back to around 109 C.E., when the emperor Trajan ordered the construction of the Via Traiana [29–33], a road that provided an alternative to the Via Appia in order to connect Rome with the strategic port of *Brundisium* (Brindisi) via Benevento.

Fig. 1 The bridge over the Ofanto river near Canosa. View from the South

That said, the presence of an earlier bridge on the site cannot be excluded, since the Via Traiana largely followed the course of the earlier Via Minucia (110 BCE) [34].

In the Middle Ages, the Via Traiana made up part of the Via Francigena, the route by which Christian pilgrims from northern and western Europe reached the ports of Apulia, from which they took ship to the Holy Land; later its route was followed by *tratturi* (drovers' roads), travelled by shepherds bringing their flocks from central Italy to pasture in the Apulian plains [35].

Although numerous restorations are attested during the reigns of Septimius Severus, Caracalla, the Tetrarchs and Constantine [36], in the fourteenth century [37] and perhaps in the sixteenth century [38], until the eighteenth century the bridge was preserved as a structure with a large central arch, as shown in some early modern cadastral representations preserved in the *Archivio di Stato di Foggia* (National Archive of Foggia) [39].

In another document, a drawing made in 1749 by a local surveyor, Francesco Delfino [40], the bridge is clearly represented with three arches, composed of a large central arch and two smaller lateral ones supporting a structure made of large stone blocks (Fig. 2).

In 1751, the main arch collapsed and a new pier was built in the middle of the span in order to provide a greater structural support. With this addition, two new, smaller arches took the place of the single central one, greatly altering the shape of the previous structure.

The current shape of the structure, which consists of five arches of different sizes, is the direct result of the mentioned series of modifications and additions made over several eras. The only remains of the original structure are probably the piers, the abutments, and the foundation *platea* [41].

In twentieth century the *platea* was covered by a concrete walkway, and during the Second World War the bridge was further damaged by retreating German troops who mined it [37, 42] and opened a breach in the flood containment walls.

Fig. 2 The bridge before the collapse of 1751, as drawn by Francesco Delfino in 1749. Archivio di Stato di Foggia, Italy

Years after the last reconstruction of the bridge, vehicular traffic was moved a few hundred metres further north to a new road while a new bridge was built to adapt to contemporary standards. As it was no longer in use, the decline of the ancient bridge accelerated and very little maintenance has been carried out in recent decades.

A post-graduate thesis in architectural heritage and landscapes, defended by the author in 2019, has been an opportunity to focus on various scattered data from different fields of research and put them together to achieve a scientific result that could reconstruct the history and the possible original shape of this monument.

3 Method

Once the research goal had been set up, the following steps were taken:

- archival research, through published sources and unstudied documents,
- metrological analysis according to ancient units,
- comparison between historical and contemporary maps and photos,
- on-site direct and indirect surveys (i.e., triangulation and drone survey),
- photogrammetry operations,
- identification and recording of the different Stratigraphic Units of Masonry,
- cross-checking of the collected data.

3.1 Interpretation of the Monument Through Metrological Analysis

While archival graphic sources, such as cadastral plans, often only outlined the three-arch layout of the bridge, written sources mention important dimensional data that allow for the verification of what survived of the original structure.

The first known written mention of the bridge dates back to 1541, when a Polish traveller described it as *altissimum pontem muratum* ("a very high masonry bridge") [43], but the first description of the dimensions of the bridge is not recorded until 1584, when another traveller, from Italy, described the bridge as *ponte bellissimo fatto* ("a very well-made bridge") and reported the size of the central arch as 128 palms long and 40 palms high [38].

A further crucial record dates to 1749, when an expert surveyor, Francesco Delfino, due to the need for maintenance work, drew up a report with the exact measurements of the arches, in which he warned of dangers to its stability: [...] *l'arco maestro, di larghezza palmi 112, d'altezza dal piano sino alla sommità palmi 44, e di fronte palmi 5, gli due archi laterali di larghezza palmi 50 per ciascuno, e di altezza palmi 25* ("...the main arch is 112 palms wide, 44 palms high from the base to the top, with a front of 5 palms, while the two lateral arches are 50 palms wide each, and 25 palms high") [40].

Table 1 Ancient units and their mutual correspondence

Unit	Roman feet	Neapolitan palms	Metres
Roman Feet	1	1.1225	0.2960
Neapolitan Palms	0.8909	1	0.2637
Meters	3.3784	3.7922	1

Since the two documents report the dimensions in palms, it is possible to assume that they were referring to the Neapolitan palm, i.e., the unit of measurement in use in the Kingdom of Naples, corresponding to 0.26367 m [44] (Table 1).

Although the conversion between the units of measurement of the different construction periods (Roman feet, Neapolitan palms, metres) may seem a mere mathematical operation, the cross-checking of survey data with ancient metrology has proved to be an efficient method [45, 46] that can explain the complexity of all the elements involved and aid in developing reliable hypotheses about the bridge construction phases through the verification against current dimensions. Although a certain degree of approximation is to be expected in the aforementioned documents, the margin of error seems to be small enough so as not to affect the consistency of the data taken into account for the purpose of the research.

Converting the numerical data reported in Delfino's report, the most recent, into Roman feet (Table 2) reveals the presence of round numbers in the major elements of the bridge, such as the arches. Such dimensions are highly plausible for a new building, e.g., in order to facilitate construction operations. In particular, the data reported for the central arch would confirm the dating of the project to the Trajanic period, due to the recurrence of dimensions of 100 Roman feet in monuments of that period, like Trajan's Column in Rome, also called *centenaria* (hundred-feet-high) for the same reason [47]. Since these dimensions correspond to about 30 m in the metric system (100 feet = 29.6 m), it is possible to suggest that the central arch of the original bridge of Canosa was among the widest of the imperial period, if we take into account that in the Italian peninsula, the major arch of the Augustus Bridge at Narni (104 feet = 30.8 m) [6] would be the only one to exceed it, albeit slightly.

Table 2 Bridge measurements conversion chart

Element of the bridge	Neapolitan palms	Metres	Roman feet
Main arch (span)	112	29.53	99.78 $\simeq$ **100**
Main arch (height)	44	11.60	39.20 $\simeq$ **40**
Lateral arches (span)	50	13.19	44.54 $\simeq$ **45**
Lateral arches (height)	25	6.59	22.27
Front (parapet)	5	1.31	4.45

3.2 From the Large-Scale to Detail: On-Site Survey and Compilation of SUM Context Sheets

At a later point in the research, a classical ground survey was conducted in order to provide all the necessary architectural information. Due to the size of the bridge, the difficulty of accessing the river and the dense vegetation that usually covers a large part of the structure, the successive use of drone technology and aerial photogrammetry [48] proved to be fundamental to verify the previous argumentation on ancient measures and provide a basis for the virtual reconstruction [49] (Fig. 3).

After the large-scale survey of the monument, the building's virtual construction sequence was read according to a method that belongs to a dedicated discipline called Archaeology of Architecture or Building Archaeology [50–55] developed within the study of medieval architectural contexts but now commonly applied in archaeological contexts relating to architectural structures. This methodology is characterized by a detailed analysis of the different parts of the structure carried out through the identification of the different *USM*s (*Unità Stratigrafiche Murarie*) [56, 57], Italian for Stratigraphic Units of Masonry [58, 59], Structural Stratigraphic Units [60] or Wall Stratigraphic Units [61]. Since each SUM represents the result of an individual action, construction process (positive units) or removal (negative units) that characterize a building, their classification and the individuation of direct relations of anteriority and posteriority between them is important to reconstruct the history of the building [62]. From the analysis of the different SUMs it is possible to observe the diachronic evolution of material preferences in the use of large, regular, hard limestone elements during the Roman period, smaller rusticated elements in the post-1751 phase, while smaller, not always regular, soft tuff blocks are used in later restorations. This qualitative difference affects enormously not only the stylistic but also the mechanical characteristics of the structure, especially for the parts in contact with water such as the paving stones of the *platea*. Due to the action of the river the restored sections of the platea are in fact much more worn than the original, Roman sections. After the numbering of each SUM and the establishment of their stratigraphic relationships, forms were elaborated and compiled for each SUM (Fig. 4), following the layout and the content of the context record sheets used in archaeology [63]. The different units were then represented graphically in the elevation drawing of the monument to provide a preliminary overview of the different phases.

4 The Virtual Reconstruction

One of the most challenging parts of the study is the reconstruction, which can be based on consistency criteria like objectivity of data, deduction from formal characteristics of the buildings, comparisons with similar structures, analogy with theoretical models such as Roman modules, hypotheses with different values of reliability [64].

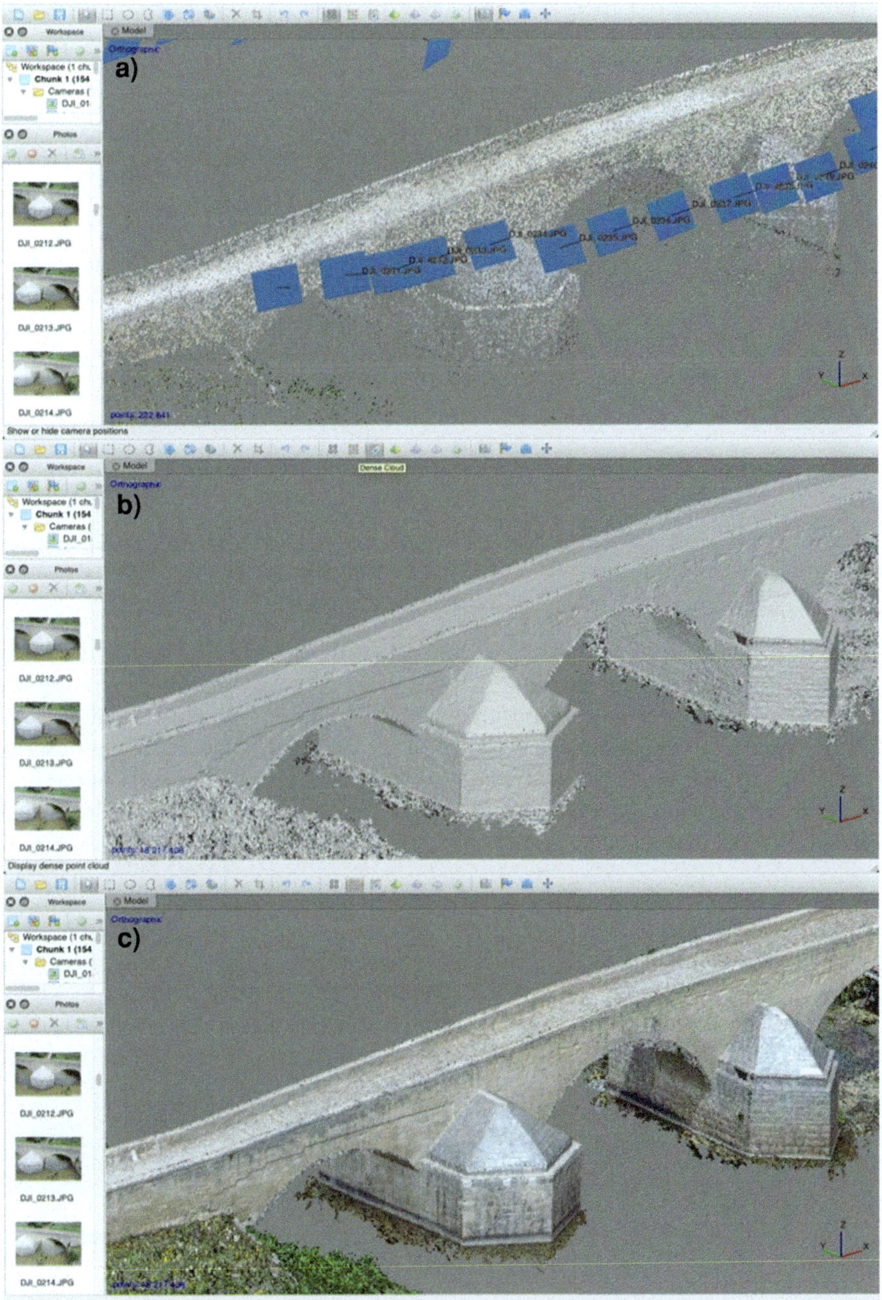

Fig. 3 The steps performed by Photoscan software to obtain the reality-based model of the bridge. **a** estimation of the camera information and positions and point cloud generation; **b** triangulation and mesh generation; **c** texture creation from oriented cameras

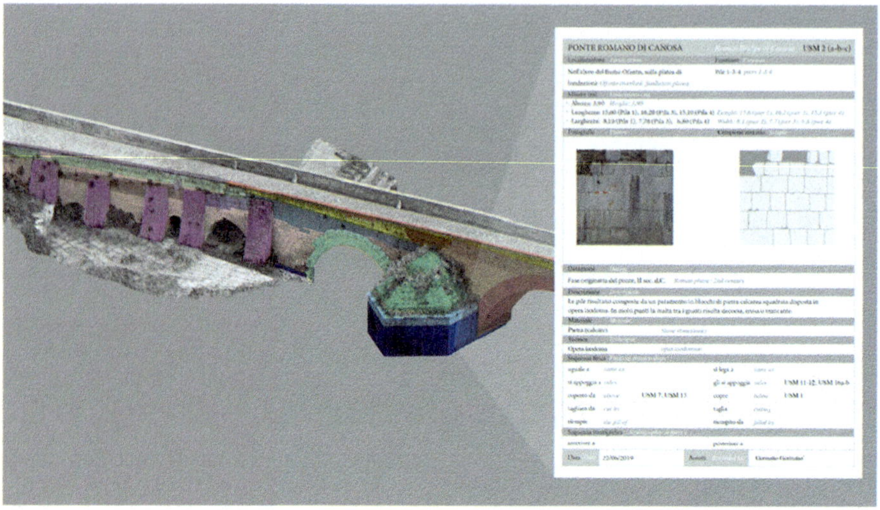

Fig. 4 N-E side of the bridge. The different Stratigraphic Units of Masonry shown in different colors on the photogrammetric model. On the right, a context record sheet for SUM

In order for the representation to be comprehensive and understandable, it is therefore necessary to recognize the geometry and shape of the elements to be represented, as well as their reciprocal relationships [65]. In fact, after the individuation of the SUMs, further investigations of construction details led to some important observations to set up the virtual reconstruction workflow. First, the inclination of the rows of blocks of the west side (Fig. 5), visibly different from the angle of the current gradient, seems to suggest an earlier and higher inclination of the slopes.

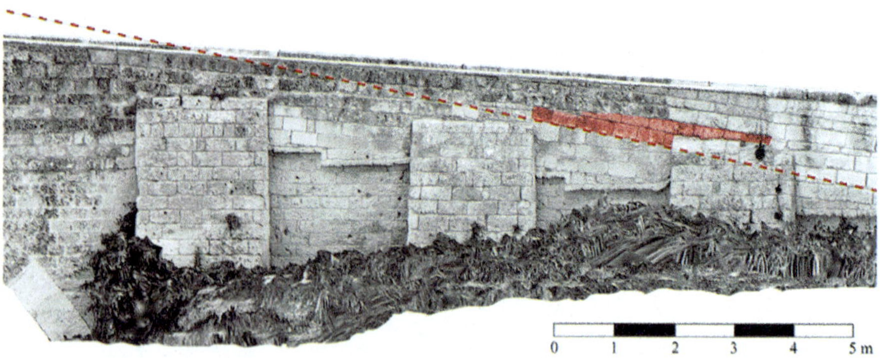

Fig. 5 Northwest face of the bridge, abutment. The red area shows the ancient inclination still identifiable in the grade of the rows of blocks

Although it is not possible to affirm beyond doubt that it belongs to the original structure, by using this inclination to reconstruct the outline of the bridge, the dimensions of the arches reported in the mentioned historical documents match with the graphic hypothesis proposed.

The resulting typology, with segmental arches and double-inclined slopes, finds comparisons in different aspects in other Roman bridges, such as the one in Ascoli Satriano in Apulia (Italy), the Pont-Saint-Martin in Valle d'Aosta (Italy) [66, pp. 199–201], the Pont Julien in Bonnieux (France) [66, pp. 264–265], the bridge of Cangas de Onìs in Asturia (Spain) [66, pp. 333–334], the Puente la Reina in Navarra (Spain) and the Mantible bridge in Assa (Spain) [66, p.367], just to name a few.

Subsequently, it was noted that in the first elevation drawings after the post-collapse reconstruction, carried out in 1756 by Amato Poulet (Fig. 6), the top of the starlings had a stepped form, just as in the nearby Roman bridge of Ascoli Satriano, while currently they have a sharpened or curved extreme, sometimes called the nose, an expedient to lighten the mass. In the same drawing, in correspondence with the first spandrel from the East, there is also an arched opening which is probably evidence of a flood opening, commonly present in Roman bridges [6].

Lastly, the comparison of the parapet dimensions reported by Delfino in 1749 (5 Neapolitan palms, that is, 131 cm) with a Roman inscribed slab found in Cerignola and assumed to belong to one of the bridges of the Via Traiana [67] shows that the heights match perfectly. Since these slabs, which celebrated the imperial building program, were set along the balustrade on the bridge this element was also inserted within the illustrated reconstruction.

The logical process required for the phase of representation [68], based on processing and elaboration of raw data obtained from the previous survey, allows the reconstruction of a complete and accurate 3D model.

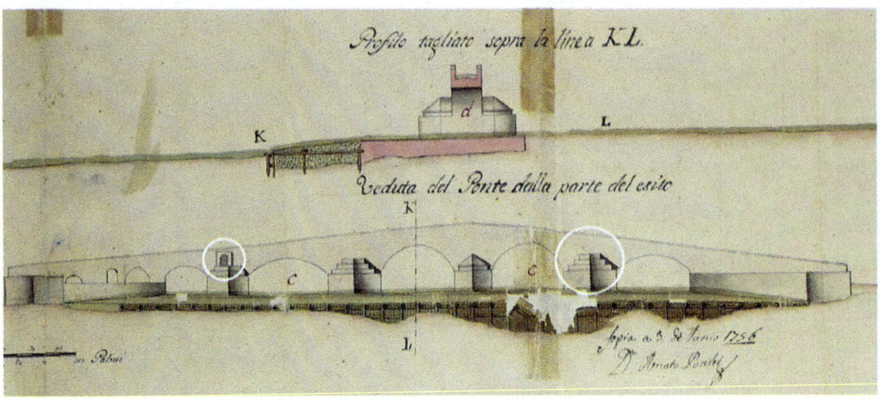

Fig. 6 Detail from the survey of Amato Poulet (1756) showing the North face of the bridge. On the left, an *oculus* or flood opening; on the right, the step-shaped top of the starlings. Archivio di Stato di Foggia (National Archive of Foggia, Italy)

In order to confirm and develop these interpretative hypotheses, a schematic geometric prototype of the three-dimensional model was created from the collected information and clusters of building elements from different construction phases were grouped by colour (Fig. 7). The reconstruction workflow was developed using Rhinoceros, a versatile and powerful 3D computer graphic and CAD software based on the NURBS mathematical model.

Thanks to the connections between the different parts, it was possible to develop reconstructive hypotheses of the three-arch Roman configuration. In addition, the entire diachronic evolution of the site was illustrated at different time scales from the laying of the foundation *platea* to the heavy alterations that occurred during the Second World War (Fig. 8).

Since incisive two-dimensional reconstructions are often best made through a three-dimensional phase [69], this final operation was carried out by merging different representation techniques, from the three-dimensional model generated with

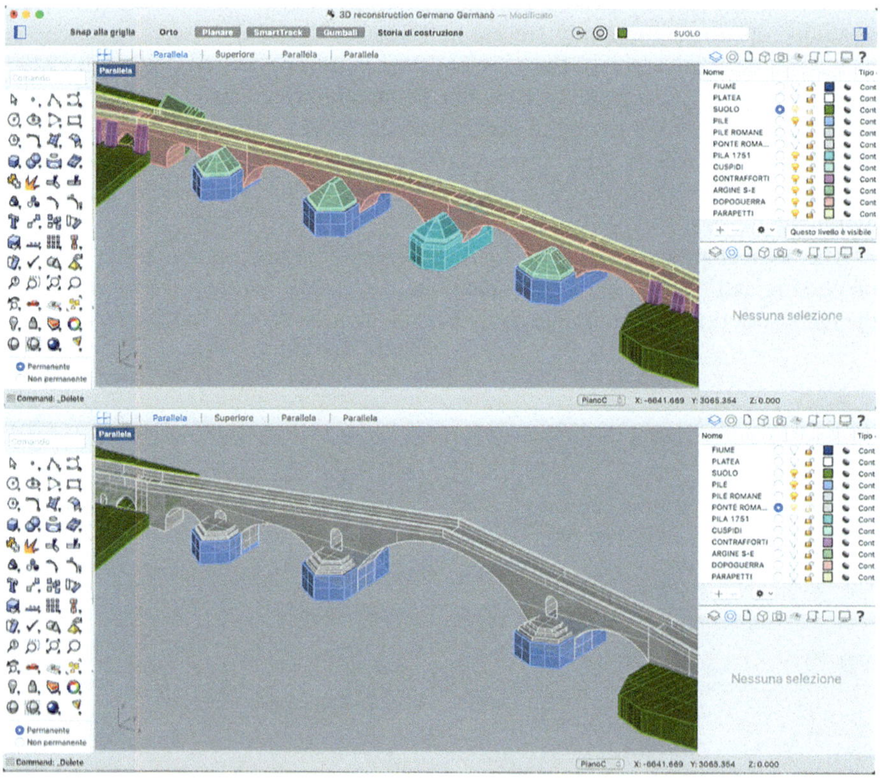

Fig. 7 Three-dimensional geometric model created with Rhinoceros software. Above, the bridge nowadays with elements relating to the different construction phases differentiated by colour (blue for Roman piers, turquoise for the 1751 pier, light green for the "noses" above starlings, fuchsia for lateral buttresses, light red for the main structure and the arches, and yellow for the parapets. Below, in grey, the hypothesised structure during the Roman phase

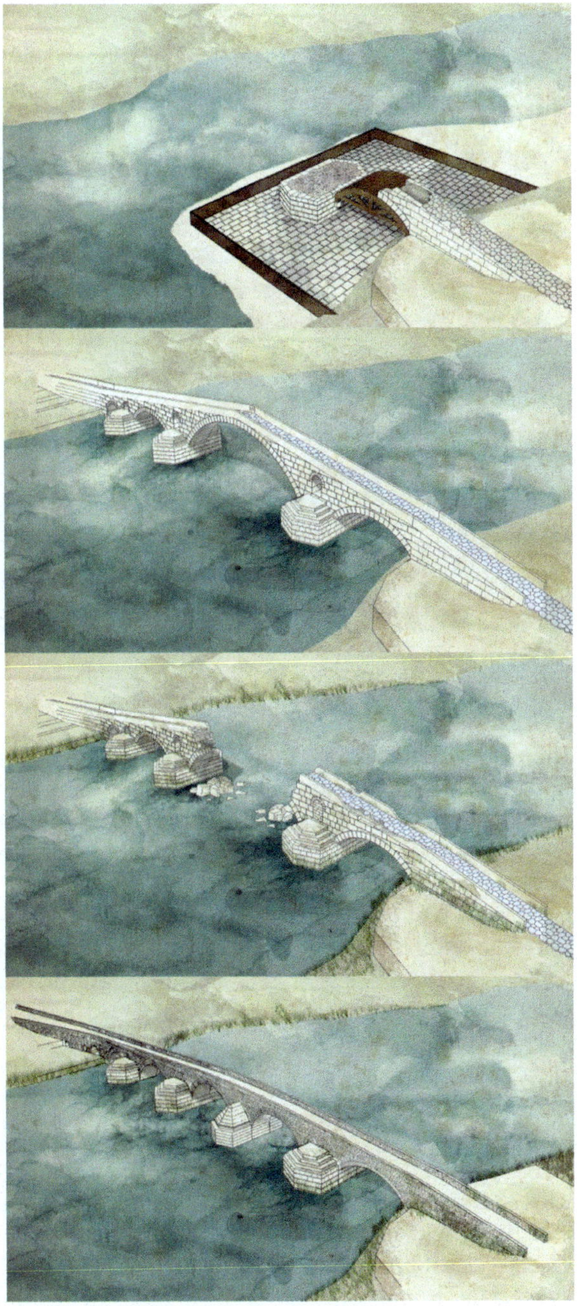

Fig. 8 Illustrations showing the diachronic evolution of the site with the main events. In order from the top: the building site, the Roman bridge, the 1751 collapse, the reconstruction

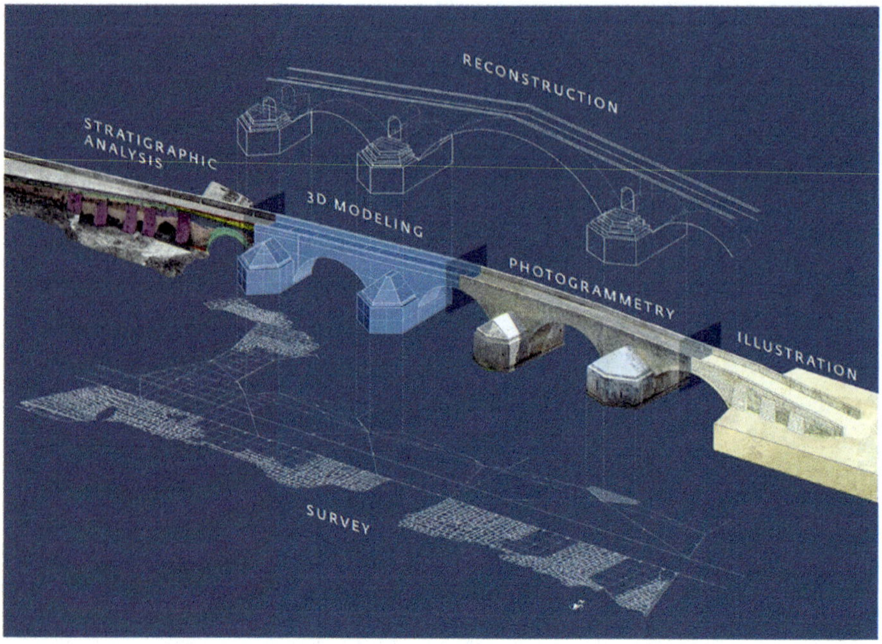

Fig. 9 Synoptic table showing the different stages of the workflow and representation methods

Rhinoceros to the watercolour characterisation in Photoshop. The virtual reconstruction is both the final result of the investigation and the basis from which further interpretations and visualisations of the monument can be developed, an important tool for understanding its history and ancient construction techniques, for both research and scientific dissemination [70].

Starting from the archaeological and architectural evidences, the reconstruction relied on different levels of accuracy (Fig. 9), including interpolation, extrapolation, and speculation [71]. As researchers, we have indeed a responsibility to document demonstrable theories as well as conjectures, to encourage constructive criticism.

5 Conclusions

This study brought back to light an element of cultural heritage surprisingly ignored and partly misunderstood until now. Through the virtual reconstruction of the diachronic evolution of the Roman bridge of Canosa, the aim of this study has been to analyse the complex history of the monument and the construction techniques used in the various eras, providing a method of analysis that is open to future updates. Making available a digital replica of a monumental bridge is also useful to makes accessible a context otherwise inaccessible such as a large river. As a critical product of a creative-interpretative operation, the 3D model is a valuable contribution to knowledge and it becomes itself a new document [72]. For these reasons, the model can be subjected to further analysis and interpretations, but also be historicized and susceptible to the risk of graphic obsolescence, because no reconstruction drawing of the past can do more than illustrate the state of archaeological knowledge at the time that it was created [73]. The medium of representation is very important for the transmission of cultural heritage, but it needs to avoid being characterised by either technical or artistic excesses. Considering a suggestive representation of the past as exempt from an accurate scientific study and sufficient for a general public of non-experts contributes to the perception of virtual reconstruction as an aesthetic tool more than a scientific tool [74]. The democratization of knowledge cannot allow the scientific value of research to be undermined. In this scenario, it is important that multidisciplinarity coincides with interdisciplinarity, so that the skills of archaeologists and architects are shared mutually, together with the mastering of graphic skills, because representation, just like architecture, to quote Vitruvius, is also *venustas*.

References

1. McCarthy, J. (2014). Multi-image photogrammetry as a practical tool for cultural heritage survey and community engagement. *Journal of Archaeological Science, 43*, 175–185. https://doi.org/10.1016/j.jas.2014.01.010.
2. Liarokapis, F., Voulodimos, A., Doulamis, N., & Doulamis, A. (Eds.) (2020). *Visual computing for cultural heritage*. Springer Series on Cultural Computing. https://doi.org/10.1007/978-3-030-37191-3_5.
3. Stylianidis, E. (2020). *Photogrammetric survey for the recording and documentation of historic buildings*. Springer.
4. Gazzola, P. (1963). *Ponti romani*. Leo Olschki, Firenze.
5. O'Connor, C. (1993). *Roman bridges*. Cambridge University Press.
6. Galliazzo, V. (1995). *I ponti romani*. Canova, Treviso.
7. Rafiee, A., & Vinches, M. (2013). Mechanical Behaviour of a stone masonry bridge assessed using an implicit discrete element method. *Engineering Structures, 48*, 739–749. https://doi.org/10.1016/j.engstruct.2012.11.035.
8. Elyamani, A., & Roca, P. (2018). A review on the study of historical structures using integrated investigation activities for seismic safety assessment. Part II: Model updating and seismic analysis. *Science Culture, 4*, 29–51. https://doi.org/10.5281/zenodo.1048241.
9. Addessi, D., Liberatore, D., & Nocera, M. (2020). damaging behavior of masonry arch bridges: Analysis of 'Ponte delle Torri' in Spoleto, Italy. *Journal Earthquake Engineering*, 1–26.
10. Vegas, F., & Mileto, C. (2008). La restauración del puente medieval sobre el río de las Truchas. *Loggia. Arquit. Restauración, 21*, 106–123.
11. D'Amelio, M. G., De Cesaris, F., & Mataloni, G. (2018). Il ponte della Badia di Vulci: Storia, rilievo, progetto e intervento di restauro strutturale. *ARKOS, 5*(21–22), 25–54.
12. Fiorini, A. (2018). An urban archeological project in Rimini. Preliminary Report (2017–2018). In *The Contribution of Building Archaeology to Research and Conservation*. Groma 3, 5–18 (2018). https://doi.org/10.12977/groma.
13. Bertocci, S., Bigognari, M., & Ricciarini, M. (2019). Digital survey for landscape knowledge: Garfagnana case studies for tourism development and versilia case study for hydraulic analysis. *DisegnareCon, 12*(22).

14. Giannelos, C., & Vintzileou, E. (2022). The rehabilitation of the historical bridge of Konitsa: A documentation-based investigation of structural behaviour. In I. Vayas, & F. M. Mazzolani (Eds.), *Protection of Historical Constructions, Proceedings of PROHITECH 2021*, pp. 711–726. Springer, Cham. https://doi.org/10.1007/978-3-030-90788-4_54.
15. Balletti, C., Guerra, F., Pillon, M., & Sartorelli, L. (2007). The Grumentum's Arena: measure, geometry and shape. In *Proceeding of CIPA 2007 XXI International Symposium 'Anticipating the Future of the Cultural Past'* (Athens, 1–6 October 2007). Athens.
16. Lubowiecka, I., Arias, P., Riveiro, B., & Solla, M. (2011). Multidisciplinary approach to the assessment of historic structures based on the case of a masonry bridge in Galicia (Spain). *Computers & Structures, 89*(17–18), 1615–1627. https://doi.org/10.1016/j.compstruc.2011.04.016.
17. Solla, M., Caama C., Riveiro, B., & Lorenzo, H. (2011). GPR analysis of a masonry arch for structural assessment. In *Proceedings of the 2011 6th International Workshop on Advanced Ground Penetrating Radar (IWAGPR)* 22–24 June, pp. 1–6. Aachen, Germany. https://doi.org/10.1109/IWAGPR.2011.5963873.
18. Omer, M., Margetts, L., Mosleh, M. H., Hewitt, S., & Parwaiz, M. (2019). Use of gaming technology to bring bridge inspection to the office. *Structure and Infrastructure Engineering, 15*(10), 1292–1307. https://doi.org/10.1080/15732479.2019.1615962.
19. Alvaro, C. (2015). Il Ponte di Statua sulla Via Aurelia. In E. De Minicis (Ed.), Archeologia delle Strade: La Viabilità in Età Medievale; Metodologie ed Esempi di Studio a Confronto, Atti del I Convegno Nazionale di Studi, Viterbo-Roma, 3–4 dicembre 2009, pp. 55–62. Aracne, Roma.
20. Ferrari, I. (2012). The Roman Bridges of the Via Traiana: An Innovative Building System. Culture, Technology and Society, Proceedings of the 4th International Congress on Construction History, Paris 3–7/7/2012In R. Carvais, A. Guillerme, V. Nègre, & J. Sakarovitch (Eds.), *Nuts and Bolts of Construction History* (Vol. 3, pp. 573–378). Picard.
21. Rodriguez-Navarro, P., & Verdiani, G. (2013). Digital survey and interpretation of a fortification fragment: The Cadi bridge at the feet of the Alhambra hill, Granada, Spain. In *Proceedings of the 2013 Digital Heritage International Congress (Digital Heritage), Marseille*, France, 28 October–1 November 2013, pp. 363–366. IEEE, Marseille.
22. Uygun Gençer, F., & Hamamcıoğlu Turan, M. (2017). The masonry techniques of a historical bridge. *Hypokremnos (İçmeler) METU Journal of the Faculty of Architecture, 34*(1), 187–207. https://hdl.handle.net/11511/50655.
23. Savini, F., Rainieri, C., Fabbrocino, G., & Trizio, I. (2021). Applications of stratigraphic analysis to enhance the inspection and structural characterization of historic bridges. *Infrastructures, 6*, 7. https://doi.org/10.3390/infrastructures6010007.
24. Haselberger, L., & Humphrey, J. (Eds.) (2006). Imaging ancient Rome. *Documentation, Visualization, Imagination, Proceedings of the third Williams Symposium on Classical Architecture*, May 20–23, 2004 (Suppl. JRA, 61). Rome.
25. Gabellone, F., Ferrari, I., Giuri, F., & Chiffi, M. (2015). The contribution of the 3D study for new reconstructive proposals of Lecce in roman age. In *IMEKO 1st International Conference on Metrology for Archaeology*. Athena Srl, Roma.
26. Piccoli, C. (2020). *Visualizing cityscapes of Classical antiquity: From early modern reconstruction drawings to digital 3D models, with a case study from the ancient town of Koroneia in Boeotia*. Greece. Archeopress.
27. Demetrescu, E. (2015). Archaeological stratigraphy as a formal language for virtual reconstruction. Theory and practice. *Journal of Archaeological Science 57*, 42–55.
28. Cassano, R. (1986). Recenti indagini archeologiche a Canosa. In *Profili della Daunia antica, 2°Ciclo di conferenze sulle più recenti campagne di scavo*, pp. 64–75. Regione Puglia, Foggia.
29. Pratilli, F. (1745). *Della Via Appia riconosciuta e descritta da Roma a Brindisi*. Arnaldo Forni: Napoli.
30. Ashby, Th., & Gardner, R. (1916). The Via Traiana. *BSR, 8*, 104–171.
31. Alvisi, G. (1970). *La viabilità romana della Daunia*. Società Storia Patria, Bari.
32. Uggeri, G. (1983). *La viabilità romana nel Salento*. Schena, Fasano.

33. Ceraudo, G. (2003). Via Traiana. In S. Le Pera Buranelli, & R. Turchetti (Eds.), *Sulla via Appia da Roma a Brindisi*. Le fotografie di Thomas Ashby 1891–1925, pp. 152–155. L'Erma di Bretschneider, Roma.
34. Ceraudo, G. (2008). Via Gellia: Una strada 'fantasma' in Puglia centrale. *Studi di Antichità, 12*, 187–203.
35. Russo, S. (Ed.) (2015). *Tratturi di Puglia—risorsa per il futuro*. Claudio Grenzi, Foggia.
36. Chelotti, M., Morizio, V., & Silvestrini, M. (1990). *Le epigrafi romane di Canosa 2*. Edipuglia, Bari.
37. Morea, G. (1969). *Arte e monumenti a Canosa*. Canosa.
38. Ieva, P. (2015). La sepultura di Re Boamundo in una inedita brieve descrittione tardo-cinquecentesca. In C. D. Fonseca, & P. Ieva (Eds.), Unde boat mundus quanti fuerit Boamundus: Boemondo I di Altavilla, un normanno tra Occidente e Oriente: atti del Convegno internazionale di studio per il IX centenario della morte, Canosa di Puglia, 5–6–7 maggio 2011, pp. 301–335. Società di storia patria per la Apulia, Bari.
39. Germano', G. (2020). The role of archives in the graphic restitution of monuments: the case of the Roman bridge over the Ofanto river in Canosa di Puglia, Italy. *Bitàcora Arquitectura, 45*, 4–11. https://doi.org/10.22201/fa.14058901p.2020.45.77630.
40. Tritto, M. R. (2005). I restauri settecenteschi del ponte romano di Canosa di Puglia. In L. Bertoldi Lenoci (ed.) *Canosa*. Ricerche Storiche 2004, pp. 71–100. Schena, Fasano.
41. Cassano, R. (1992). Il ponte sull'Ofanto. In R. Cassano (Ed.) *Principi, Imperatori, Vescovi*. Duemila anni di storia a Canosa, pp. 708–711. Marsilio, Venezia.
42. Germano', G. (2023 *i.p.*). Bridges at war and conservation of cultural heritage: a case study from Italy. In A. Crisà (Ed.), *Antiquities, Sites and Museums Under Threat: Cultural Heritage and Communities in a State of War (1939–45)*, Proceedings of the International Conference at Ghent University (Gent, Belgium, 15–16/10/2020), Brill, Leiden.
43. Bilinski, B. (1987). La Puglia e Bari nel "Diario di viaggio" di Jan Ocieski, ambasciatore polacco nel 1541. In: AA.VV. (Eds.), *La regina Bona Sforza tra Puglia e Polonia*, Atti del Convegno (Bari 27/4/1980), pp. 16–40. Leopoli.
44. Salvati, C. (1970) Misure e pesi nella documentazione storica dell'Italia del Mezzogiorno, pp. 34–35. L'arte tipografica, Napoli.
45. Neal, J. (2020). The structure and function of ancient metrology. In N. Sekunda (Ed.). *Wonders Lost and Found*. A celebration of the archaeological work of Professor Michael Vickers, pp. 33–47. Archaeopress, Oxford.
46. Germano', G. (2022). Ancient metrology in architecture: a new approach in the study of the Roman bridge of Canosa di Puglia (Italy). *Acta IMEKO, 11 (1)*, 1–7. https://doi.org/10.21014/acta_imeko.v11i1.1092.
47. Frova, A. (1961). *L'arte di Roma antica*. UTET.
48. Remondino, F. (2011). Heritage recording and 3D modeling with photogrammetry and 3D scanning. *Remote Sens., 3*, 1104–1138. https://doi.org/10.3390/rs3061104.
49. Germano', G. (2021). 3D Reconstruction of a Roman Bridge in Canosa Di Puglia (Italy). *Studies in Digital Heritage, 1*(5), 75–87. https://doi.org/10.14434/sdh.v5i1.30812.
50. Mannoni, T. (1984). Metodi di datazione dell'edilizia storica. *Archeologia dell'Archittura, 9*(11), 396–403.
51. Doglioni, F. (1988). La Ricerca sulle strutture edilizie tra archeologia stratigrafica e restauro architettonico. In R. Francovich, & R. Parenti (Eds.), Archeologia e Restauro dei monumenti, pp. 223–247. All'Insegna del Giglio, Firenze.
52. Parenti, R. (2000). Archeologia dell'Architettura. In R. Francovich, & D. Manacorda (Eds.), *Dizionario di archeologia*. Temi, concetti, metodi, pp. 39–43. Laterza, Roma-Bari.
53. Redi, F. (2007). L'Archeologia del costruito: un bilancio di trenta anni di ricerche e di discussioni metodologiche. In *Archeologia del paesaggio medievale: studi in memoria di Riccardo Francovich*. Quaderni di archeologia medievale 9, 9–48.
54. Brogiolo, G. P., & Cagnana, A. (Eds.) (2012). Archeologia dell'architettura. Metodi e interpretazioni. Collana Metodi e temi dell'archeologia medioevale. All'insegna del Giglio, Firenze.

55. Azkarate, A. (2020). La Arqueologìa de la Arquitectura a Revisiòn. *Arqueologìa de la Arquitectura, 17*, 101.
56. Parenti, R. (1985). La lettura stratigrafica delle murature in contesti archeologici e di restauro architettonico. *Restauro & Città, 1*(2), 55–68.
57. Brogiolo, G. P. (2008). Procedure di documentazione e processi interpretativi dell'edilizia storica alla luce delle Linee Guida per la valutazione e riduzione del rischio sismico del patrimonio culturale. *Archeologia dell'architettura, 13*, 9–13.
58. Fiandaca, O., & Salvo, G. (2019). ArcheoGIS for the roman thermal bath of Sofiana: Informative support for a digital restoration. *Heritage, 2*(2), 1530–1550. https://doi.org/10.3390/heritage2020096.
59. Trizio, I., & Savini, F. (2020). Archaeology of buildings and HBIM methodology: Integrated tools for documentation and knowledge management of architectural heritage. In *IMEKO International Conference on Metrology for Archaeology and Cultural Heritage*, MetroArchaeo, pp. 84–89. Athena Srl, Roma.
60. Fenthress, E., & Goodson, C. J. (2003). Patricians, monks, and nuns: the abbey of S. Sebastiano, Alatri, During the Middle Ages. *Archeologia Medievale, 30*, 67–105.
61. Trizio, I., Brusaporci, S., Savini, F., Maiezza, P., Tata, A., & Giannangeli, A.: advanced digital processes for the survey: analysis and conservation of built heritage. In C. M. Bolognesi, & C. Santagati (Eds.), *Impact of Industry 4.0 on Architecture and Cultural Heritage*, pp. 76–110. IGI Global, Hershey.
62. Parenti, R. (1988). Le tecniche di documentazione per una lettura stratigrafica dell'elevato. In R. Francovich, & R. Parenti (Eds.), *Archeologia e Restauro dei monumenti*, pp. 280–302. All'Insegna del Giglio, Firenze.
63. Pavel, C. (2010). *Describing and interpreting the past—European and American approaches to the written record of the excavation*. University of Bucharest Press.
64. Dell'Unto, N., Leander, A. M., Dellepiane, M., Callieri, M., Ferdani, D., & Lindgren, S. (2013). Digital reconstruction and visualization in archaeology: Case-study drawn from the work of the Swedish Pompeii Project. In *Proceedings of the 2013 Digital Heritage International Congress (DigitalHeritage)*, Marseille, France, 28 October–1 November 2013, vol. 1, pp. 621–628. IEEE: Piscatway. https://doi.org/10.1109/DigitalHeritage.2013.6743804.
65. De Luca, L. (2014). Methods, formalisms and tools for the semantic-based surveying and representation of architectural heritage. *Applied Geomatics, 6*, 115–139. https://doi.org/10.1007/s12518-011-0076-7.
66. Galliazzo, V. (1995). *I ponti romani*, vol. II, Catalogo Generale. 2nd edn. Canova, Treviso.
67. Ceraudo, G. (2013). A proposito delle lastre iscritte dei ponti della Via Traiana. *ATTA, 22*, 143–153.
68. Croce, V., Caroti, G., Piemonte, A., & Bevilacqua, M. G. (2021). From survey to semantic representation for Cultural Heritage: The 3D modeling of recurring architectural elements. *Acta IMEKO, 10*(1), 98–108.
69. James, S. (1997). Drawing inferences: Visual reconstructions in theory and practice. In *The Cultural Life of Images: Visual Representation in Archaeology*, pp. 22–48. Routledge, London, New York.
70. Bruseker, G., Guillem, A., & Carboni, N. (2015). Semantically documenting virtual reconstruction: building a path to knowledge provenance. ISPRS Ann Photogramm Remote Sens. Spat. Inf. Sci. II-5/W3, 33–40. https://doi.org/10.5194/isprsannals-II-5-W3-33-2015.
71. De Kramer, M. (2020). Relict–interpolated–extrapolated–speculative: An approach to degrees of accuracy in virtual heritage reconstruction. In F. Liarokapis, A. Voulodimos, N. Doulamis, & A. Doulamis (Eds.), *Visual Computing for Cultural Heritage Visual Computing for Cultural Heritage*. Springer Series on Cultural Computing, pp. 409–422. https://doi.org/10.1007/978-3-030-37191-3_21.
72. Brusaporci, S. (Ed.) (2015). *Handbook of research on emerging digital tools for architectural surveying, modeling, and representation*. IGI Global: Hershey, PA. https://doi.org/10.4018/978-1-4666-8379-2.

73. Greaney, S. (2013). Reconstruction drawings: Illustrating the evidence. In N. Mills (Ed.), *Presenting the Romans. Interpreting the Frontiers of the Roman Empire World Heritage Site*, pp. 31–39. The Boydell Press, Woodbridge.
74. Demetrescu, E. (2018). Virtual Reconstruction as a Scientific Tool: The Extended Matrix and Source-Based Modelling Approach. In S. Münster, K. Friedrichs, F. Niebling, & A. Seidel-Grzesinska (Eds.), *Digital Research and Education in Architectural Heritage* (pp. 102–116). Springer.

Digital Restoration

Multidisciplinary Approach for the Knowledge of Historical Built: Digital Tools for the Virtual Restoration

Adriana Marra, Ilaria Trizio, and Francesca Savini

Abstract The identification of strategies to be implemented for the restoration of cultural heritage, considered in its broad meaning, results from complex knowledge path. This process involves different disciplines collecting information for the cultural heritage safeguard and identification of innovative solutions that support the professionals and administrators in the conservation actions. At the same time, the advances in digital technologies provided novel solutions capable of supporting the knowledge, conservation and valorisation processes of built heritage. Based on these assumptions, the paper deals with a solution based on Building Information Modelling (BIM) and Virtual Reality (VR) techniques for addressing the issues concerning the processes of cultural heritage. The digital environment is suitable for managing an operative procedure that integrates the methods traditionally used by the restoration and archaeology of buildings disciplines to converge in the virtual restoration of a historical artefact. The HBIM model of the church of *Santa Maria della Vittoria in Fontecchio* (AQ) is developed starting from the in-situ investigation of the artefact. The model integrates multidisciplinary data with the aim of planning and designing proper restoration interventions for the examined asset, making these visible within the digital model and Virtual Tour in the perspective of virtual restoration. The strength of the proposed solutions is the virtual intervention of restoration actions on the artefact replica that is not a simple reproduction of the real but a digital object that communicates with a wider audience and facilitates the design and the comparison between alternative solutions, promoting the e-conservation paradigm.

Keywords Multidisciplinary approach · Built heritage · Data management · Integrated digital tools for conservation · Virtual reality

---

A. Marra (✉) · I. Trizio · F. Savini
ITC-CNR, Institute for Construction Technologies, Italian National Research Council, 67100 L'Aquila, Italy
e-mail: Marra@itc.cnr.it

I. Trizio
e-mail: ilaria.trizio@itc.cnr.it

F. Savini
e-mail: Savini@itc.cnr.it

© The Author(s), under exclusive license to Springer Nature Switzerland AG 2023

I. Trizio et al. (eds.), *Digital Restoration and Virtual Reconstructions*, Digital Innovations in Architecture, Engineering and Construction,
https://doi.org/10.1007/978-3-031-15321-1_13

1 Introduction

The identification of strategies and actions to be implemented for restoration and conservation of cultural heritage, considered in its broad meaning, namely the whole of movable and immovable assets with historical, artistic and cultural value, is the result of a complex knowledge path. This process involves different disciplines to collect and process information for the cultural heritage safeguard and identification of innovative solutions that support the professionals and administrators in the conservation actions. At the same time, the advances in digital technologies and the field of representation sciences provided novel solutions, such as the digital semantic models, capable of supporting the knowledge, conservation and valorisation processes of built heritage [1–3].

To develop these research activities, the Institute for Construction Technology of the National Research Council has founded the InnResLab, a laboratory for the documentation, analysis, conservation, valorisation and regeneration of the built environment distributed throughout Italy. The main activities of the laboratory focus on the conservation and protection of historical and architectural buildings, starting from the survey, documentation and analysis of artefacts with a multidisciplinary approach. The digitalisation of the process of knowledge and management of the built and infrastructural heritage is one of the issues that L'Aquila research unit is facing, developing procedures for the documentation and digital modelling of the built heritage and defining integrated methodologies for its conservation. This encourages the development of the novel e-conservation paradigm [4], which foresees the implementation of digital tools and best practices that can facilitate and support public administrations in the phases of management, maintenance and conservation of the existing heritage. This scenario reveals new issues related to the different digital skills of a multi-target audience, such as public administrations, both in large cities and in small towns in the inner areas. Therefore, the research must contribute to developing digital procedures that align the conservation of cultural heritage issues with the techniques of e-governance and must examine the methods to be implemented for supporting the authorities in charge of the management and protection of the built heritage towards digital transition.

Starting from these assumptions, the collaboration with the municipal administration of Fontecchio, a historical village located in the inner area of the Abruzzo Apennines, has led to the testing of operational procedures to validate the proposed methodological principles. The municipality of Fontecchio is particularly careful in the management of the extensive environmental and cultural heritage, both tangible and intangible, located in its territories, such as historical buildings, churches and monasteries, highland settlements, historical bridges, museum collections, and natural reserves. Many monuments are currently involved in the difficult post-earthquake reconstruction process, while others need to be renovated from the perspective of maintenance and conservation, as well as use, fruition and enhancement. Among these, the *Santa Maria della Vittoria* church, in a state of neglect and acquired by the administration, has been identified to be converted into an auditorium capable of hosting cultural events and welcoming the local and non-local community. The church is in a poor state of preservation and needs to be restored, so it represented a valuable example to validate an operational workflow for a virtual restoration resulting from multidisciplinary investigations, which start with the integrated survey and thematic analysis of the artefact. The final goal is to achieve a solution capable of addressing the decision-making processes for the conservation of the built heritage and developed by integrating the phases of knowledge with the restoration design within Building Information Modeling (BIM) (Fig. 1).

The contribution illustrates some of the research outcomes starting from the critical analysis of the methods currently available, concerning the application of new technologies to historical artefacts in compliance with the principles of the restoration discipline, to the definition of an HBIM model, capable of integrating and relating the data resulting from the multidisciplinary analyses performed on the artefact and useful for the planning and design of restoration work, until to the use of the three-dimensional model virtually restored in a digital environment.

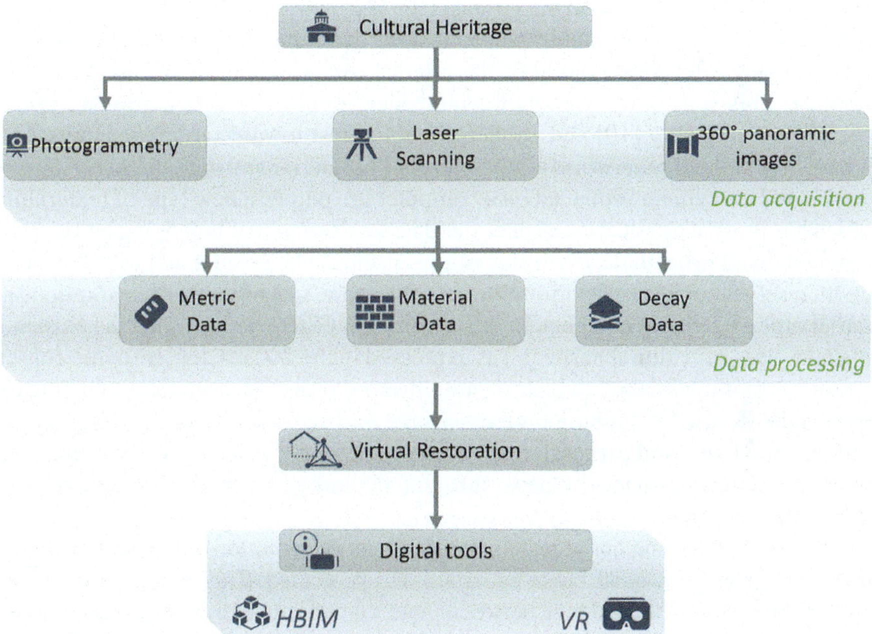

Fig. 1 Procedure flowchart for the virtual restoration

2 Background and Applied Method

The growing attention paid in recent years to cultural heritage and landscape, in its broadest sense, has contributed to decreasing the risk of abandonment and degradation of the historic building, especially in the inner and most disadvantaged areas of our country, on the one hand. On the other, it has favoured operational methods and conservation interventions often oriented in various ways [5], especially during the reconstruction interventions due to the seismic events that have struck the Italian territory in recent decades. At the same time, the increasing use of Information and Communication Technology (ICT) in the field of documentation, analysis and management of cultural heritage has led to the development of new methods of acquiring, archiving, relating and interpreting data, as well as use and enhancement through the creation of digital replicas of real objects, easily usable through different software and applications [6–9]. The digital tools widely used also in the field of conservation and restoration currently seem to offer possibilities to be explored and strengthened, which have oriented many researchers interested in the problems of translatability from the real to the virtual [10]. The main problem with the conversion of the real object to its digital twin would seem to be that the purpose of restoration is not so much the conservation of the artefact over time but its use in the present moment. This shift in the paradigm would seem to be related to a growing loss of interest in materiality [10] and an increasing interest toward virtuality. However, it is known that the digital world is not entirely immaterial and that numbers, images, sounds and everything produced on a computer are part of a new type of materiality, which has its own physical consistency and its own domain of belonging and circulation [11]. In addition to these considerations, it should be noted that these new media are themselves subject to deterioration and, therefore, to a problem of restoration and conservation of virtual contents. In this regard, UNESCO has affirmed the existence of a new class of 'cultural heritage', as expressed in the 2003 "Charter for the conservation of digital heritage", which identified some founding principles and actually opened the debate by defining digital heritage as a set of irreplaceable resources of human knowledge and expression. To this type belong all digital materials made up of texts, databases, images, videos, infographic and audio, in a wide and growing variety of formats.

This ever-expanding heritage, which represents an evolution of human knowledge and expression [12], must be safeguarded and protected. The definition of virtual restoration is also in the balance between material and immaterial, a concept that has been contested since its beginning, as it would seem to be a real oxymoron [12] since, in the most classic definition, there could be no restoration without intervention on the material [13]. The meaning of virtual restoration, which has been consolidated on a theoretical and practical level in the last decades and has recently found opportunities for further investigation [14], orients all the potentialities of virtual reality applied to Cultural Heritage within methods, rules and principles of conservation and restoration, from which it directly derives [15–17].

The main purpose of the virtual intervention is therefore the simulated restitution of the formal unity of a work, whether architectural, pictorial or sculptural, as it would appear following a physical restoration. In addition to this, the possibility of reproducing the different intervention phases on the digital twin offers useful tools for foreseeing the final result, providing all the functional cognitive elements for planning the real intervention [17]. Furthermore, according to the state of preservation of the artefacts, different methods can be used in the digital intervention: one of "virtual restoration", working directly on what remains of the original, or rather on its digital representation, and the other of "virtual reconstruction", which aims to restore what is no longer visible but which is documented by indirect sources and can be reconstructed on an analogue basis. These interventions aim to reconstruct the initial condition of the artefact using the tools of ICTs and Computer Vision [17].

The use of digital technologies for the study of architecture and built landscape is not in itself an innovative aspect of research but rather the dynamic association (in input and output) of images and alphanumeric data. The different combination of such information generates new data—appropriately georeferenced—releasing the digitisation of cultural heritage from a strictly instrumental role to make it in itself a new methodological approach [18]. Among the many digital tools available, the parametric environment is suitable for managing the fundamental steps of an operational procedure that integrates the methods traditionally adopted by the disciplines of restoration and archaeology of architecture to merge into the virtual restoration of a historical artefact [19–22].

In addition to this, a database acquired through an accurate integrated survey campaign, capable of recording and visualising the state of damage of the original artefact [21], represents a solid starting point on which to make virtual hypotheses about a possible intervention [23]. Predictive tools regarding the restoration actions to be undertaken and possible results of the interventions can be made available to the authorities responsible for the protection of the artefacts [24], which are useful tools for sharing the proposals with the public to ensure a participatory approach the protection actions [25]. Finally, in the context of virtual restoration, activities for digital reconstruction of destroyed works are included [26] to make digitally accessible lost artefacts. And this purpose seems to remember once again that the use of digital is particularly suited to the "narration" and fruition of heritage, while the potential of digital procedures in the context of restoration and conservation is still being validated.

3 The HBIM for the Restoration of the *Santa Maria della Vittoria* Church

3.1 *The Knowledge Phase: From the Survey to Thematic Analyses*

The church of *Santa Maria della Vittoria* (Fig. 2) is a small structure located outside the urban centre of the Fontecchio municipality along the valley floor road. The building stands on the remains of an ancient temple, dating to the first century AD. [27], whose traces are still visible on the western front [28]. Sources document that the church was originally dedicated to St. Peter and was already part of the Abbey of Farfa in the eleventh century [29]. The name change was probably due to the town's triumph over Captain Quinzi's troops, who occupied the village and caused considerable damage to the church. Today, the church has a single nave, with walls marked by paired pilasters and barely visible round-arched niches, and with the presbytery area slightly higher than the hall. Valuable elements of this building are present on the external facades. The Baroque-style window at the top and the portal and two openings with stone cornices at the bottom can be seen in the main façade, which ends with a horizontal crowning, and historical masonries with different construction techniques can be noticed on the side walls. Several traces can be found in the masonry, testifying to the changes and interventions that the structure has undergone over the centuries, not least that of 1995 [30] that led to the restoration of the roofs after the church had lost its original function and had been improperly used as a sheepfold [31].

Different investigations and thematic analyses were carried out to develop a project for the virtual restoration of the ancient building that would be respectful of its

Fig. 2 Church of *Santa Maria della Vittoria* in Fontecchio: general and detail views

material and cultural authenticity and in line with the characteristics of the structure and its history.

In the first phase, a digital survey of the church was carried out using and integrating laser scanning procedures (TLS and SLAM) with photogrammetric ones (terrestrial and aerial). The artefact's expeditious survey was integrated with the acquisition of spherical panoramas linked together in a Virtual Tour (Fig. 3). The data processing made it possible to obtain a reliable survey from a metric point of view and high-quality photorealistic orthomosaics that supported the development of the digital model and several thematic analyses. Indeed, these latter were carried out by correlating the in-situ investigations with those obtained from a careful analysis of the digital renderings resulting from the survey process, including the Virtual Tour of the current state achieved from the 360° images. In particular, it was possible to understand the materials, construction techniques, masonry textures and, finally, the different degradation forms of the building.

The analyses showed that the vertical structures are masonries made of local limestone, with different manufacturing techniques. Indeed, stones worked in regular blocks and arranged in horizontal rows are visible in the lower part, while in the upper part, the rows lose their horizontality, and the stone elements are roughly shaped. The several masonry textures detected are further evidence of the different construction actions that occurred over time following destructive events or extensions to the original structure.

A square brick pavement, with a perimeter and central connecting strip, is visible inside the church; it is deteriorated and has many missing parts. Furthermore, an

Fig. 3 Survey of the *Santa Maria della Vittoria* church

excavation at the main entrance shows a tomb found during the excavation and restoration works of the church. The analysis of the degradation forms showed that the structure is affected by several pathologies whose causes are related to different actions. In particular, the lack of continuous maintenance and neglect has provoked the development of superficial deposits, such as guano, soil and other debris, on the horizontal and vertical structures, as well as the lack and detachment of stone elements that characterize the architectural and finishing elements inside the church (Fig. 4). These phenomena are also combined with the detachment and missing portions of plaster, chromatic alterations, patinas and efflorescence caused by the action of rising damp. The lack of external doors and windows facilitated the rainwater infiltration into the structure, causing flow traces and infesting vegetation on the wall surfaces. Finally, stains, which can be traced back to anthropic actions, were found inside the niches and the perimeter walls.

The state of degradation and neglect of the building highlights the need for appropriate recovery actions. Therefore, measures to recover and preserve the degraded

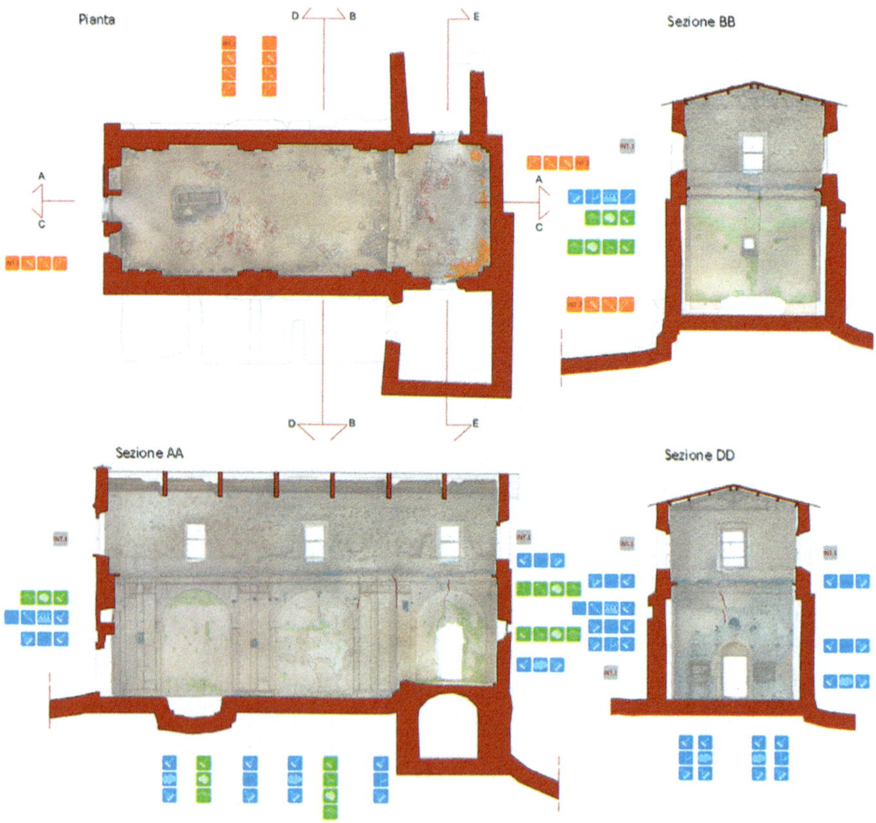

Fig. 4 Analysis of the current state and conservation actions

surfaces, in compliance with the Italian legislation concerning the protection, were defined according to the collected information (Fig. 4). Low-impact mechanical cleaning actions for both vertical and horizontal structures, restoration and consolidation of finishes and stone elements, with the elimination of those strongly compromised, have been considered. Finally, protection, integration and new construction interventions with materials and techniques compatible with possible use of the building have been evaluated.

Specifically, for the masonry structures in view, the planned interventions concern the cleaning of the surfaces through the use of low-impact mechanical tools, for removing deposits and crusts more or less adherent to the surface, solvent solutions in the most compact parts and their subsequent washing. The joints should be consolidated with mortar compatible with the original, while missing parts (in correspondence with the window frames and cornices) should be filled with materials similar but still distinguishable to the original, ensuring that the connection between existing and new masonry.

The plastered surfaces should be consolidated, and the missing plasters should be integrated. In particular, the mechanical cleaning of the surfaces with low-impact tools and solvent solutions in the presence of more compact deposits should be considered. The restoration of plaster, and its cohesion with masonry support, should be ensured through micro-injections of solutions or mortars compatible with the support, the filling of cracks, even small ones, and the application of a new layer of plaster.

3.2 *The HBIM Model for Virtual Restoration*

The digital modelling of the *Santa Maria della Vittoria* church was carried out to integrate all the information collected within a single model. The aim of this digital replica is to provide support in the knowledge of the building's historical-constructive events, as well as in the analyses of the current conditions of the asset to define the interventions to be carried out for the conservation and enhancement [32–34]. Therefore, a parametric model was implemented since the BIM offers great advantages both in the planning of the actions to be implemented after the knowledge process and in the management and decision-making processes [35, 36].

The parametric model of the church was developed after a critical analysis of the acquired data, importing the point cloud obtained during the survey process into Autodesk's Revit software and using procedures already used and tested (Fig. 5). New parametric families were developed to correctly represent the peculiarities identified within the building, such as the portal, by defining geometric and descriptive parameters for all the features of the architectural elements [37, 38].

At the same time, the orthomosaics obtained from the photogrammetric survey were imported into the parametric project to correctly represent the current condition of the building and the degradation forms detected in the analysis phase. The orthomosaics were edited for the specific thematic analysis to represent, using an

Fig. 5 Development of the HBIM model of the *Santa Maria della Vittoria* church: from the point cloud to the final model in realistic view

adaptive family, the pathologies afflicting the building surfaces [39, 40]. For each degradation form represented within the model, shared parameters were defined to identify and describe the pathologies detected, namely *ID-Defect*, *Defect* and *Image* parameters. In addition, a hyperlink was defined to relate the modelled element to the relative survey module of the degradation and alteration form, which was designed using online formats (Fig. 6).

The representation in the parametric environment of the information acquired during the knowledge and analysis processes supports the identification of restoration and conservation strategies to be implemented [32, 33, 39]. In this perspective, specific labels were modelled to facilitate the identification of interventions to be implemented. In particular, new parameters were added to identify the required interventions according to the type of support, as briefly mentioned in the previous section, the information on the actions to be implemented, and to connect the label to the corresponding intervention form through a hyperlink (Fig. 6).

The resulting model provided the reference point for developing the restoration and reuse project for the church. The creation of an auditorium was proposed, taking into account the building's features and the territorial context in which it is located, to meet the social and cultural needs of the area (Fig. 7).

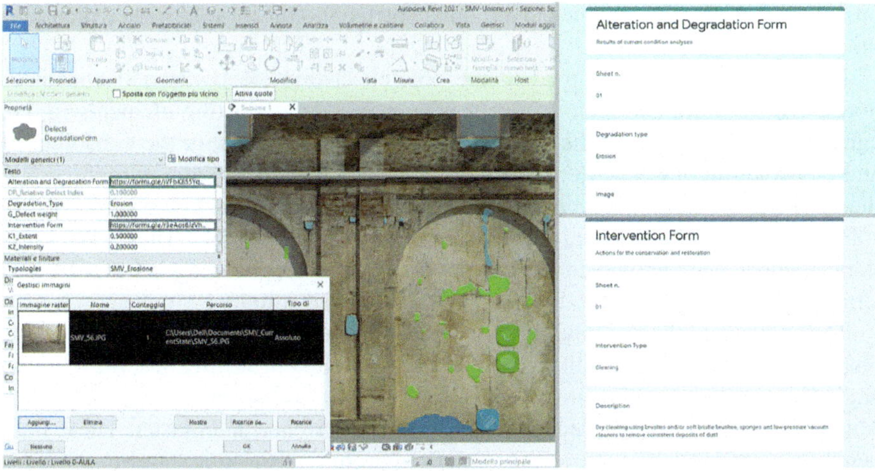

Fig. 6 Modelling of degradation forms and restoration interventions in the parametric environment. Implementation of new parameters dedicated to the modelled families and correlation to the degradation and intervention forms shared in a storage cloud

Fig. 7 Virtual reconstruction of the *Santa Maria della Vittoria* church after the proposed restoration and renovation works

The digital renovation project was implemented by creating a new project phase in the HBIM model in which all the interventions considered after the multidisciplinary analysis were represented. In particular, the interventions of integration with new materials were carried out directly in the parametric environment.

The current lack of windows and doors was solved by adding specific parametric families to the model, which removed the degradation forms related to the infiltration of rainwater and the development of surface deposits on both vertical and horizontal surfaces. The new families included in the project are made of natural essences, which integrate aesthetically with the old building, and technological solutions, which solve the problems of humidity.

Particular attention was paid to the possible actions to be carried out on the horizontal structure and stairs: from the recovery of the existing elements in their current configuration with the integration of the missing ones, to the consolidation and construction of a new floor, after the recovery, consolidation and protection of the original parts. The potential of digital restoration permitted to test of the second solution, which best satisfied the requirements of the new use. Therefore, a new floor element was defined in the parametric model, representing all the different stratigraphies that can correctly describe the proposed solution.

The analysis of the artificial lighting inside the church, favoured by the rendering phase of the digital model both in a parametric environment and in a specific real-time engine, oriented the choice of materials to be used and proposed for real restoration. This process validated the decision of actions on masonry and plaster. These interventions were implemented in the digital environment through careful editing of high-resolution orthomosaics, which made it possible to simulate the cleaning of the masonry, the restoration of the original colour of the plasterwork and the integration of missing or damaged coatings.

Finally, some choices have been oriented to highlight the traces of the past through the re-proposition of the baroque altar and functionalising it in a conference table, or the video mapping system that projects the archaeological evidence and the original finishes of the church on the floor surface (Fig. 8). This solution enriches the architecture with visual content that can change over time thanks to new knowledge resulting from further investigations, on the one hand, and, on the other, actively involves the public, who will approach the history of the place innovatively.

Fig. 8 Virtual restoration of *Santa Maria della Vittoria* church

The created HBIM model, representative of both the current conditions of the building and the possible future ones, confirms itself as a support tool for sharing information and design proposals, especially for technicians and specialists in the field. Therefore, to fully benefit from the possibilities offered by BIM processes and to involve a wider audience, it is necessary to exploit the advantages of integrating the obtained results within different solutions.

4 Digital Tools for the E-conservation

To facilitate the dissemination of data and results within the HBIM model, in the perspective of the development of the Heritage Community [41], it is necessary to make the digital replica of the artefact available to a wide audience, from public authorities to tourists. The Virtual Reality techniques, applied to many domains [42, 43], satisfy well these needs and increase the accessibility of data to the public administration. Indeed, the Virtual, Augmented and Mixed Reality show advantages, in the cultural heritage field, for more technical aspects such as knowledge and monitoring of historical buildings [44, 45] and for enhancement and enjoyment tourists [46].

Therefore, a digital tool that, starting from Virtual Reality from spherical panoramas, can integrate technical aspects with those related to valorisation, connecting multi-data (textual, geometrical/three-dimensional, images) to the digital representation to transmit them to a variety of users, was developed. The resulting interactive tool can be used with PCs, tablets and smartphones, or an immersive mode through best-performing devices such as the Oculus Quest. This digital tool provides users with easy access to the three-dimensional model and related information.

The spherical images acquired during the survey phase were used to create a Virtual Tour of the Church, which allows the user to navigate the space, view the current condition of the structure and access technical information and data forms on the deterioration and the work to be carried out during the restoration phase (Fig. 9).

From the digital environment of the real Church, a digitally restored reconstruction of the building can be viewed using specific hotspots (Fig. 10). This step was made possible by integrating modelling and visualisation techniques and importing the HBIM model into Twinmotion, a 3D immersion software for the real-time rendering. The spherical panoramas of the digitally restored model were extrapolated from this environment (Fig. 11). The digital tour allows users to visit the restored church in a virtual environment where they can directly perceive the balance between spaces, colours and the effect of natural or artificial light on the proposed materials.

Within this VR environment, thanks to a specific hotspot, a button that link to web site, it is possible to view the HBIM model and query it to access geometric data and technical information, as well as to view the restoration work to be carried out (Fig. 12). The proposed solution thus becomes a tool for validating the techniques to be adopted, which enriches the BIM project.

The digital platform is structured as a hub capable of linking different ways of representing architectural heritage. This is able to integrate multiple models and link

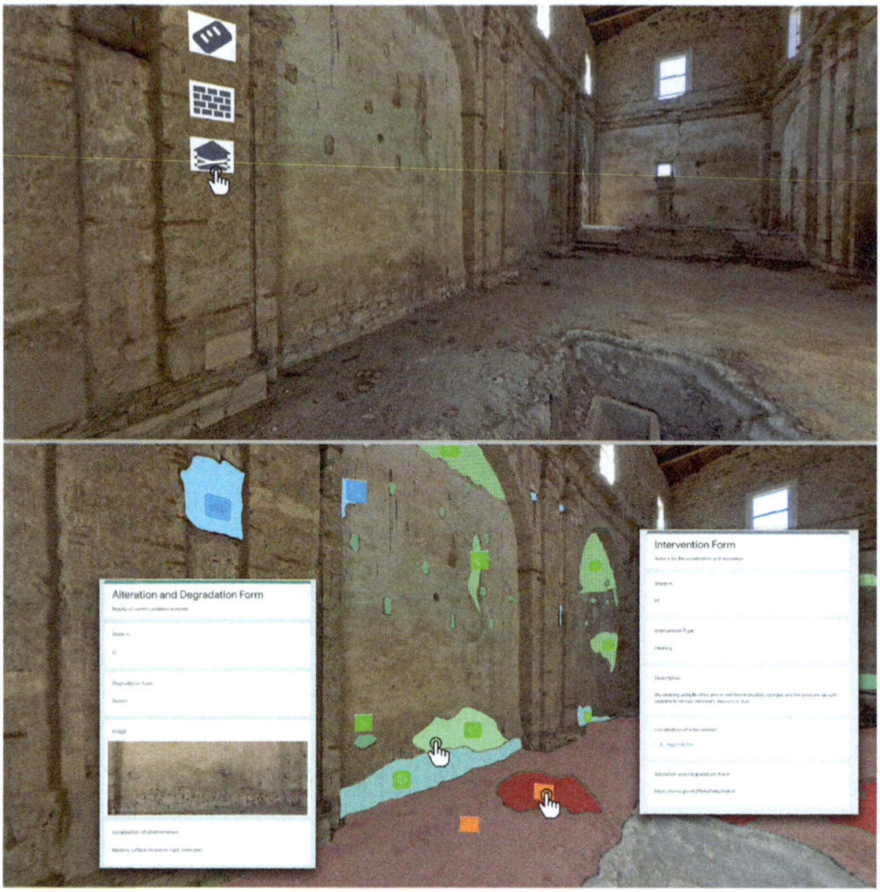

Fig. 9 Digital Tool for the accessibility of data: current state of the building with hotspots to activate additional information (top); mapping of deterioration with linked sheets and symbols referring to conservation interventions activating specific forms (bottom)

information, such as those derived from actions over time, a before and an after. The integrated system, accessible through the web, can be easily made available to the public administration, thus favouring the e-conservation process. It becomes a digital tool that supports the decision-making phases of management and conservation of the historic built heritage.

The documentation and management of these data, which can be considered dynamic because acquired on different time scales, is more important for the artefacts conservation. Furthermore, the custom of a digital environment makes it possible to communicate this technical information to multiple users. The simplicity of use and the intuitive nature of the digital operations to be carried out for the navigation of the space, cleared by the most common online engines for the navigation of places on a cartographic basis, i.e. Google Maps with its Street View system, allows translating

Fig. 10 An integrated digital system that allows to navigate the church in its current phase (left) and restored state through digital technology (right)

complex data with a prevalent graphic and visual language. In order to permit the use of semantic contents, the virtual tour has been planned in a multilevel way, which can be selected from a drop-down menu on the homepage (Fig. 13).

An *Information tour,* accessible by a button in the menu, has been developed. It provides general data on the history and evolution of the church and is enriched by texts written communicatively and historical photos that are able to involve a large audience made up of the local population and the many visitors who frequent this area and who, before embarking on the real visit, plan their trip online. In this way, it is possible to transform technical data into tourist information and improve the enhancement of the building and the tourist offer of territory.

5 Conclusions

The analyses and tests performed on the case study demonstrated how digital models resulting from a critical knowledge process can become a valuable tool to manage the restoration actions, from their design to the virtual hypothesis of implementation, also thanks to their integration within different digital applications.

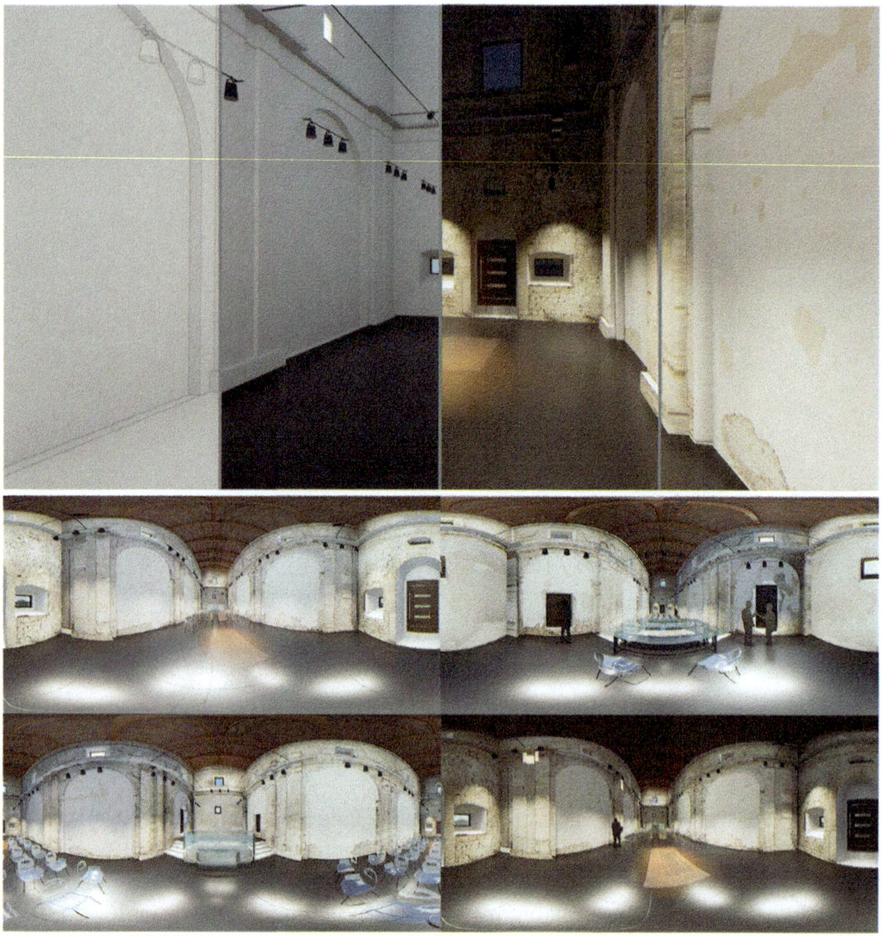

Fig. 11 Rendering process of the three-dimensional model (top) and equirectangular images of the reconstruction used to create the Virtual Tour (bottom)

The strength of the BIM process application is the virtual intervention of restoration actions on the artefact replica that is not a simple reproduction of the real but a digital object that facilitates the design and the comparison between alternative solutions, promoting the dialogue on the most appropriate choices. Indeed, interoperability in a virtual environment makes the model a hub to cooperate for proper preservation by evaluating the strategies to be implemented and validating the expected results.

The new paradigm of the e-conservation of cultural heritage is strengthened from the approach and solution proposed. The e-conservation will facilitate the management and preservation of cultural heritage by using innovative digital technologies for

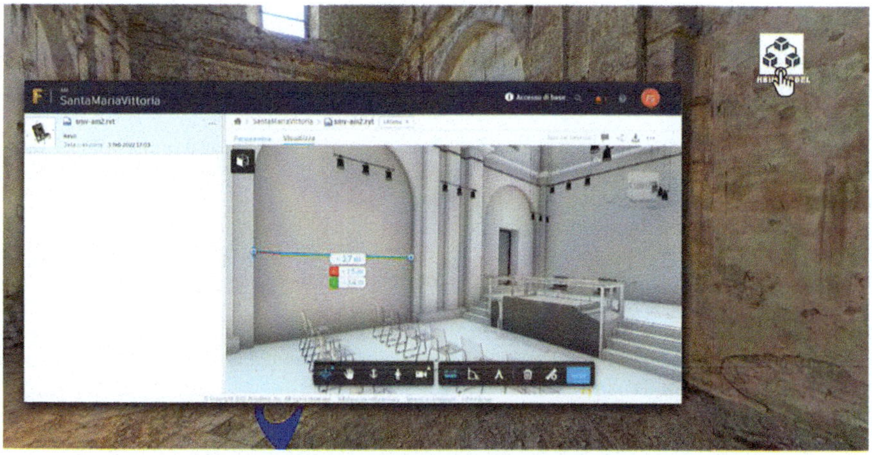

Fig. 12 Accessibility of HBIM within the digital platform

Fig. 13 Multilevel structure of the Virtual Tour with general data for tourism information

the processes of knowledge and analyses of assets, encouraging community involvement and the development of heritage community according to the Faro Convention indications and e-government principles.

Acknowledgements The research is part of the activities of InnResLab, a laboratory of innovation and research within the Institute for Construction Technology of Italian National Research Council and directed by Ilaria Trizio.

The authors are grateful to the municipal administration of Fontecchio (AQ), in particular, the mayor Dr Sabrina Ciancone, for granting the experimentation. They thank Andrea Ruggieri for the laser scanner survey, Marco Giallonardo for the photogrammetric survey, and Alessio Cordisco for implementation the spherical panorama from the HBIM model. Finally, the authors also thank Giulia Ciotti, Leonardo del Vecchio, and Berardino Scafati, students of the course of "Rilievo e modellazione digitale dell'architettura" at the University of L'Aquila held by Ilaria Trizio, for the geometric restitution of the church.

Credits Although the research presented is the result of the authors' collective work and continuous comparison, it is credited to Ilaria Trizio the Sect. 2, Adriana Marra the Sect. 3, and Francesca Savini the Sect. 4. All authors wrote Sects. 1 and 5.

References

1. Biagini, C., Capone, P., Donato, V., & Facchini, N. (2016). Towards the BIM implementation for historical building restoration sites. *Automation in Construction, 71*, 74–86.
2. Messaoudi, T., Véron, P., Halin, G., & De Luca, L. (2018). An ontological model for the reality-based 3D annotation of heritage building conservation state. *Journal of Cultural Heritage, 29*, 100–112.
3. Godinho, M., Machete, R., Ponte, M., Falcão, A. P., Gonçalves, A. B., & Bento, R. (2020). BIM as a resource in heritage management. An application for the National Palace of Sintra, Portugal. *Journal of Cultural Heritage, 43*, 153–162.
4. Trizio, I., Marra, A., Savini, F., & Fabbrocino, G. (2021). Survey methodologies and 3d modelling for conservation of historical masonry bridges. *ISPRS Annals of the Photogrammetry, Remote Sensing and Spatial Information Sciences, 8*, 163–170.
5. Fiorani, D. (2020). Editoriale. Progetto restauro: questioni (ri)aperte. *Materiali e strutture. Problemi di conservazione, Nuova serie,* anno IX, *17*, 5–9.
6. Limoncelli, M. (2012). *Il restauro virtuale in archeologia.* Roma: Carocci editore.
7. Barone, Z., & Nuccio, G. (2017). For a Consciuos Fruition of the Cultural Heritage of Ancient Noto (Sicily). EFIAN Project as Opportunity for Valorisation. *The International Archives of the Photogrammetry, Remote Sensing and Spatial Information Sciences, XLII-5/W1*, 89–97.
8. Banfi, F., Brumana, R., & Stanga, C. (2019). Extended Reality and informative models for the architectural heritage: From Scan-To-BIM process to virtual and augmented reality. *Virtual Archaeology Review, 10*(21), 14–30.
9. Pietroni, E., & Ferdani, D. (2021). Virtual restoration and virtual reconstruction in cultural heritage: Terminology, methodologies, visual representation techniques and cognitive models. *Information, 12*, 167.
10. Fiorani, D. (2015). Materiale/immateriale: frontiere del restauro. *Materiali e strutture. Problemi di conservazione. Nuova serie*, anno III, *5/6*, 9–23.
11. Gabellone, F. (2014). Approcci metodologici per una fruizione virtuale e arricchita dei Beni Culturali. *Arkos. Scienza Restauro Valorizzazione, 5/6*, 7–17.

12. Buzzanca, G. (2008). Si restaurano i byte? Dal restauro virtuale al restauro del virtuale, ovvero dagli 'smanettoni' all'applicazione della critica del restauro nella conservazione del digitale. *OPD Restauro. Rivista dell'Opificio delle Pietre Dure e Laboratori di Restauro di Firenze, 20*, 131–138.
13. Brandi, C. (1963). *Teoria del restauro*. Torino: Giulio Einaudi Editore.
14. Trizio, I., Demetrescu, E., & Ferdani, D. (Eds.). (2021). Virtual reconstruction and restoration. Comparing methodologies, practices, and experiences. *DisegnareCon, 14*(27).
15. Limoncelli, M. (2011). Applicazioni digitali per l'archeologia: Il restauro Virtuale. *DigItalia, 1*, 42–59.
16. Limoncelli, M. (2017). *Virtual Restoration. Paintings and Mosaics*. Roma: L'Erma di Bretschneider.
17. Coralini, A., & Limoncelli, M. (2020). Pittura antica e restauro virtuale. In P. Giulierini, A. Coralini, & V. Sampaolo (Eds.), *Picta Fragmenta. La pittura vesuviana: una rilettura*, pp. 377–390. Cinisello Balsamo: Silvana editoriale.
18. Fiorani, D., Acierno, M., Cutarelli, S., & Donatelli, A. (2021). Transformation of tools and conservation of architecture: some researches on the use of digital systems for the intervention on the historical buildings. *Építés-Építészettudomány, 49*(1–2), 97–131.
19. Demetrescu, E. (2015). Archaeological stratigraphy as a formal language for virtual reconstruction. Theory and practice. *Journal of Archaeological Science, 57*, 42–55.
20. Apollonio, F. I., Ballabeni, M., Bertacchi, S., Fallavollita, F., Foschi, R., & Gaiani, M. (2017). From documentation images to restauration support tools: A path following the Neptune Fountain in Bologna design process. *The International Archives of the Photogrammetry, Remote Sensing and Spatial Information Sciences, XLII-5/W1*, 329–336.
21. Chen, S., Yang, H., Wang, S., & Hu, Q. (2018). Surveying and Digital Restoration of Towering Architectural Heritage in Harsh Environments: A Case Study of the Millennium Ancient Watchtower in Tibet. *Sustainability, 10*, 3138.
22. Biagini, C., Capparelli, F., & Verdiani, G. (2021). BIM design lead for restoration of shipwreck museum in Kyrenia castle in Cyprus. In *Proceedings of the joint international event 9th ARQUEOLÓGICA 2.0 & 3rd GEORES* (Valencia, Spain, 26–28 April 2021), pp. 63–70. Valencia: Editorial Universitat Politècnica de València.
23. Barba, S., De Feo, E., D'Auria, S., & Guerriero, L. (2012). Survey and virtual restoration: The Castle of Magacela (Spain). In *2012 18th International Conference on Virtual Systems and Multimedia*, pp. 641–644. IEEE.
24. Di Paola, F., Vitella, M., & Antoci, F. (2014). Tecnologie digitali per la conservazione dei Beni Culturali. Un'esperienza di studio/Digital technologies for Cultural Heritage preservation. A case study. In P. Giandebiaggi, & C. Vernizzi (Eds.), *Italian survey & international experience. 36° Convegno Internazionale dei docenti della Rappresentazione*, pp. 271–280. Roma: Gangemi Editore.
25. Bilis, T., Kouimtzoglou, T., Magnisali, M., & Tokmakidis, P. (2017). The use of 3d scanning and photogrammetry techniques in the case study of the roman theatre of Nikopolis. Surveying, virtual reconstruction and restoration study. *The International Archives of Photogrammetry, Remote Sensing and Spatial Information Sciences, 42*, 97–103.
26. Dong, Y. Q., Hou, M. L., Qiao, Y. F., Wang, R. L., Li, Z. F., & Hua, W. (2020). A virtual restoration framework for the "Moon Gate" based on a modified scale of evidence. *International Archives of the Photogrammetry, Remote Sensing & Spatial Information Sciences, 43*, 1451–1454.
27. La Regina, A., & Coarelli, F. (1984). *Abruzzo Molise* (vol. 9). Bari: Laterza.
28. Mattiocco, E. (2002). *Antonio De Nino. Scoperte Archeologiche comunicate all'Accademia Nazionale dei Lincei e pubblicate nelle notizie degli scavi di antichità*. L'Aquila: Edizioni Libreria Colacci.
29. Paolini, G. (2003). *La vita transecolare del contado aquilano. Villa San Pio, Fontecchio e familia Paolini di Aquila*. Colledara: Andromeda Editrice.
30. Console, C., & Marsili, C. (1995). *Fontecchio nel Parco Naturale Regionale Sirente-Velino. Storia, Territorio, Ambiente*. L'Aquila:Comune di Fontecchio.

31. Gasbarrini, A., Donati, G., & Soldani, F. (1980). *Album per un paese. Fontecchio estate 1979*. L'Aquila: Marcello Ferri editore.
32. Oreni, D., Brumana, R., Della Torre, S., & Banfi, F. (2017). Survey, HBIM and conservation plan of a monumental building damaged by earthquake. *The International Archives of the Photogrammetry, Remote Sensing and Spatial Information Sciences, XLII-5/W1*, 337–342.
33. Bacci, G., Bertolini, F., Bevilacqua, M. G., Caroti, G., MartínezEspejo Zaragoza, I., Martino, M., & Piemonte, A. (2019). H-BIM methodologies for the architectural restoration: the case of the ex-church of San Quirico all'Olivo in Lucca, Tuscany. *The International Archives of the Photogrammetry, Remote Sensing and Spatial Information Sciences, XLII-2/W11*, 121–126.
34. Croce, V., Caroti, G., De Luca, L., Piemonte, A., & Véron, P. (2020). Semantic annotations on heritage models: 2D/3D approaches and future research challenges. *The International Archives of the Photogrammetry, Remote Sensing and Spatial Information Sciences, XLIII-B2–2020*, 829–836.
35. Jordan-Palomar, I., Tzortzopoulos, P., Gracia-Valldecabres, J., & Pellicer, E. (2018). Protocol to manage heritage-building interventions using heritage building information modelling (HBIM). *Sustainability, 10*(4), 908.
36. Bruno, N., & Roncella, R. (2019). HBIM for conservation: A new proposal for information modeling. *Remote Sensing, 11*(15), 1751.
37. Santagati, C., & Lo Turco, M. (2016). From structure from motion to historical building information modeling: Populating a semantic-aware library of architectural elements. *Journal of Electronic Imaging, 26*(1), 011007.
38. Savini, F., Fabbrocino, G., & Marra, A. (2021). Digitization of historical architectural elements, a workflow for knowledge of minor centers in inner areas. *DisegnareCon, 14*(26), 9.1–9.15.
39. Lo Turco, M., Mattone, M., & Rinaudo, F. (2017). Metric survey and BIM technologies to record decay condition. *The International Archives of the Photogrammetry, Remote Sensing and Spatial Information Sciences, XLII-5/W1*, 261–268.
40. Marra, A., Trizio, I., Fabbrocino, G., & Savini, F. (2021). Digital Models for e-conservation: The HBrIM of a bridge along the Aterno river. *SCIRES-IT, 11*(2), 83–96.
41. CoE—Council of Europe. *Framework Convention on the Value of Cultural Heritage for Society*. Council of Europe Treaty Series—No. 199, (2005), https://rm.coe.int/090000168071aabc. Accessed 26 November 2021.
42. Dieck, M., & Jung, T. (Eds.). (2019). *Augmented reality and virtual reality*. Cham: Springer.
43. De Paolis, L. T., Arpaia, P., & Bourdot, P. (Eds.). (2021). *Augmented reality, virtual reality, and computer graphics: 8th International Conference, AVR 2021, Virtual Event, September 7–10, 2021, Proceedings. Lecture Notes in Computer Science,* vol. 12980. Cham: Springer.
44. Bacco, M., Barsocchi, P., Cassará, P., Germanese, D., Gotta, A., Leone, G. R., Moroni, D. Pascali A. M., & Tampucci, M. (2020). Monitoring ancient buildings: real deployment of an IoT system enhanced by UAVs and virtual reality. *IEEE, 8*, 50131–50148.
45. Fabbrocino, G., Savini, F., Marra, A., & Trizio, I.: Virtual investigation of masonry arch bridges: Digital procedures for inspection, diagnostics, and data management. In C. Pellegrino, F. Faleschini, M. A. Zanini, J. C. Matos, J. R. Casas, & A. Strauss (Eds.), *Proceedings of the 1st Conference of the European Association on Quality Control of Bridges and Structures. EUROSTRUCT 2021. Lecture Notes in Civil Engineering*, vol. 200, pp. 979–987. Springer, Cham.
46. Bekele, M. K., Pierdicca, R., Frontoni, E., Malinverni, E. S., & Gain, J. (2018). A survey of augmented, virtual, and mixed reality for cultural heritage. *Journal on Computing and Cultural Heritage (JOCCH), 11*(2), 1–36.

Virtual Restoration and Persuasive Storytelling for a Virtual Visit to Palazzo Rosso, Genoa

Francesco Gabellone

Abstract The take-off of 5G technology in Italy coincided with a series of initiatives aimed at demonstrating the potential and possible applications of this network in the areas of smart technologies, internet of things, smart cities and next-generation smart homes. The use of this new high-speed connection will also positively involve the enjoyment of cultural heritage, proposing to offer new and more powerful solutions, especially in the field of immersive virtual visualization. With these premises is born a project promoted by Ericsson, Fastweb, Genoa Municipality and Strada Nuova Museums (UNESCO heritage), which aims to virtualize some spaces of the Museum of Palazzo Rosso through innovative solutions, immersive, emotional and persuasive. With the virtualization of the Palazzo Rosso museum we intend to test some features of the 5G network, investigating response times, streaming capabilities and portability of audio, video and 3D resources in real time. The main outputs concerned the development of immersive stereoscopic panoramas to be visualized on VR headset and, within these, stereoscopic movies. The realization of narratives including reconstructive films, immersive navigation in the paintings and virtual restoration of the ceilings destroyed during the Second World War, are made possible by the use of 3D-based technologies and the potential of methods based on Camera Mapping. The paper describes the main techniques used and exposes the results of the case study as a guideline for possible applications in other similar contexts.

Keywords 5G network · Virtual visit · Storytelling · Camera mapping · Stereoscopy · Virtual restauration

1 The 5G Project at Palazzo Rosso

The launch of 5G technology in Italy coincided with a series of initiatives aimed at demonstrating the potential and possible applications in the fields of smart technologies, internet of things, smart cities and next-generation smart homes. The use of

F. Gabellone (✉)
National Research Council—Nanotec, 73100 Lecce, Italy
e-mail: francesco.gabellone@cnr.it

© The Author(s), under exclusive license to Springer Nature Switzerland AG 2023

I. Trizio et al. (eds.), *Digital Restoration and Virtual Reconstructions*, Digital Innovations in Architecture, Engineering and Construction,
https://doi.org/10.1007/978-3-031-15321-1_14

this new high-speed connection will positively involve also the fruition of cultural heritage, proposing to offer new and more powerful solutions, especially in the field of immersive virtual visualization. With these premises is born a project promoted by Ericsson, Fastweb, Genoa Municipality and Strada Nuova Museums, which aims to virtualize some spaces of the Palazzo Rosso Museum (Fig. 1), through innovative, immersive, emotional and persuasive solutions. Along today's Via Garibaldi, the magnificent Renaissance and Baroque Strada Nuova is home to an original museum itinerary that connects three important Genoese palaces: Palazzo Rosso, Palazzo Bianco and Palazzo Doria Tursi.

The entire street axis of Strada Nuova, residential district of the Genoese aristocracy, is unique in the world for its urban and architectural quality. For the painter Pietro Paolo Rubens, the buildings located along this road even became a model of housing civilization to be proposed to his contemporaries. Built between 1671 and 1677, Palazzo Rosso is now home to a prestigious museum that houses important paintings, rich polychrome marble floors and parts of the furnishings of the original noble residence. It was built by Ridolfo Maria, the firstborn, together with his brother Giovanni Francesco I. After the death of Ridolfo, Giovanni Francesco I inherited the entire palace and in the following years his nephew, Giovanni Francesco II, commissioned the architect Francesco Cantone for the decoration of the facade. John Francis II adopted for the palace the style "Regency" then prevailing, and wanted to renew environments and furnishings according to the new fashion. This policy of artistic magnificence was crowned in 1746 by the election of Giovanni Francesco II as Doge of the Republic of Genoa. But it is to the Duchess of Galliera, Maria Brignole Sale De

Fig. 1 The Palazzo Rosso museum

Ferrari, that we owe the birth of the first Genoese museums. With the donation to the City of Palazzo Rosso (1874) and the hereditary bequest of Palazzo Bianco (1889), these important residences became the main seats of the Strada Nuova Museums, declared a UNESCO World Heritage Site. In Palazzo Rosso is exhibited a rich picture gallery which includes paintings collected over more than two centuries by the noble Brignole-Sale family. Among the artists exhibited in the gallery of the aristocratic Genoese palace are Dürer, Veronese, Guercino, Strozzi, Grechetto, Van Dyck, and many others. With the virtualization of the Palazzo Rosso museum we intend to test some features of the 5G network, investigating response times, streaming capabilities and portability of audio, video and 3D resources in real time.

2 Technologies and Objectives

One of the main limitations in the development of VR applications is given by the computational burden required to properly display the content and the network bandwidth, i.e. the maximum data transfer rate of a network at a given time over a specific connection. This project therefore aims to verify the overall performance of immersive applications capable of stressing the current networks and compare them with the potential of the 5G network, the fifth generation of mobile technology. It was created with the aim of implementing mobile network quality and increasing speed, improving wireless services and reducing latency, i.e. the response time of a device to the network. More than an evolution of the previous existing networks, the 5G network is a completely different way of conceiving the mobile network. It is based on a different communication technology than the current 4G network. 5G provides for better coverage, with special 5G antennas, different frequencies and data transmission modes [1]. These improvements have been conceived and developed to meet the new needs of users, who increasingly need to connect to a stable and performing network the so-called smart objects (Internet of Things, such as appliances or cars). Such a network, which is based on a different system of decomposition into cells and receivers, allows to:

- manage a greater quantity of connected devices at the same time: currently, in fact, the 4G network can manage from 1,000 to 100,000 connected devices per km^2, while 5G networks will come to manage more than a million simultaneously;
- achieve higher data transmission speeds: it is estimated, in fact, that 5G mobile networks can theoretically transmit data up to 10 Gigabits per second (Gbit/s), a speed 10 times faster than the 4G network;
- optimize and reduce energy consumption of devices connected to the network by about 90%: 5G, in fact, does not use the hardware of smartphones or devices to operate but a special system of 5G antennas, which results in significant energy savings for each device.

As regards, instead, the possible harmful effects on health, the 5G network has a system of organization in cells more articulated and complex than the current one, but uses a lower signal power, which reduces the risk of damage in case of excessive exposure to the network itself. What is certain, however, is that the benefits of 5G will be found in different areas of application. It's not just normal user activity that will benefit, but also and above all the use of connections in the business and public sectors, which will lead to be able to offer better services in less time (source: https://www.orakom.it/rete-5g/). In order to test the benefits of 5G band in the context of the project exposed here, some strategies have been identified, mostly oriented to the realization of cultural contents that have required a considerable commitment of the network [2]. The main outputs involved the development of immersive stereoscopic panoramas to be displayed on VR headsets of the Oculus Go type and, within these, stereoscopic movies. Both are notoriously more onerous from the computational point of view and therefore, in terms of weight in Mb, more challenging for wi-fi streaming. The project wants therefore, on the one hand, to test the new potential of 5G in reference to the bandwidth only. The data rate of 5G is up to 100 times higher than 4G, with a maximum potential speed of 20 Gbps (Giga bits per second). In the second instance, however, the goal is also to demonstrate that the benefits of the new network will also involve the virtual world and social media involving the use of VR headset. In brief, the test concerned: the creation of stereoscopic VR panoramas; the realization of a fruition platform based on headset (Oculus Go, currently out of production); the realization of narratives including reconstructive movies; 3D navigation in paintings [3]. The management of all these media is therefore based on a 3D approach, which will be described in the next chapter.

3 Virtual Restoration Techniques at Palazzo Rosso

Following the bombings of the Second World War, Palazzo Rosso underwent two successive restorations, respectively in 1953 and 1961, coordinated by the then director of the Office of Fine Arts and History, Caterina Marcenaro and conducted by architect Franco Albini [4]. The restorations of these years brought to light some parts of the frescoes painted in 1692 by Bartolomeo Guidobono. The entire figurative program of the Gioventù in cimento was reconstructed on the basis of the remains still preserved. The intent of the restorers was to restore the original volumes which, through the insertion of marble floors and door frames, would give the whole a prestigious appearance, where the sober elegance of the architectural lines could be related without excessive contrasts to the system of arrangement of the works of art. Thanks to digital technologies, it is possible to understand the nature of the transformations of the building following the restoration and to faithfully reconstruct some of the spaces destroyed by the bombings of the Second World War.

Two rooms have been examined, the north loggia of the second piano nobile and the ceiling of the main hall of the same floor. In the first case, the reconstructive technique uses the methods of virtual restoration, in particular an approach based on the projection of a two-dimensional image on the 3D restitution of the real environment. To this purpose, a 3D model based on digital photogrammetry has been created, on which some period photos have been attached (Fig. 2), taken from different points of view. The technique used is quite simple, on Maxon's BodyPaint 3D software we visualize the UV map of the real space, automatically generated by Agisoft Metshape software, used for the restitution of the loggia (Fig. 3). The points of the photo are then anchored with the 3D model, with a technique known as texture mapping, called camera mapping or projection mapping. This technique is well known and has been discussed in depth in tutorials and papers in the past [5–7]. An alternative automated approach is described in the literature [8] through the extraction of corresponding features between the 3D model and the reference image.

With a matching feature set, the orientation of the image to the point cloud, generated by laser scanner or photogrammetry, can be obtained by spatial resection. The basic idea of this approach is to use the SIFT operator described by Böhm and Becker [9] in 2007 to detect matching points in the camera image and an intensity image of the laser scanner data. The determination of matches consists of four steps: detection of salient features, description of features, matching of descriptions in both images, and evaluation of correct matches. From an operational point of view, techniques of this type, although extremely refined and oriented towards automating the process, require specialized computer skills. In our case, which is very different from the one just described, the projection mapping method has provided excellent results, mainly oriented to the narration by images of a stage prior to the current one, without the need to apply complex algorithms and specialized computer procedures.

Fig. 2 Two old photos show the appearance of the North Loggia before its partial destruction

Fig. 3 On the left the 3D restitution of the North Loggia in its current state. On the right the virtual restoration obtained with the Camera Mapping techniques

We have therefore sought to refine this old, widely known mapping method in an attempt to provide even the least experienced with the tools necessary to produce textured 3D models with good accuracy. Using high quality, high resolution images, it is possible to obtain extremely realistic models, almost indistinguishable from the real thing, to be used productively in documentation for the knowledge of historical buildings, in restoration operations, in faithful representations of the state of affairs, in video games, etc. [10]. The usefulness of this method is even more appreciated in mapping operations performed from historical photographs. It is in fact known that UVW projections, whether cylindrical, spherical or cubic, are applicable only in particular cases and can never be used to project photos directly onto a 3D object. Planar mapping, for example, projects the texture onto the object according to the normal to a plane. The location of the plane, and therefore the direction of projection, is chosen according to the needs and the type of object to be textured, but the type of projection can never be coincident with the camera shot. In a large number of cases similar to ours, this technique is mistakenly used to map complex objects, assigning small patches of the 3D object to specific photographic textures, with poor results and obvious signs of stretching in areas with different angle to the direction of projection [11].

This means that, strictly speaking, planar mapping could only be applied to a planar object with an orthorectified texture, according to the laws of orthogonal projections. On the other hand, every photographic image is a perspective view, with a point of view, a field of view, and deformations dependent on the quality and nature of the lenses. In theory then, knowing exactly these four parameters, and mapping the photographic image according to the rules of perspective, i.e. the same method of camera mapping, it is possible to obtain a mapping with almost perfect 3D texture-model superposition.

The first step of this process, extremely important for the success of the method, is the elimination of the distortions induced by the lenses from the texture. "Barrel" or "bearing" distortions are always present, despite the use of professional aspherical lenses generally advertised as having no distortions. These can be easily removed using specific software, usually provided by the manufacturers. For the recognition of the point of view and the aiming point, you can use the rules of descriptive geometry, but also specific software, for example TAG Camera Calibration. This simple function allows to solve the position of the focal points and of the projection plane. Consequently, through small eventual adjustments it will be possible to obtain an almost perfect collimation of the photo with the model. Knowing the camera position, the focal point and the aiming point, it will be possible to apply effectively the camera mapping (Fig. 4). The application of the described technique to the virtual restoration of the ceiling of the salon (second piano nobile) is particularly interesting. This ceiling was also destroyed during the Second World War. Its original decoration was centred on the myth of Phaeton, son of Helios, a very recurrent theme in the aristocratic residences of the 1500 and 1600s. In the representation, Phaethon (Fig. 5), inexperienced in handling the reins and keeping the horses at bay, lost control and the chariot came too close to the Earth, causing destruction everywhere. It is the allegorical representation, always current, of human inexperience; of a young man who, measuring himself arrogantly with the forces of nature, ends up burning the forests and drying up the seas. A teaching that comes from ancient mythology, which invites us to shun arrogance, pursuing rationality. Thanks to the virtual restoration, based on Camera Mapping, Phaeton lives again at the wheel of his solar chariot. A historical hole and a sketch of the painting, offer us the basic elements to remap the ceiling and reproduce the original colors. Currently, the ceiling of the hall is treated with a neutral-colored paint, so in the next restoration works, scheduled starting from the end of last year, the image obtained from the virtual restoration can be projected on the ceiling, using the same techniques of projection mapping. Operationally, the real projector can be positioned anywhere in the room, precisely because the virtual restoration operation provides a fully mapped 3D model, optimally visible from different points of view.

Fig. 4 The complete restoration of the North Loggia within the movie

Fig. 5 Actual state and virtual restoration of the Phaethon ceiling

4 Immersive Clips, Animated Paintings and Digital Photogrammetry

A similar approach to the one just described was used for the three-dimensional reconstruction of old photos, useful to represent the different restorations carried out on the building. Simplified three-dimensional models, based on photos taken before the post-war restoration, have been obtained from camera mapping techniques. From these models, reconstructed from one or two photos, it is possible to derive simple camera movements [12], very interesting from the point of view of narration, but even more effective for the use of stereoscopy. As mentioned at the beginning, stereoscopic animated clips have been integrated into the visit platform according to a choice oriented to the testing of the 5G network, but also to the pursuit of a persuasive communication, obtained in immersive mode [13]. The clips have been realized in native stereoscopy and integrated within the tour with the Krpano software (https://krpano.com/home/).

```
“ <image stereo="true" stereoformat="SBS">
  <sphere url="plugin:video" />
  </image>
```

With this setting you can use MJPEG HTTP video streams (mjpegstream) as a source for the pano image. The supported image types are: <sphere>, <cylinder> and <flat> and <fisheye>, in the case of a movie the choice will be the flat format. A second possibility of using the stereoscopic movie is possible with the identification of the movie resource as a “hotspot” that can be activated with a simple trigger. In this case my XML script is the following:

```
<hotspot name="video1"
         visible="false"
      pausedonstart="true"
      url="plugins/videoplayer.swf"
         alturl="plugins/videoplayer.js"
         videourl="clipvideo.mp4"
         altvideourl="clipvideo.mp4"
      stereo="SBS"
      posterurl="cats-poster.jpg"
      antialias="false"
         ath="0" atv="0"
         distorted="true"
         depth="off"
      scale="0.5"
         rx="0.0"
         ry="0.0"
         rz="0.0"
         volume="1.0"
         directionalsound="true"
         range="90.0"
         rangefalloff="1.0"
         outofrangevolume="0.0"
         onvideoready=""
         onvideopaused=""
         onvideocomplete=""
         onunsupported=""
      handcursor="true"
      onclick="togglepause(); tween(hotspot[vr_cursor].scale,0,0);"
         />
```

With this expedient, the group of programmers has developed a stable application, with stereoscopic views of the entire museum, in which are integrated animated clips, in stereoscopy, useful to deepen some topics, such as the history of restoration and the critical reading of some paintings.

In the latter case, both the classic techniques of decomposition of the paintings in 2.5D were used, but alongside these, some characters were reconstructed completely in 3D (Fig. 7). The technique is a useful communicative device to generate interest in unobservant or uninterested visitors. The decomposition of a painting in three-dimensional space is in any case interesting, because it allows to critically isolate some elements of the scene, allowing the visitor to visualize in the foreground or background the narrative themes, told by the speaker.

The recreated spatiality establishes interesting new perspectives and different reading keys of the work, sometimes necessary to underline geometries, compositional rules, palettes and hidden details. The same communicative purpose can be found in the need to tell the stylistic and morphological connotations of the ceilings dedicated to the seasons. The complexity of the decorations and the chromatic richness of these ceilings cannot be fully appreciated from the usual points of view, those of a normal visitor. For this reason, the four rooms dedicated to the seasons have been restored with digital photogrammetry techniques, in order to obtain 3D models from which to derive stereoscopic clips obtained with camera movements only possible with high-end cinematographic equipment [2]. The availability of the

Fig. 6 The Autumn room. 3D restitution obtained by digital photogrammetry. The availability of the 3D model allows to tell and make more visible details barely visible by the visitor

Fig. 7 Full 3D painting. On the left the frame in the animation, on the right the 3D model

3D model gives us access to unusual shots (Fig. 6), a few centimetres from the stuccoes and paintings, to offer the visitor an emotional story to discover the allegories represented in these rooms. The three-dimensional models have a resolution of about 3 mm/mesh, in the most detailed areas. The technique used is widespread and well known. In our case we used a Fujifilm X-T30 camera with 26 Mpx resolution, Fujifilm 16–55 F2.8 lens. For each environment were taken about 400 frames, which allowed to generate UVW textures of 20,000 × 20,000 pixels.

5 Conclusions

This case study has allowed to expose different techniques on a 3D basis to test the efficiency of the 5G network and to highlight its potential also in the fields of multimedia and edutainment, applied to communication for cultural heritage. The impact of the platform on the audience has been very positive, also with reference to the feedback obtained on press releases [14]. The streaming vision did not show any slowdown, with performances similar to the off-line reproduction. It was not possible to test the use with a very high number of devices, due to their limited availability during the experiment. The applications on cultural heritage, in particular Virtual Reality with VR Headset will certainly benefit from the wide bandwidth guaranteed by 5G.

A sincere thanks to Elena Marchetto (Fastweb s.p.a.), Marco Giacomobono (Ericsson s.p.a.), Raffella Besta (Municipality of Genoa, Museo di Palazzo Rosso) and Piera Castagnacci (Municipality of Genoa), for having strongly supported this project. The VR platform was developed by Digimat s.p.a., thanks to Angelo Donvito and Francesco Carbone.

References

1. Roussou, M., & Efraimoglou, D. (1999). High-end interactive media in the museum. In *International Conference on Computer Graphics and Interactive Techniques: ACM SIGGRAPH 99 Conference Abstracts and applications* (Vol. 8, No. 13, pp. 59–62)
2. Jain, A. K., Acharya, R., Jakhar, S., & Mishra, T. (2018). *Fifth generation (5G) wireless technology revolution in telecommunication, 1867–1872.* https://doi.org/10.1109/ICICCT.2018.8473011
3. Gabellone, F. (2014). Approcci metodologici per una fruizione virtuale e arricchita dei Beni Culturali. In *Arkos Scienza Restauro e Valorizzazione*, ISSN 1974-7950, edItinera Thesauron, Gennaio-Giugno
4. Besta, R., Boccardo, P., & Priarone, M. (2017). Genova - Musei di Strada Nuova, Palazzo Rosso, Palazzo Bianco e Palazzo Tursi, Guida alla visita, Silvana Editoriale
5. Meierhold, N., Spehrb, M., Schilling, A., Gumholdb, S., & Maasa, H. G. (2010). Automatic feature matching between digital images and 2D representations of a 3d laser scanner point cloud. isprs.org
6. Gabellone, F. (2015, April). *Digital technologies and communication: prospects and expectations, open archaeology* (Vol. 1, no. 1). ISSN (Online) 2300-6560. https://doi.org/10.1515/opar-20150005
7. Gabellone, F. (2019). *Archeologia Virtuale. Teoria, tecniche e casi di studio* (pp. 13–34). Edizioni Grifo
8. Yastikli, N. (2007, September–December). Documentation of cultural heritage using digital photogrammetry and laser scanning. *Journal of Cultural Heritage, 8*(4), 423–427. https://doi.org/10.1016/j.culher.2007.06.003
9. Böhm, J., & Becker, S. (2007). Automatic marker-free registration of terrestrial laser scans using reflectance features. In *8th Conference on Optical 3D Measurement Techniques* (pp. 338–344)
10. Haddad, N. A. (2016). Multimedia and cultural heritage: A discussion for the community involved in children's heritage edutainment and serious games in the 21st century. *Virtual Archaeology Review, 7*(14), 61–73. https://doi.org/10.4995/var.2016.4191
11. Becker, S., & Haala, N. (2007). Combined feature extraction for facade reconstruction. *ISPRS Workshop on Laser Scanning and SilviLaser, 36*, 44–49.
12. Boccardo, P., Dequal, S., Lingua, A., & Rinaudo, F. (2001). True digital orthophoto for architectural and archaeological applications. *ISPRS Archives, XXXIV–5/W1*, 50–55

13. El-Hakim, S. F., Beraldin, J.A., & Blais, F. (2003). Critical factors and configurations for practical 3D image-based modeling. In *6th Conference on 3D Measurement Techniques* (Vol. II, pp. 159–167), Zürich, Switzerland
14. https://www.youtube.com/watch?v=9Zt0TxR3tnY

Methodological Practice for the Physical and Virtual Reconstruction of "Absent" Museum Goods: Hypotheses for Their Protection, Valorization and Inclusive Dissemination

Rita Valenti, Fernanda Cantone, and Emanuela Paternò

Abstract The present study focuses on the adoption of a methodological approach which can bring enjoyment, in the context of an inclusive society, from "invisible" museum works. Reference is made to the invisible museum heritage, stored in basements not accessible to the public and therefore "hidden". The specific action involves an meaningful dialogue between the University and the local museums both of whom have entered into formal institutional agreements. The collaboration is based on an interdisciplinary activity where survey, prototyping and technological processes are bound together to bring back to light the different artistic fragments, silent sometimes, but part of the museum collections. The purpose of the research is to develop an appropriate action plan for intervention to boost audience engagement implementing different multidisciplinary strategies. It is necessary then to create with the support of digital technology multi-sensory 3D models designed for a wider audience which includes both non-disabled people who can enjoy the full display (not only the fragments on view in the museum's rooms also those off display kept in storage) and visually impaired visitors who can have an interactive tactile experience. The architectural-sculptural fragments studied in the present research are kept in the Museo Civico in Noto and belong to the main entrance and to the apse coronation found in the excavation site of the ruins of Ancient Noto annexed to thestructure of San Michele "al Castello". Hopefully, the applied methodology can contribute to the sustainable protection of those artifacts whose storage in the museum basements often causes further damages.

Keywords Cultural heritage · 3D virtual reconstruction · Prototyping · Cultural dissemination

---

R. Valenti (✉) · F. Cantone · E. Paternò
Special Educational Structure of Architecture, DICAR, University of Catania, Syracuse, Italy
e-mail: rita.valenti@unict.it

F. Cantone
e-mail: fernanda.cantone@unict.it

E. Paternò
e-mail: emanuela.paterno@unict.it

© The Author(s), under exclusive license to Springer Nature Switzerland AG 2023
I. Trizio et al. (eds.), *Digital Restoration and Virtual Reconstructions*, Digital Innovations in Architecture, Engineering and Construction,
https://doi.org/10.1007/978-3-031-15321-1_15

1 Introduction

1.1 Digital Communication of Cultural Heritage

The "Code of the Cultural and Landscape Heritage" defines the **protection** of cultural heritage as the exercise of the functions and the regulation of the activities aimed at **identifying** the properties constituting the cultural heritage and at ensuring the protection and conservation of the aforesaid heritage for purposes of public enjoyment. This fundamental principle, enshrined in Art. 6 Par. 2, states that enhancement is to be carried out in forms which are compatible with **protection** and which are such as not to prejudice its exigencies.[1] Museums which keep extensive collections of cultural heritage have to meet the requirements of protection and enhancement of the works they own fostering and promoting a type of knowledge accessible to everybody. In particular, according to the most recent regulations, museums have both a cognitive, social and an inclusive educational role. The Faro Convention ("framework Convention" of the Council of Europe on the value of cultural heritage for society), ratified in September 2020 by the Italian Government, establishes that all the involved parties take steps to improve access to the heritage, especially for young people and the disadvantaged, in order to raise awareness about its intrinsic value. The present study focuses on the adoption of a methodological approach which can bring enjoyment, in the context of an inclusive society, from "invisible" museum works. Intangible cultural heritage, synthesis of traditions and knowledge, is a part of the museum identity, rich in new values and distinctiveness [1]. Reference is made to the invisible museum heritage, stored in basements not accessible to the public and therefore "hidden". Stored in the museum basements it actually prevents the community from the complete interpretation of its own past, becoming "invisible" and precluding sometimes researchers as well, from a full comprehension of the relation of the artefacts with the original architectural structures and from the possible reconstruction of the decontextualised pieces or fragments, often stored without any systematic catalogue.

Modern society with all the issues connected to the emergence of COVID-19 has contributed to the research of new operating methods for the enhancement and scientific dissemination. The use of digital technologies is considered, by the scientific community itself, "the driving force behind the development of new paradigms for cultural heritage" [2]. The data derived from the survey conducted by the ICOM with a set of questions asked to Museums from 23 to 30 April 2020 turned up that a high percentage of museums have created specific digital contents after the lockdown. However, only 27% of the surveyed museums have published contents regarding undisplayed objects stored in basements (the "hidden" collections).[2]

---

[1] Legislative Decree 22 January 2004, n. 42 "Code of the Cultural and Landscape Heritage, art. 10 Law 6 July 2002, n. 137", published in the Official Gazette n. 45, 24 February 2004—Ordinary Supplement n. 28.

[2] International Council of Museums Italia, http://www.icom-italia.org/comunicazione-digitale-dei-musei-ai-tempi-del-covid-19/.

The temporary closing of museums has given rise to digital platforms with sometimes not so high-quality contents. Consequently, all the issues concerned with the Digital Cultural Heritage (DCH) represent today a challenging field of study in relation to the epochal change in exploring and interacting with virtual museums. In particular, one of the four macro-tasks, identified by Clini and Quattrini, in the process of creating a new chain of digital, sustainable, proactive awareness of cultural heritage deals with the digitalisation of 2D and 3D cultural models with the contribution of "a network of relations between scientific institutions (Universities and Public Administrations)" [2]. Thus, virtual modelling is suitable for the documentation, analysis and reconstruction of visible and invisible museum artefacts, to determine the best communicative performances of artistic, decontextualised or undisplayed products in order to enhance and promote them, creating a single "democratic" tool for better access and understanding. The specific action involves an open and meaningful dialogue between the University and the local museums both of whom have entered into formal institutional agreements. The dense collaboration is based on an interdisciplinary activity where survey, prototyping and technological processes are bound together to bring back to light the different artistic fragments, silent sometimes, but part of the museum collections. The desire to tangibly display a selection of "absent" works of art responds to the need to reconstruct and recontextualise archaeological and architectural fragments which are housed for limited display space in basements, often inappropriate and, as previously said, lacking complete data for their correct position in time and space (Fig. 1).

The purpose of the research is to develop an appropriate action plan for intervention to boost audience engagement implementing different multidisciplinary strategies.

It is necessary then to create with the support of digital technology multi-sensory 3D models designed for a wider audience which includes both non-disabled people who can enjoy the full display (not only the fragments on view in the museum's

Fig. 1 Archaeo-architectural fragments stored in the local museum basements

rooms also those off display kept in storage) and visually impaired visitors who can have an interactive tactile experience.

People with visual impairments are provided greater accessibility to some artistic objects on view in traditional museums, real semiophores to enjoy the visit touching and experiencing masterpieces through dedicated paths.

Thus, displaying the "absent" cultural heritage through rapid prototyping techniques gives the chance to create multisensory paths usually found in museums which are catering to people with a wide range of needs and at the same time producing simulations for the non-disabled so to have a personal understanding of living with visual impairment [3].

1.2 Tactile Perception of Hidden Cultural Heritage

The process of digitalization of the fragments, conducted with non-invasive and precise survey methods, the following creation of copies with rapid prototyping techniques, perfectly matching the originals is functional to formal-stylistic studies for the reconstruction of the architectural structures they belong to, through the single elements.

The geometry investigated with the support of 3D restitution allows the restoration of fragmented artefacts giving them their original unity.

The purpose is to reorganize their fragmented storage and their decontextualisation from the excavation sites. 3D printing can build more inclusive cultural paths for a wider audience.

The architectural-sculptural fragments studied in the present research are kept in the Museo Civico in Noto and belong to the main entrance and to the apse coronation found in the excavation site of the ruins of Ancient Noto annexed to the structure of San Michele "al Castello"[3] (Fig. 2).

In particular, the fragments of the entrance are partly kept in museum (lunette) and partly, those regarding the apse coronation, are stored in a warehouse adjacent to the museum lacking any systematic catalogue (Fig. 3).

Given the circumstances, the first operations dealt with a close inspection of the places to recover the fragments with the support of the published photos and archive

[3] The remains of Ancient Noto are located in the site of Monte Alveira. The medieval town and the whole Val di Noto were destroyed, as it is known, by an earthquake in 1693. Despite its position which made Ancient Noto "invincible" it was completely abandoned. "Here, the archaeological excavations conducted in 2007 in the area of the Royal Castle gave back the structure of San Michele "al Castello", interpreted as a probable palatine chapel of Norman times" [4]. Cf. also Gazzè [5].

The compositive and morphological reconstruction of historical architecture through survey. An example of study and methodological approach to the representation of the emerged elements. Master's Degree Thesis in Architecture-SDS Architecture, Siracusa; Bares [6]. La cappella reale di San Michele nel castello di Noto Antica (XII-XVI secolo). Palermo: Edizioni Caracol.

Fig. 2 Archaeological site of San Michele al Castello in Ancient Noto

materials. The comparison showed that not all the pieces represented in the photos could be recovered.

Of course, this contributes to the gradual decline of data useful for the implementation of an approach aiming at the investigation of the structure further aggravated by the possibility that some fragments could definitely become "absent" and not merely "hidden". Actually, the ruins of the glorious Norman monument, object of the research, were studied with the support of reverse engineering which uses 3D technology particularly appropriate for the present purposes.

Being sculptural objects with no colored parts they are suitable for the realization of scale prototypes which can transmit information through an exclusively haptic form of exploration.

Fig. 3 Fragments on view (lunette); Fragments stored in the warehouses adjacent to the museum

2 Materials and Methods

2.1 State of the Art

The ongoing evolution of survey technology and instruments offers a considerable number of advantages in terms of 3D documentation of museum artefacts gradually transforming the museum system [7, 8].

In particular, 3D reality-based technology is applied to protect, disseminate and restore cultural heritage.

3D models represent effective tools for the protection and enjoyment of cultural heritage since they can be used for the virtual creation of museum archives, for cultural dissemination on the web, for the contextualization of artistic objects, for 3D printing design [9].

Digitalisation and 3D printing have generated a new field of application regarding the museum-going experience [10].

Scientific literature has confirmed that 3D models and the architecture of 3D installations offer a great variety of materials which meet the expectations of people with special needs [3, 11–13].

Tactile reproduction provides an opportunity for visitors to explore the objects from different points of view and with the use of a different number of senses.

There are many rapid prototyping techniques whose difference depends on methods and materials used (powder, liquid, solid).

Unlike CNC machines (Computer Numerical Control) which operate subtracting material from a block in order to produce an end product, additive manufacturing models are created depositing material, layer upon layer.

In FDM technology (Fused Deposition Modeling) implemented in the present research, a plastic filament pours out from an extrusion nozzle.

The nozzle is heated to melt the material which is deposited onto a base in layers one strand at a time.

2.2 *Data Acquisition*

As already said, the conducted research aims at the digitalisation and 3D printing of architectural and sculptural objects stored in the basements of the Museo Civico in Noto. Projects such as the present one have difficulties in the acquisition and analysis of digital data for the dimension and complexity of the investigated cultural objects and for the difficulty of survey operations.

It was necessary, then, to find out the ideal solution which could allow high resolution, metric quality and the acquisition of the greatest amount of data for digital and tactile reproduction.

When surveying museum objects, it is necessary to take into account all the possible situations and the different problems which might arise: dimension and position of objects, complexity of surfaces to be surveyed, reflectance of the surface of materials, lighting conditions.

All these factors may cause problems in data acquisition and in the 3D model construction determining, for example, mistaken surface reconstructions. In the present case study two different methods for the acquisition of data were used: range-based techniques using the FARO Laser Scanner CAM2 Focus3D and the LeicaC10 Laser Scanner coupled with image-based techniques which rely on the potential of Structure from Motion (SfM) employing the Nikon D5300 digital camera.

3D Laser Scanner was extremely important for its accurate precision. Different scans were obtained from different survey stations and "chess" targets were placed for the subsequent automatic connection to the PC.

Considering that there isn't any terrestrial laser scanner with self-adaptability to all surveying procedures and taking into account the previously mentioned difficulties, in order to achieve a positive result and to avoid shadows (inaccessible areas) data were integrated with terrestrial photogrammetry (Fig. 4).

Images were acquired in JPEG format at a variable distance of 0.5–2 m. Shots were taken maintaining the camera at three different heights with a recommended overlap of 60% between flying tracks. Anyway, the real difficulties in the data acquisition processing were caused by the position of some fragments in the basements.

In particular, the fragments pertaining to the main entrance of the structure of San Michele "al Castello" in Ancient Noto were adjacent to the walls and it was not

Fig. 4 Survey operations with Faro CAM2 Focus and Nikon D5300

possible to move them. It was impossible to survey the whole surface of these objects, consequently, the final mesh generated with the combination of the two previously mentioned techniques presents different holes. In the case of flat objects (canvas, for example) it would be possible to reconstruct the back creating a plane. But dealing with sculptural objects with particular decorations, the reconstruction of extensive portions would have led to a model quite different from the original [14].

Images were acquired using no specific source of light, rather using diffused ambient light and respecting the traditional adjustments of SfM, that is, disabling flash and auto white balance and setting equal exposure. In the case of the fragments of the apse coronation of San Michele in Ancient Noto, it was possible to have a complete survey, combining the two techniques, because they were placed far from the walls and in a position which allowed the rotation around them.

2.3 *Data Processing*

The two data sets were processed in single units during both the alignment and the cleanup phases. The scans, obtained with a 3D laser scanner, were linked together using Cyclone software.

Photos extracted from the digital camera were imported into 3DF Zephyr able to generate a sparse point cloud, searching for the homologous points, and then a dense point cloud. The two sets were subsequently georeferenced within the same frame of reference using the function "Point Cloud Registration" (ICP). In particular, the point cloud from Cyclone was imported into 3DF Zephyr and was manipulated as a reference object, conversely, the photogrammetric model was considered as a moving object and as such it was scaled and placed onto the other. The dense point cloud, generated by overlapping and matching the two previous point clouds, went

Fig. 5 Lunette of San Michele al Castello: from point cloud to texturized model

through the mesh generation process. The obtained polygonal model, improved with Geomagic Wrap, went through a long and careful retopology process (Fig. 5).

The same procedure was followed for all the surveyed fragments. It allows appropriate operations of digital recomposition and reconstruction aiming at a more extensive and engaging cultural communication especially when it is not possible, as is the case, to appreciate these artefacts on site simply because they are not on view (Fig. 6).

Before 3D printing, all 3D processed models required a hole-filling review, especially for the parts which make up the apse and which didn't present big problems during survey operations. On the contrary, the models referring to the entrance of San Michele al Castello had large empty areas which prevented 3D printing but made it possible, though, to test the reconstructive hypothesis, to carry out geometric studies and to allow visualization (Fig. 7).

Digital models ready for 3D printing were exported into .STL format for their subsequent 3D printing. 3D printer PowerWasp was used to produce the physical object. Cura was the software which worked with the printer's parameters while Pronterface managed the printing process.

Fig. 6 Fragments of the apse coronation of San Michele al Castello in Ancient Noto: 3D model

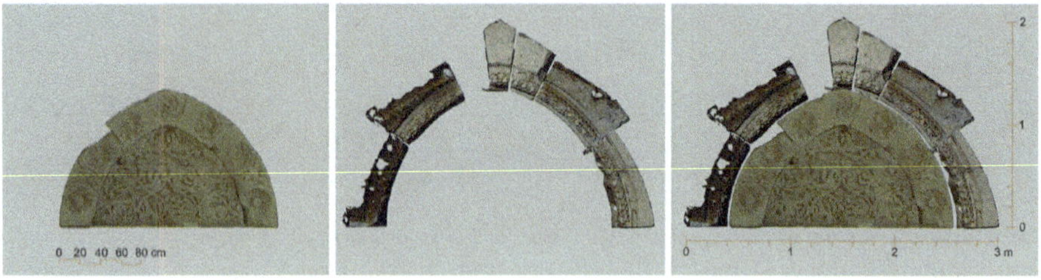

Fig. 7 The models referring to the entrance of San Michele al Castello: virtual reconstruction

Printing times depend on various factors, the size of the object for example. In the present case it took about 11 h to print the copies of the architectural fragments on a scale 1:8. Once printed, objects were separated from the build platform and finished removing supports and smoothing surfaces (Fig. 8).

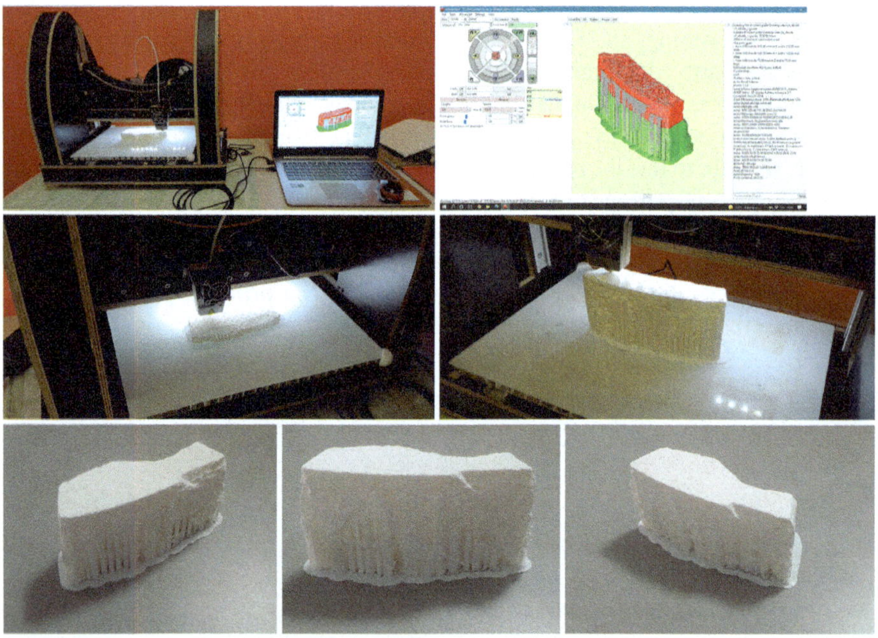

Fig. 8 3D printing operations on an architectural fragment

3 Exhibition Spaces and Routes

3.1 Documentation and Design

The exhibition space is the place of vision of objects and works. It is an architectural space where references can be found in the shape of the space, in its material and functional consistency and then in the characteristics of the objects to be exhibited. In the past, the rules of the exhibition space were mainly simplicity and cleanliness.

In particular, Le Corbusier, Mies van der Rohe and F. L. Wright were promoters of a fluid and easily passable space; the phase of contemplation and focus of the exhibited objects is linked to their influence.

Since the 1960s, it has been noted that society is beginning to prefer messages for images, fast and immediate tools that are easy to understand and disseminate.

Today, contemporary society requires an exhibition space in which the levels of contact between object and public can be identified in the collection and selection of problems mainly related to communication and then to the characteristics of physical space. It is a question of identifying the right image that can affect the public, in relation to the objects on display and the architectural space.

Contemporary culture considers perception as an optimal manifestation of the artistic and semantic qualities of the objects on display [15].

In addition, the relationship between objects and architectural space is closely linked to the image that you want to create, the ability to manage the space and its objects.

In general, ancient art is exhibited in highly textured and colorful places, made of backgrounds and depths that exalt the relationship with the frame of the paintings, contemporary art is instead characterized by white spaces, the white cube of O'Doherty [16], where there is no perception of the outside but a strong appeal to the religious space persists.

In the field of the design and selection of the objects to be exhibited, the succession of exhibitions can follow several methods that change according to the themes, objectives and possibilities of the architectural space.

It is a matter of choosing between chronological, systematic, thematic, geographical, historical-architectural (environment), didactic or mixed exposition. Succession choices influence the overall organization of the exhibition.

Today, the objectives of a contemporary exhibition are based on the concept of semiophore. It indicates an object bearing particular meanings and exposed in such a way as to enhance its meaning and its function. In other words, any object highlights its meanings when it has only representative function.

The path thus organized can give "social context" to objects, trying to highlight meanings and values, in a narrative plot understandable to everyone. In this perspective, objects take on the value of what is inherited and what they leave behind. The history of objects attracts the public and creates empathy.

An object must not only be observed but understood in its entirety, it must send an immediate message to everyone. The so-called "communication to the masses" is the goal to be achieved.

The construction of the exhibition, so rich in meanings, is based on the visual-verbal system and tends to build a metaphorical image, perceptive, conceptual, almost a theatrical transposition where objects are to recite.

The spectator perceives the atmosphere, the objects, the colors and the materials, in a space-architecture relationship that must abandon the concept of traditional exposure and turn towards figurative and graphic experiences.

In a context of this kind, the path is characterized by dynamism and different levels of information.

An exhibition must communicate the most significant messages, historically founded and understandable to a heterogeneous audience, both in age and culture. It must allow vision to any person, even with movement limitations or visual impairments. The latter, in particular, could see objects through touch.

3.2 The Integrated Fruition Model

Today, the technological possibilities of the exhibition are expressed through physical paths, designed to arouse direct visual emotions, and multisensory paths, bearers of an innovative and engaging experience.

The physical paths provide fluid passage solutions through the spaces and a comfortable observation. They are made with materials chosen for their characteristics of practicability, safety and usability.

Designing according to the security requirement class means identifying all the conditions related to the safety of users, as well as the defense and prevention of damage dependent on accidental factors.

In this sense the paths must be free of visual and physical obstacles, provide surfaces that allow a safe practicability, not too slippery, not too rough, have dimensions that allow maneuvers and displacements.

The exhibition supports must help to communicate and interpret the exhibited objects in all phases of the exhibition: of perception, of approach, of observation, of contemplation and of distance. The multisensory paths produce a complete perception of the exposed object, both physical and virtual.

They are the new frontier of the museum exhibition and are addressed to several categories of users, including also disadvantaged people such as the visually impaired.

They have the experience of tactile knowledge that involves both ergonomics, design and technology. In terms of usability, the project of multisensory paths provides spaces without obstacles, lights that guide the path, writing, banners and explanatory tables clear and simple (Fig. 9).

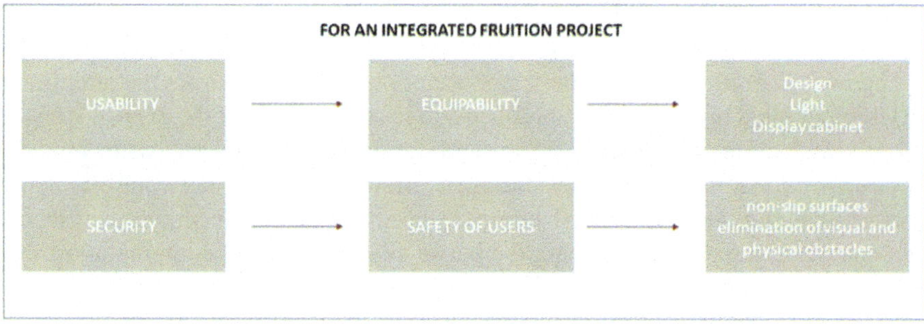

Fig. 9 Classes of needs involved in the project

The model proposed here is set on a thematic exhibition sequence, with environmental values (where possible) that leads the visitor along a path of reconstruction of the past and visibility of artifacts and forgotten objects.

The path becomes multisensory, with great potential for communication and opens to the visually impaired who can finally see-through touch. The image of the route tells the story of the pieces and ruins of buildings symbol of a city that no longer exists. The levels of information are therefore distributed both in the visual and tactile spheres, proposing a direct and comprehensible communication.

From the performance point of view, the path proposes reconstructed objects to be exposed and touched; they must be placed in particular spaces easily accessible, at appropriate heights and must be made in order to be easily touched, identifying the characters of the various parts.

The paths reserved for the disabled and in particular for the visually impaired are of particular importance. For this category of subjects, the exhibition includes a tactile phase on a perfectly reproduced copy of the object on display. To allow the visually impaired to touch the reproduced copy of the object, it is necessary to provide specific exhibition criteria:

- form and size of the object,
- height of the object,
- distance from the route.

Each object, according to its shape and size, will need a different display cabinet, not standardized, which will facilitate the approach and the tactile phase, without hindrance.

Furthermore, in order to define the characteristics of this multisensory model, the existential performance approach directs towards the classes of safety requirements (users' safety against risks) and usability of spaces (size, equipment, accessibility, possibility of interrelation with other spaces, etc.).

In particular, the characteristics of the floors, in the ambit of the regulations on the exhibition spaces in force, must prevent falling or slipping, must not be too rough and make it easy to transit. Lights, colors, physical filters guide the path and guide in all five phases of exposure.

The initial perception is the most important phase, it must strike the attention of the visitor, it must induce him to approach and to stop in front of the exposed objects.

Lighting, colors and attractive shapes are fundamental elements in this phase, as well as the two subsequent phases that must allow the user to stay comfortably and observe, contemplate, let himself be involved in the history of the object. In fact, light can guide the path, suggesting the pause, observation and contemplation.

In conclusion, the schematic indications introduced by this model of path, characterized by a strong multidisciplinarity, lead to the drafting of an innovative and expanded multisensory exhibition project both functionally and communicatively.

4 Conclusions

3D printing technology, using reverse engineering process, enabled the reconstruction of architectural "invisible" remains whose traces were found in the archaeological sites after careful excavations.

This reconstruction represents the final product which makes visible the "hidden" artifacts in a sustainable and inclusive way (for non-disabled and visually impaired people).

From a broader perspective, virtual and/or tactile reconstruction regarding, in particular, the remains of the chapel of San Michele "al Castello", has generated geometric and morphological analysis for the composition of their unity.

Geometric studies provide all the possible relations between the parts and the scale printed models remove the present fragmented situation in data acquisition (Fig. 10).

The applied methodology gives a whole new perspective on the implementation of digital technology with the goal to build more inclusive museum exhibitions.

Visually impaired people will have a chance to enjoy cultural heritage on view in traditional museums following tours not specifically dedicated to them but accessible and enjoyable to them, too.

Multisensory exhibitions make visible the "invisible" archaeological remains reaching out the visitors' other senses leading to a kind of integration between non disabled and visually impaired people.

Besides looking at a work of art, they can experience it through a sense of touch, that instinctive gesture which is necessarily and categorically forbidden with the original. Hopefully, the applied methodology for the realization, with a backward process, of digital twins can contribute to the sustainable protection of those artifacts whose storage in the museum basements, waiting for a more appropriate placement, often causes further damages.

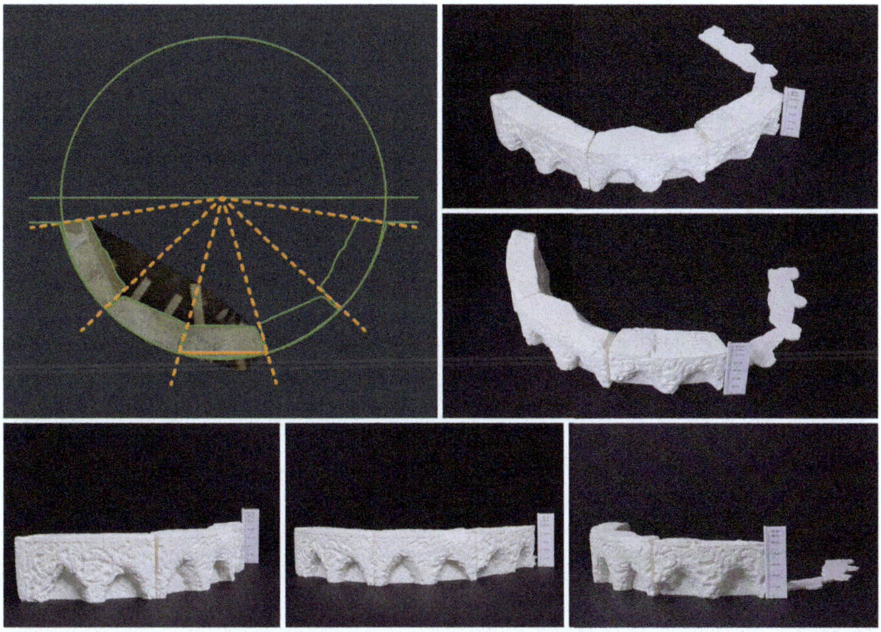

Fig. 10 Geometric studies and tactile reconstruction of the architectural remains

Acknowledgements The authors would like to thank Dr. Giuseppina Giurdanella, responsible manager 8th sector Tourism and Culture and architect Raffaele Di Dio, general director service 3° municipal library, Unesco office, technical support for museum areas of the municipality of Noto, who made possible the research about the dissemination of knowledge of sculptural heritage of Museo Civico di Noto. Translation by Dr. Melania Grancagnolo. This research was partially supported by the research program named "Visibile e Invisibile: percorsi interdisciplinari per una fruibilità diffusa dei beni Museali. Ricerca-Azione per una didattica inclusiva" (University research PIA.CE.RI 2020–2022) is the result of the collaboration between the authors. In particular, Rita Valenti wrote the following paragraphs 1, Conclusions; Emanuela Paternò wrote the paragraph 2; Fernanda Cantone wrote the paragraph 3.

References

1. Sammarro, D. (2021). La valorizzazione del bene culturale e il quadro normativo di riferimento: brevi osservazioni, 2019, Ratio Iuris, n. LXV. Retrieved March 24, 2021, from https://www.ratioiuris.it/la-valorizzazione-del-bene-culturale-e-il-quadro-normativo-di-riferimento-brevi-osservazioni/
2. Clini, P., & Quattrini, R. (2021). Digital Cultural Heritage, arts reproduction and Museums systems. Languages and techniques in a Covid and post-Covid scenario for new forms of heritage against the silence of a fragile culture. *SCIRES-IT—SCIentific RESearch and Information Technology, 11*(1), 1–10

3. Riavis, V. (2019). Discovering architectural artistic heritage through the experience of tactile representation: state of the art and new development. DISEGNARECON (Vol. 12, No. 23, pp. 10.1–10.9). Retrieved from http://disegnarecon.univaq.it
4. Valenti, R. M. F., Aliano, C. L., Paternò, E. M., & Gazzè, E. (2021). Silent cultural heritage: Digitalisation of the archaeological artefacts of Santa Lucia di Mendola for an interdisciplinary reconstructions. *SCIRES-IT—SCIentific RESearch and Information Technology, 11*(1), 179–194.
5. Gazzè. (2020). *The compositive and morphological reconstruction of historical architecture through survey. An example of study and methodological approach to the representation of the emerged elements.* Master's Degree Thesis in Architecture-SDS Architecture.
6. Bares. (2012). *La cappella reale di San Michele nel castello di Noto Antica (XII-XVI secolo).* Palermo: Edizioni Caracol.
7. Calisi, D., & Molinari, M. (2019). Towards a digital archaeological archive: the case study of the artefacts of the area of Fori Imperiali. *The International Archives of the Photogrammetry, Remote Sensing and Spatial Information Sciences*, XLII-2/W9, 189–196.
8. Koehl, M., & Fuchs, M. (2019). 3D modelling of architectural blocks and antique sculptures for the conservation and the promotion of archaeological heritage—experiments in Alsace. *The International Archives of the Photogrammetry, Remote Sensing and Spatial Information Sciences*, XLII-2/W15, 625–632.
9. Adami, A., Balletti, C., Fassi, F., Fregonese, L., Guerra, F., Taffurelli, L., & Vernier, P. (2015). The bust of Francesco II Gonzaga: from digital documentation to 3D printing. *The International Archives of the Photogrammetry, Remote Sensing and Spatial Information Sciences*, II-5/W3, 9–15
10. Clini, P., El Mehtedi, M., Nespeca, R., Ruggeri, L., & Raffaelli, E. (2017). A digital reconstruction procedure from laser scanner survey to 3D printing: The theoretical model of the Arch of Trajan (Ancona). *SCIRES-IT—SCIentific RESearch and Information Technology, 7*(2), 1–12.
11. Cardaci, A., Versaci, A., & Fauzia, L. R. (2015). 3D documentation for archaeological conservation: Some case studies in central Sicily. *SCIRES-IT—SCIentific RESearch and Information Technology, 5*(2), 49–70.
12. D'Agnano, F., Balletti, C., Guerra, F., & Vernier, P. (2015). Tooteko: a case study of augmented reality for an accessible cultural heritage. Digitization, 3D printing and sensors for an audio-tactile experience. *The International Archives of the Photogrammetry, Remote Sensing and Spatial Information Sciences*, XL-5/W4, 207–213
13. Fregonese, L., Giordani, N., Adami, A., Bachinsky, G., Taffurelli, L., Rosignoli, O., & Helder, J. (2019). Physical and virtual reconstruction for an integrated archaeological model: 3D print and Maquette. *The International Archives of the Photogrammetry, Remote Sensing and Spatial Information Sciences*, XLII-2/W15, 481–487
14. Gonizzi Barsanti, S., & Guidi, G. (2013). 3D digitization of museum content within the 3dicons project. *The International Archives of the Photogrammetry, Remote Sensing and Spatial Information Sciences*, II-5/W1, 151–156
15. Zolotova, A. S., Likholet, E. V., & Fedotova, N. N. (2019). Current trends in the museums and exhibitions exhibition space formation. In *CATPID-2019, IOP Conference Series: Materials Science and Engineering*, 698 033021. https://doi.org/10.1088/1757-899X/698/3/033021
16. O'Doherty, B. (1999). *Inside the white cube: The ideology of the gallery space.* Berkeley: University of California.
17. Ciancicotto, L. M. (2019). Public space, common space and the space in-between. *City & Community, Wiley Online Library, 19*(3), 676–703.

Structural Investigation on 3D Reality Based Models for Cultural Heritage Conservation and Virtual Restoration

Sara Gonizzi Barsanti

Abstract Conservation of Cultural Heritage is a key topic and structural changes and damages can affect the structural behaviour of Cultural Heritage artefacts and buildings. Considering that the application of technologies can help in preserving, conserving and restoring ancient structures, it is mandatory to find the best pipeline to produce the correct analysis that can be provided with the use of Finite Element Analysis, a well-known technique used in modelling stress behaviour of objects and structures. The necessary workflow invests and involves design experts enabled to mathematically master the potential of models that pass from interpretation to design control. Following the principles of good experimental practice, the typical workflow involves the use of CAD 3D models made by Non-Uniform Rational B-splines (NURBS) surfaces, representing the ideal shape of the object to be simulated. Conversely, when applied to 3D models of Cultural Heritage, often altered by the time passed since their original creation, the representation with a schematic CAD model may introduce an excessive level of approximation leading to wrong simulation results. The focus of this work is to present a method based on retopology, hence on quadrangular element instead of triangles. Such process aims at generating the most accurate 3D representation of a real artefact from highly accurate 3D models derived from image-based and range—based techniques, maintaining the accuracy of the high—resolution polygonal models in the solid ones to be created for FEA. A proper pipeline can help to carry out the necessary interventions and perform virtual restoration prior to the physical one.

Keywords Retopology · FEA · Restoration · Conservation · Cultural heritage

S. G. Barsanti (✉)
Department of Engineering, Università degli Studi della Campania Luigi Vanvitelli, Aversa (CE), Italy
e-mail: sara.gonizzibarsanti@unicampania.it

© The Author(s), under exclusive license to Springer Nature Switzerland AG 2023

I. Trizio et al. (eds.), *Digital Restoration and Virtual Reconstructions*, Digital Innovations in Architecture, Engineering and Construction,
https://doi.org/10.1007/978-3-031-15321-1_16

1 Introduction

For research purposes, documentation needs, preservation and conservation issues in archaeology and Cultural Heritage field, it is very important to appropriately document and survey artefacts and sites: indeed, an accurate and complete 3D digital documentation is a prerequisite for further analyses and interpretations. In modern times, atmospheric agents, the growing of the urban drift and of the density of constructions, bombing during the two world wars, the negligence during the centuries, as well as the current situation of political instability and danger in certain areas have all contributed to damage and strongly influence stability of our Heritage; it is, therefore, fundamental to define a proper methodology to preserve our Cultural Heritage. In order to achieve diagnostic studies aiming at understanding the level of decay of Cultural Heritage and at selecting the appropriate preservation methods and materials, a scientific base for allowing correct interventions would be fundamental. However, it is always difficult to predict how a historical object or structure, built thousand years ago, will suffer for environmental agents such as earthquakes, pollution, wind and rain, or human factors like constructions in the surroundings, vehicular traffic, or heavy tourism. Structural changes and damages can affect the structural behaviour of Cultural Heritage artefacts and buildings. Although the use of technologies can help in preserving, conserving and restoring ancient structures, it is mandatory to find the most appropriate and effective pipeline to produce the correct analysis.

The analysis of the structural behaviour of a Cultural object can be made by using Finite Element Analysis, a well-known technique used in engineering for modelling stress behaviour of objects and structures. The typical workflow of such analysis involves the use of CAD 3D models made by Non-Uniform Rational B-splines (NURBS) surfaces, which represent the ideal shape of the object to be simulated [1]. The major FEA packages have meshing modules capable to transform a NURBS model, made only of its exterior surfaces, to a volumetric mesh. Differently from a surface mesh, typically generated with 3D capture methods, these models have nodes distributed both on the exterior surface and in the interior volume, connected each other by elementary volumes such as tetrahedron, pyramids, prisms or hexahedral. This workflow is appropriate in the mechanical field, where a physical element to be simulated is very close to its ideal drawing within strict tolerances. Conversely, when applied to 3D models of Cultural Heritage objects or structures, often altered by the time passed since their original creation, the representation with a schematic CAD model may introduce an excessive level of approximation leading to wrong simulation results.

Nowadays, 3D documentation of CH has been widely developed through active sensors or passive approaches like photogrammetry, but the models, formed by the exterior surfaces of the objects captured at high resolution and therefore represented by millions of polygons, are not suitable for a direct use in FEA. The mesh has in fact to be converted from superficial into volumetric, and the density has to be reduced since the computational complexity of a FEA grows exponentially with the number of nodes representing the simulated object.

This paper presents a new method based on a strong simplification of the mesh associated to a topological rearrangement of it. Such process aims at generating the most accurate 3D representation of a real artefact/scenario from highly accurate 3D digital models derived from image-based and range—based techniques, maintaining the accuracy of the high—resolution polygonal models in the solid meshes that have to be created for FEA. In addition, the proposed process keeps into account the suitability of the simplified mesh to be converted in a set of NURBS surfaces through an automatic procedure. This allows producing volumetric models of a Cultural Heritage object or structure maximizing the closeness of the resulting NURBS model with the acquired one, and in the meantime minimizing the number of NURBS patches required for describing it.

The retopologized mesh is typically based on quadrangular element (quads) instead of triangles. The organization of the polygons in the retopologized models allows having a better distribution of the element, thus leading to strongly reduce the number of the final polygons of which the model is made up. The idea is that a more organized topology could be favourable for converting a polygonal mesh in a NURBS model, while maintaining a better coherence with the digitized artefact. This can be useful when dealing with reality-based models of Cultural Heritage, which are usually accurate and precise but with a complex geometry. The approach here proposed is based on a wise use of retopology procedures, coupled with a transformation of a simplified and retopologized model to a mathematical model made by NURBS surfaces, close as much as possible to the real shape, but suitable to be transformed in reasonably complex volumetric 3D models through standard FEM packages.

The aim of the paper is to provide a geometric process to create models suitable for FEA. The structural analysis presented are a completion to establish the validation of the methodology and to check if the models are suitable for being investigated using the common FEA pipeline. For this reason, the analysis presented are preliminary (Fig. 1).

A proper pipeline can help in finding the probable causes for future problems and therefore plan and carry out the necessary interventions: this will allow for a more effective conservation of the structure.

1.1 Theoretical Background: Reality-Based 3D Modelling

3D optical instruments for the 3D digitization of objects and sites [2] include imaging sensors [3] and ranging sensors like laser scanners or structured light instruments [4]. Often both the approaches are integrated in order to exploit the intrinsic advantages of each one and overcome possible problems [5–8]. Photogrammetry requires experience and images have to be properly acquired, otherwise the results are not satisfactory, in particular with fully automated black-boxes tools. One of the problems of the models produced with photogrammetry is scaling them in a proper way, in order to have an accurate and precise model. Laser scanning, on the other hand,

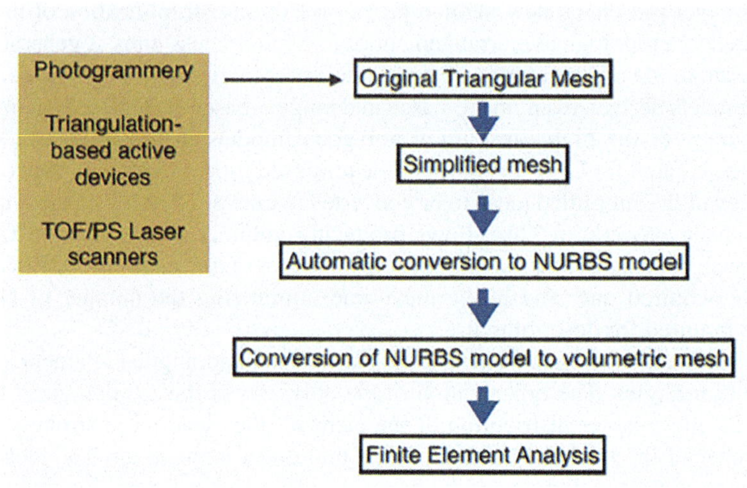

Fig. 1 The pipeline presented in this paper

is not so difficult to be used during the surveying, but it requires a lot of time and experience during the processing in the lab. The choice of the technique depends on different factors and it is often strictly related to the budget of the project. Both surveying techniques have their advantages and disadvantages (Table 1).

The documentation that can be derived from the 3D surveyed and modelled data is various: plans, sections and façades, useful for further investigations of the site and for conservation and preservation policies. Textured 3D models of archaeological sites are also useful for visualization purposes to engage the public and assist archaeologists in interpretations of past uses of space.

1.2 Creation of Volumetric Models of Cultural Heritage for FEA

By analysing the previous works, different approaches have been followed: (a) redrawing with a CAD modeller a new surface model following the superficial mesh originated by the acquired 3D cloud; (b) using directly the triangular mesh as starting point for volumetric models; (c) generating a volumetric mesh from the point cloud with no preliminary surface meshing; (d) passing from 3D reality-based models to BIM/HBIM for FEA; (e) creating a new tool.

The first approach has been used for simulating the behaviour of the Trajan's Markets [10], and in many other applications [11, 12]. In some cases, the reconstructed mathematical model was refined with the insertion of limited patches of reality-based superficial meshes [13].

Table 1 Synthesis of the main differences between photogrammetry and laser scanning [9]

	Photogrammetry (Image-based modelling)	Laser scanning (Range-based modelling)
Characteristics		
Cost of the instruments (HW and SW)	Low	High
Manageability/portability	Excellent	Sufficient
Time of data acquisition	Quite short	Generally long
Time for modelling	Short but experience required	Often long
3D information	To be derived	Direct
Distance's dependence	Independent	Dependent
Dimension's dependence	Independent	Dependent
Material's dependence	Almost independent	Dependent
Light's dependence	Dependent	Almost/totally independent
Geometry's dependence	Quite dependent	Independent
Texture's dependence	Dependent	Independent
Scale	Absent	Implicit (1:1)
Data volume	Dependent on the image resolution and on the measurements	Dense point cloud
Detail's modelling	Good/excellent	Generally excellent
Texture	Included	Absent/low resolution
Edges	Excellent	Quite problematic
Statistics	For each 3D point	Global
Open-source software	Many	A few

An option of using a 3D CAD model extrapolated from a reality-based model is to extract or draw cross-sections or profiles from the 3D model to produce the CAD one [14] used to create a model suitable for FEA This methodology was applied to the Frigidarium of the Baths of Diocletian in Rome [15] and to the monitoring of the Monastery of Jerónimos in Lisbon [16]. If this procedure, with all its accuracy limitations due to the re-draw of the model in a CAD software, can be applied to CH buildings because the geometry of the structure can be replicated through a CAD drawing using profiles, cannot be instead used for statues, whose geometry is more complex and that cannot be simplified through elements as bean, truss or shell, used for the modelling in FEA.

The second approach has a range of slightly different methods, such as bare simplification of the triangular surface mesh before converting it in a volumetric one, that may give significant deviations between the actual shape and the simulated. The simplified description of the shape in this case are discretized profiles giving a low-resolution representation of the interior and the exterior of the structure, from which produce a volumetric model [17, 18] or the fitting of the acquired model with a parametric model suitable to be converted in volumetric mesh [19]. A recent project used the triangular mesh to perform the overall stability analysis of a marble statue when placed on its new pedestal and the mechanical stresses generated on the statue and the pedestal materials [20]. This paper presented a new way of creating the geometrical model of the statue, by subtracting the mesh from a prismatic block shape, according to a boolean operation performed with an open-source CAD software, to preserve the spatial positioning of the statue. Then, another open-source software was used to create the tetrahedral mesh.

The third strategy does not even consider the mesh, generating a volumetric approximation of the shape from the raw 3D cloud of points [21], later compared by the same authors to the other approaches [22, 23]. The fourth process uses 3D reality-based model to produce BIM/HBIM models to be used for FEA [24]. This method can imply two levels of approximation, the first due to the passage from 3D reality-based model to BIM one and the second when creating the volumes for FEA. Oreni et al. [25] explored the use of BIM model as a support for sites monitoring and management, extrapolating thematic information for structural analysis from the model, even if the authors didn't provide a Finite Element Analysis on the structure.

To end with, the Masegra Castle located near the city of Sondrio in Italy, was used as a test object for a procedure called Cloud-to-BIM-to-FEM [26]. An accurate survey integrating geometrical aspects, diagnostic analysis based on destructive and non-destructive inspections, material information, element interconnections, architectural and structural considerations was the basis for the BIM creation. The BIM was then turned into a finite element model with a geometric rationalization taking into consideration the preservation of irregularities and anomalies, such as verticality deviation and variable thickness.

Finally, a new procedure for performing Finite Element Analysis directly on a 3D mesh was created [22].

All the processes have their disadvantages (Fig. 2).

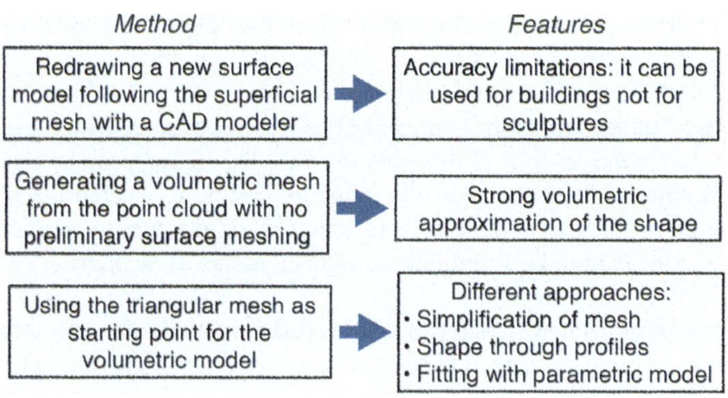

Fig. 2 Pro and cons of the different processes used for the generation of volumetric models of Cultural Heritage for FEA

There is not a unique approach to be used due to the complex field of investigation. Cultural Heritages are too complex to be treated as mechanical parts and the level of simplification that is acceptable for the common pipeline of FEA is not suitable for them. As a consequence, each post-processing of reality-based models used for the creation of volumetric models for FEA can result as the most suitable solution for one kind of CH object and not for another one. The main issue is the lack or loss of accuracy. It seems that the direct use of the mesh obtained through reality-based techniques as a starting point for the creation of the volumetric meshes can be a suitable approach.

1.3 Mesh: Triangular-Based and Quad-Dominant

Topology is referred to the study of geometrical properties and spatial relations between the polygons of a mesh, independently by continuous variation of shape and size of them. Any abrupt change in this relationship is considered a topological error, like for example the flip of the normal in two adjacent polygons. Triangular meshes are defined by a series of triangles which barycentre describes a linear surface representation. A mesh can be described by a set of vertices

$$V = \{v1, \ldots, vV\}$$

and a set of triangles connecting them

$$F = \{f1, \ldots, fF\},\ fi \in V \times V \times V$$

Quad-based topology is used by most 3D artists when creating their models and consists of rows and columns. The advantages are that it will ensure a clean mesh (especially for subdivision), it easily adds edge-loops, and is much more suitable for deforming the model for animation. A quad-based model has a topology that is simple, offers edge flow easily adjustable, and the outcome can be easily subdivided. A model with triangle-based topology can product sharp angles that can affect the design of a mesh. With quads, it's easier to add or manipulate edge loops to obtain a smoother deformation. The quadrangulation method samples the original mesh at a spatial resolution lower than the original with a degree of accuracy higher than sampling the mesh with triangular elements, because it preserves the global geometry of the original mesh, re-defining from scratch its topological structure.

Retopology is a key 3D process that has been developed for optimizing the use of polygons in describing 3D shapes in Computer Graphics (CG) animations, where the rendering time is directly related to the number of polygons of the represented 3D geometry [27]. Retopology is the creation of a new topology for a 3D model. In practical applications, it is obtained by laying down a low-polygon mesh over top of the high-density model for simplifying it and in the meantime start a brand-new polygonal organization, possibly created in order to follow the main geometrical feature of the object described by the 3D model. The retopologized mesh is typically based on quadrangular element (quads) instead of triangles [28]. This simplifies also all the processing stages following the modelling, like the rigging, or the action needed for transforming a static 3D model of a character in a dynamic entity, capable to move, walk and run.

1.4 Finite Element Analysis—FEA

The Finite Element Method (FEM) is a numerical technique used to perform Finite Element Analysis (FEA), initially developed for structural mechanics and then applied to the solution of other kind of problems, such as dynamics, thermal, electrical, fluid and biomaterial problems. The physical problem normally involves a structure or structural component subjected to certain loads. The idealization of this kind of problem to a mathematical one involves specific assumptions that lead to differential equations. The finite element analysis solves this mathematical model, and since this solution technique is a numerical procedure, it is necessary to consider the solution accuracy. If the accuracy is not met, the numerical solution has to be repeated with refined parameters (e.g., a finer meshes) until an adequate accuracy is obtained.

In FEA, is commonly used a mathematical model, an idealization of the physical object built to predict or simulate its behaviour. The analysis is then carried out on the models meshed using given elements, different if a 2D or a 3D problem is evaluated. 2D elements are the triangular and the quadrangular ones: quad elements are preferred since the triangular ones have inferior accuracy. 3D elements are the tetrahedral and the hexahedral: as for the 2D elements, the hexahedral ones are more accurate (e.g., deform in a lower strain energy state) but is more difficult to mesh a 3D volume with this kind of elements, if it is not segmented [27]. The 3D elements can be linear or quadratic: the difference is that the quadratic ones have nodes also in the mid side, varying the number from 4 nodes (linear tetrahedron) to 20 nodes (quadratic hexahedron). Studies have been done to analyse the behaviour of the different kind of elements that can be used in the meshing process.

The Finite Element Analysis is an approximation of the exact solution of the problem and the behaviour of any point inside the finite element is described through nodal displacement, that are the first result of a FEM calculation. This displacement is expressed as the combination of the nodal movement through the form functions, that describe the displacement inside an element and from the formulation of which depends the accuracy of the solution. Since the deformations are linked to the derivative of the displacement and the stresses to the deformations through the material, the accuracy of these characteristics is related to the form functions.

2 Methodology

The starting point of the methodology proposed in this paper is the consideration of the complexity of doing structural analysis on Cultural Heritage. The difficulty lies on the geometry of the objects, compromised by the time passed, on the construction techniques and on the different materials used for buildings. Starting from the previous analysis and considering the lack of information that the methodology used since now carries on in performing structural analysis, the intrinsic characteristics of the different element involved in the pipeline and the non-linear and non-symmetric geometry of the structures, the methodology proposed took into account the use of retopology for the decimation of the reality-based 3D models. The pipeline starts from the post-processing of the acquired reality-based models considering the correction of the topological errors and of the orientation of the model on a suitable reference system, usually with the XY plane corresponding to the base of the model, and the Z-axis passing from the centre of gravity of the artefact. This was considered a proper positioning for the following finite element analysis. In this way, the high-resolution models used were closed meshes properly reoriented in order to post process them to suitable volumetric models for FEA. The second step regarded the use of retopology for decimating the high-resolution models for creating 3D objects lighter that the initial one and suitable for the transformation to NURBS. The conversion of a polygonal model in a NURBS model is available on some CAD packages as automatic process, that in general tend to produce a higher number of small patches when the original mesh is topologically unorganized. As a result, by rearranging the initial topology of the mesh, a preliminary condition for minimizing the number of NURBS patches of the converted model is set. And this represents a better starting point for the volumetric mesher embedded in any standard FEM packages. After all these passages, the 3D models are finally ready for the finite element analysis. The conversion from a mesh to a NURBS is implemented in CAD software or similar (e.g., 3DMax, Blender, Rhinoceros, Maya, Grasshopper, etc.) and it transforms a mesh composed by polygons or faces to a faceted NURBS surface. In details, it creates one NURBS surface for each face of the mesh and then merge everything into a single polysurface. Depending on the mesh, the conversion works in different ways:

- If the starting point is a triangular mesh, and while, by definition, triangles are plane, the conversion creates trimmed or untrimmed planar patches. The degree of the patches is 1×1 surface trimmed in the middle to form a triangle.
- If the starting point is a quadrangular mesh, the conversion creates a 4-sided untrimmed degree 1 NURBS patches, meaning that the edges of the mesh are the same as the outer boundaries of the patches.

Considering the theory, a quadrangular mesh is more suitable to be converted in NURBS (Fig. 3). For this work, the MeshToNurb tool implemented in Rhinoceros was used.

Before creating the volumetric mesh, the retopologized models were analysed to correct any topological errors like non-manifold edges or self-intersecting faces. This

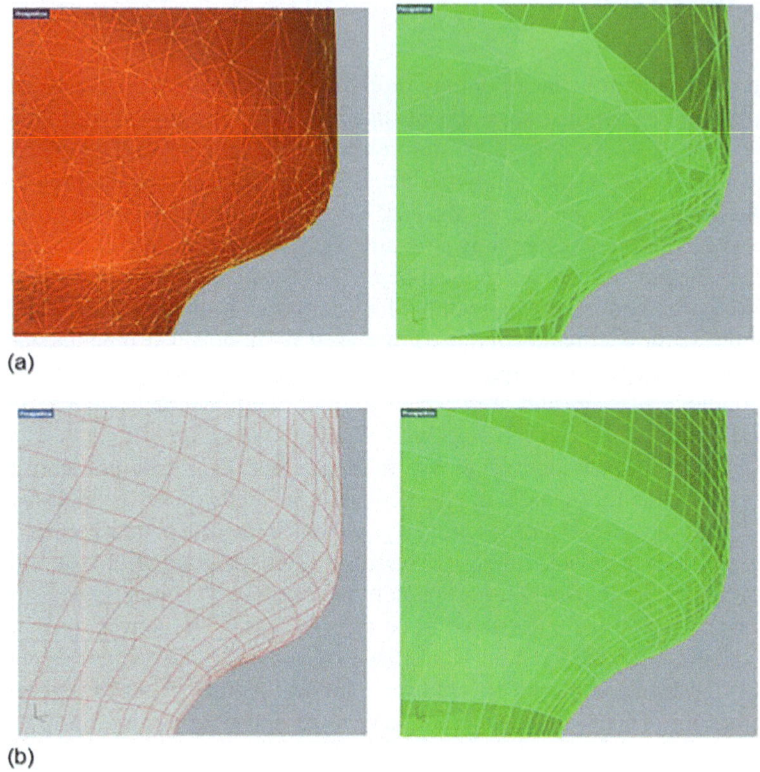

Fig. 3 The comparison between a simplified triangular mesh and its corresponding NURBS (**a**) and the retopologized mesh with its corresponding NURBS (**b**). The meshes are represented in red (left side of the images) and the NURBS in green (right side of the images)

procedure was done using Meshlab, in order to have a closed mesh to be imported in Rhino. This passage is fundamental because if the mesh is not closed, the program is not able to create a closed surface and this will affect the Finite Element Analysis. If the volumetric model is not closed, it is not possible to mesh it with tetrahedrons in the FEA software, meaning that the volume is not closed. In this way, the process in the analysis is not correct. In Meshlab, the tool "cleaning and repairing" permits to correct any topological error and to close any open vertices that could have existed.

3 Validation

The possible errors that can occur in FEA process are user errors (incorrect set up), modelling errors (incorrect representation) and discretization error (the density of the mesh is not sufficient to gain a correct result).

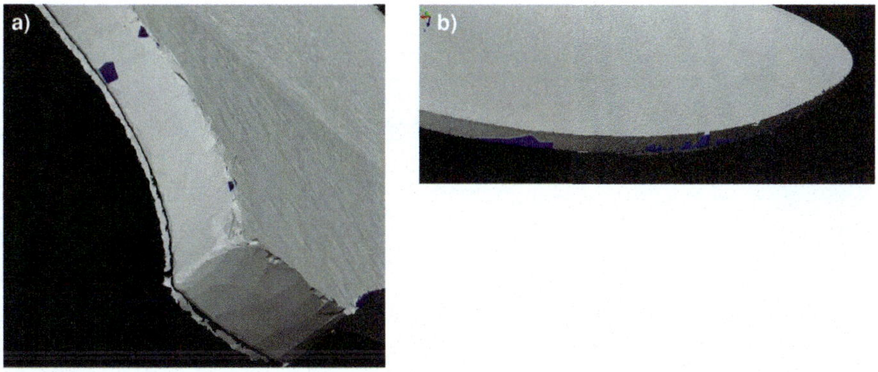

Fig. 4 The initial mesh of the violin: **a** the misalignment of the different meshes on one side of the object; **b** holes on the surface

FEA takes into account a wider number of details, taking a system assembled by differential equations and subdividing it into regions that can be individually solved with simple linear equations. This subdivision generates linear approximations that identify as much as possible the true solution. FEA solves loads and constrains on the model for the displacements of each node.

The process was applied to evaluate a vibroacoustic numerical model of the violin, based on accurate structural modelling [29]. Each side of the violin was surveyed with a 6-axis arm laser scanner with a tolerance of 0.03 mm. The post processing started from the correction of the alignment of the different meshes acquired (Fig. 4). It was not possible to correct perfectly the misalignment on one side of the model but the mesh was cleaned and completed where the missing parts or holes were visible.

After this step, the superficial mesh of each side of the violin was simplified using retopology with the adapt size parameter set to 100 to obtain a 22 K model for each part of the violin. Form this superficial mesh, two NURBS were created and exported in *.iges format (Fig. 5).

The volumetric models were then meshed in the FEA software and some modal analyses were provided for each model and then compared to the experimental ones (Fig. 6). The results gave a good accuracy compared to the laboratory ones. The slightly difference between the numbers has to be attributed to the error in the alignment of the initial mesh that caused a reduction of the thickness of one side of the violin of 0.4 mm. This test represents a solid validation of the process presented in this thesis because was applied to an object with a complex geometry and for which was possible to evaluate the results of the FE analysis with test directly performed on the physical object.

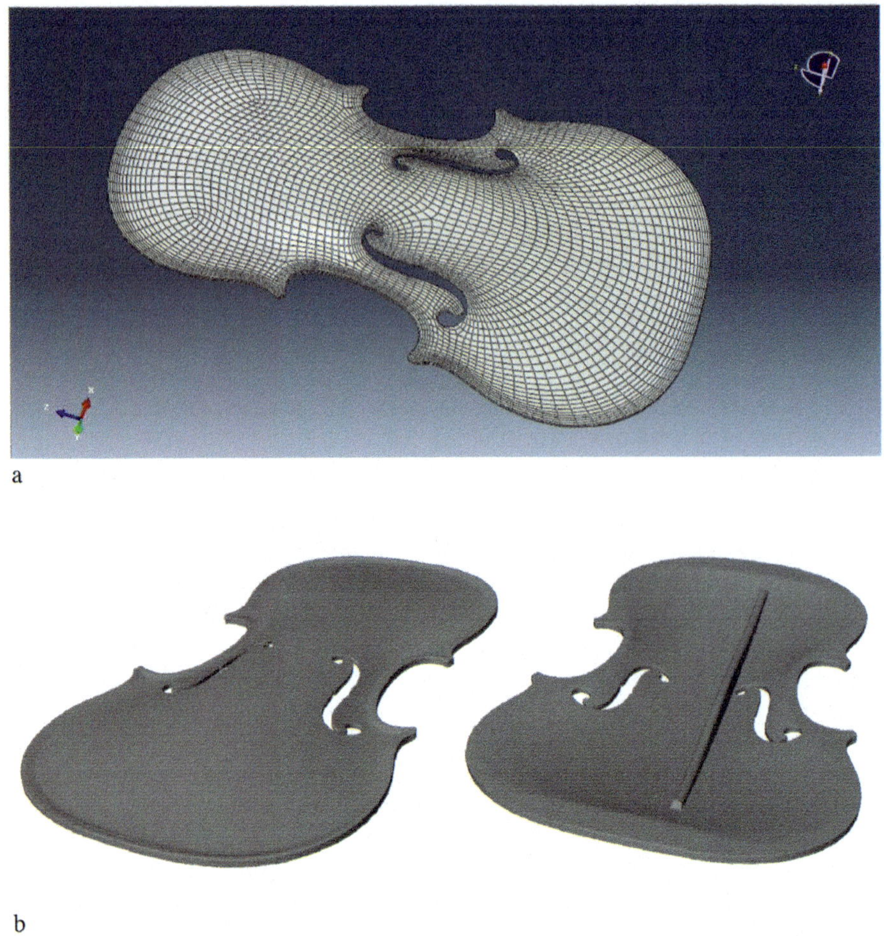

Fig. 5 The retopologized model of the of the soundboard of the violin (**a**) and the NURBS obtained of both the soundboard and the back of the violin (**b**)

4 Application

The same analysis was applied to a statue held in the Uffizi Gallery: the Aristogyton, a 2 m high marble statue, copy of the famous Greek statue made in 447 BC by Kritios and Nesiotes. It was surveyed with photogrammetry using a mirrorless camera Sony NEX 6 placed on a pole, with a 16 mm lens, ISO set to 1000 and aperture to 5.[1] The high-resolution model had 2,816,589 faces and was simplified to 32 K nodes. The

[1] The survey was performed by U. Shafqat Malik—Politecnico di Milano—for a project done in cooperation among Uffizi Gallery, Indiana University—Prof. Frisher and Politecnico di Milano—Prof. Guidi.

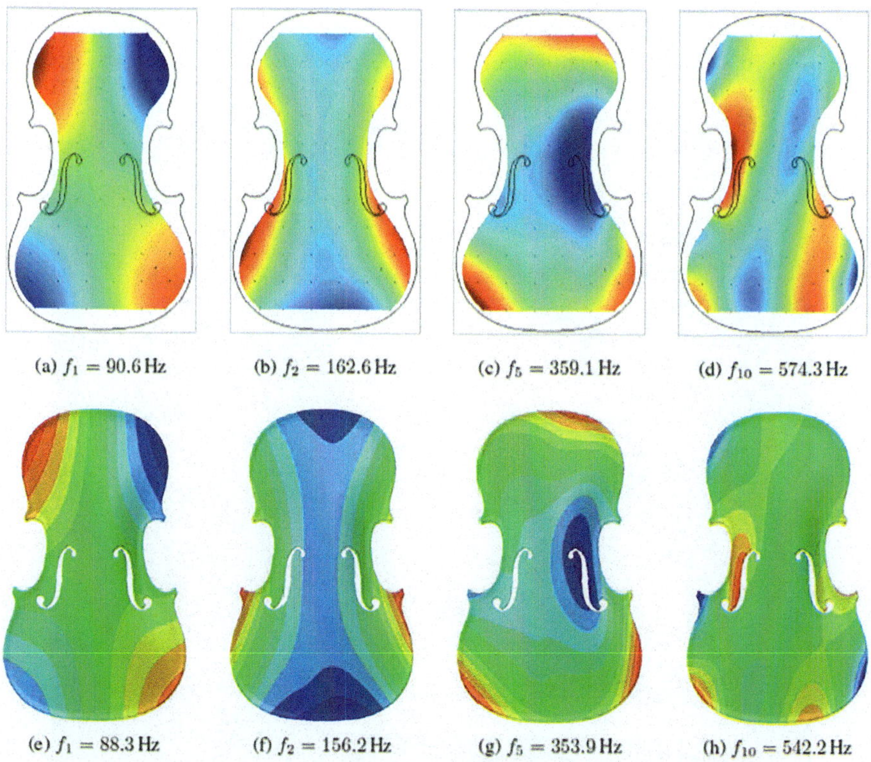

Fig. 6 1st, 2nd, 5th and 10th mode of the free soundboard from the experimental data (upper part of the image); same mode evaluated through the FEA (lower part of the image)

comparison gave 0.0001 m for the mean and 0.0008 m for the standard deviation (Fig. 7).

The mesh was then converted in NURBS and imported in the FEA software for two different analyses: a static one imposing the gravitational load and a modal one, fixing 10 analyses to determine the natural mode shapes and frequencies of the object during free vibration (Fig. 8a–b). The conditions set for the Finite element analysis were:

- Density: 2500 kg/m^3;
- Young's Modulus for marble 78,000 MPa;
- Poisson Ratio 0.3;
- Gravity on −Z axis;
- Fixed Support under the basement of the statue as boundary condition;
- Meshing element: 10-nodes Tetrahedrons;
- Element size: 14 mm, close to the mean measure of the quadrangular mesh elements composing the retopologized mesh to fit the volumetric elements to the

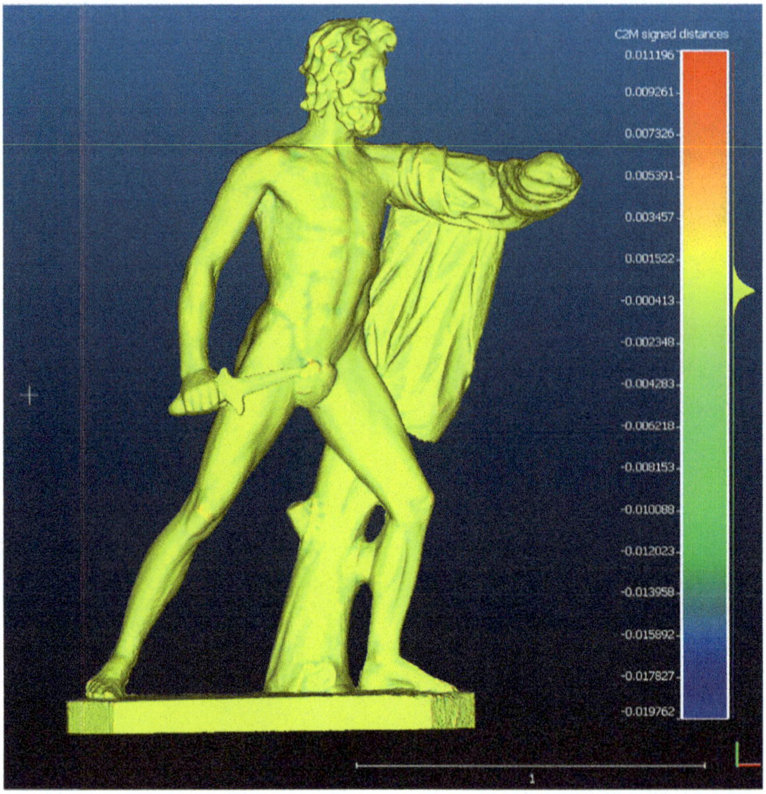

Fig. 7 The comparison of the high resolution and the retopologized model for the Aristogyton statue

subdivision of the superficial mesh in a more adequate way avoiding overlapping of nodes that could have caused errors or discrepancies.

5 Virtual Restoration

An accurate structural analysis of a Cultural Heritage artefact or building may have a strong influence of its restoration process. It is wildly known that the conservation of Cultural Heritage is a time consuming and extremely costly operation. Sometimes, due to lack of resources, buildings may be strongly damaged, a situation that can be easily avoided if the interventions would have been done in time.

In the field of restoration, the use of 3D models can detect deterioration and deformations that would otherwise be invisible. It can also be used to create a register of these weaknesses over time by capturing data at regular intervals. The possibility to predict the weakest parts of an artefact or monument can help in programming ad

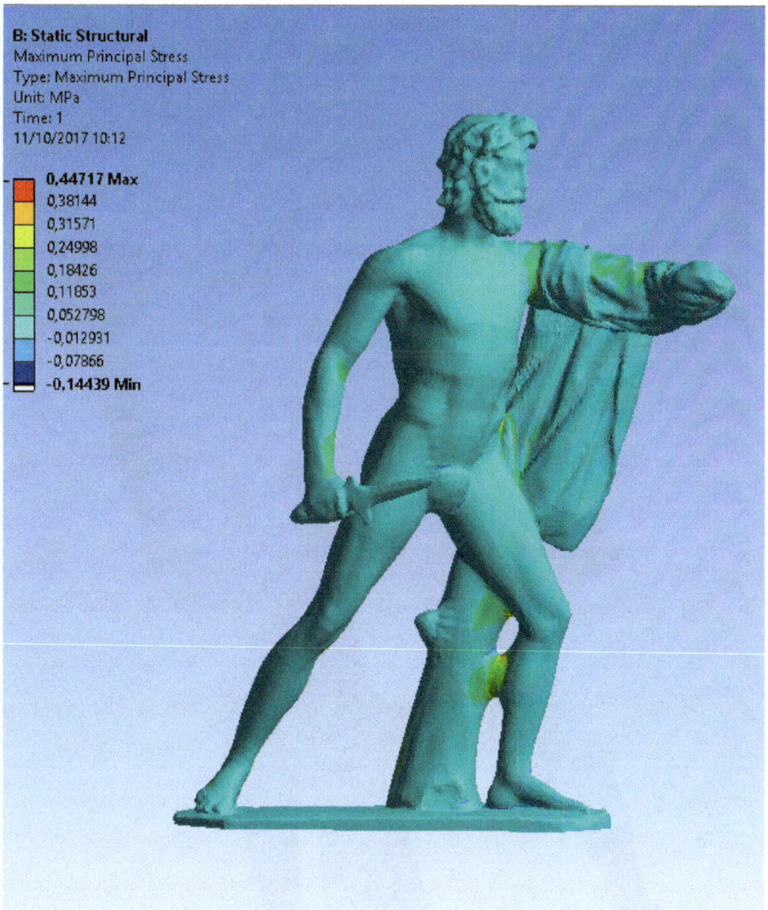

a

Fig. 8 The two structural analyses on the Aristogyton statue: **a** imposing the gravitational force; **b** the modal analysis

hoc interventions that can help in saving money and, on the other hand, intervene exclusively on the parts that really need restorations. In the case of fragmented works, it is possible to study how the fragments fit together without any contact with the Heritage. Another advantage is to create moulds of the missing pieces in order to try them on the original artefact, to analyse which best fit the missing part, or supports which are perfectly adapted to the pieces themselves. A second advantage is the possibility to test, verify and perform virtual restorations on the 3D models, in order to find the adequate procedure. In this way, several tests and interventions can be done without the risk of damaging the Heritage, leading to less expensive actions. The acquisition of a 3D model with reality-based techniques and the subsequent

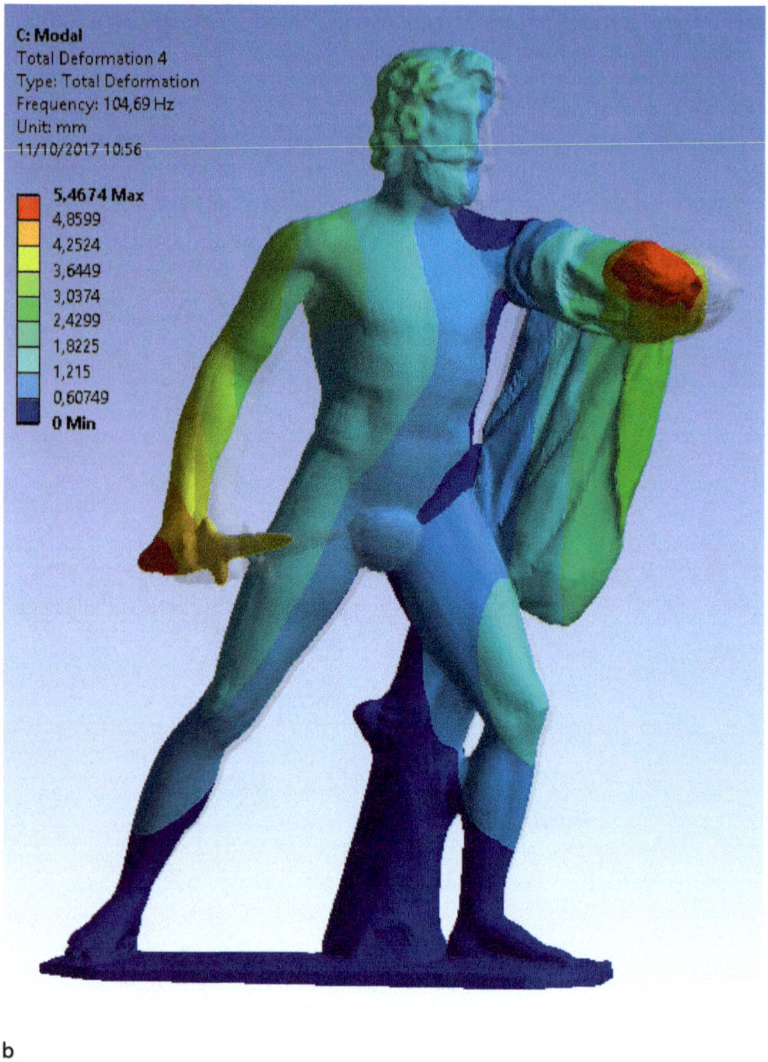

Fig. 8 (continued)

damages identification can help in the set-up of different possibilities to restore and artefact and also to the creation of mock-ups [30].

The restoration process is still tricky and faces major challenges since usually geometry of the damaged parts of artifacts is unknown or is difficult to be specified. The result is that the restored part can be made differently. The first step is then the analysis of contemporary or similar artefacts in order to restore the pieces in the truest appearance. Starting from a 3-dimensional model is helpful because it provides a virtual basis on which applying different restored parts to provide a more accurate reconstruction. Different techniques have been used as:

- Extraction and matching of surface boundaries to offer a more consistent information about the restoration;
- Shape matching of complementary three-dimensional polygonal arcs using as constraint tangent vector across boundaries and an automated restoration of fragmented artifacts;
- Spatial surface-matching techniques;
- Automated assembly of thin artifact fragments;
- Use of symmetry for the estimation of the geometry of the missing parts, in order to ensure the reliability of virtual restoration results [31].

Thanks to Virtual and Augmented reality is also possible to work in the digital space for a shared restoration phase [32] and to show to the public the process and the restored 3D models with the use of a virtual app [33].

6 Conclusions and Future Work

The use of retopology and the method proposed showed the analytical and critical potential of the procedures, obtained great performances and presented interesting properties in the process as a consistent pre-processing step to directly use a reality-based model for Finite Element Analysis. Starting from the idea to take advantage of the high resolution and of all the geometrical details of the reality-based models for the structural analysis of complex artefacts as Cultural Heritage objects and buildings, the possibility to transform all these information into a model that can be processed in FEA software is the most important result. The statement that using a size of the volumetric element close to the size of the superficial one gave an important parameter to be set, especially when dealing with models of objects that cannot be tested in the laboratory. The tests on the violin permitted to confirm the process on a more geometrically complex object, having the possibility to compare the results with the one performed in the laboratory, even if the starting point of the simplification was a mesh not perfectly aligned. No evidence of differences occurred in the post-processing phase f models generated through photogrammetry or laser scanning related to the retopology process. The simplification worked in the same way on both the models.[2] Different geometries could give different issue in the process, mostly starting from the acquisition phase: a more complex geometry is more difficult to acquire and the 3D reality-based model can present holes and missing parts. The post-processing phase in these cases is more invasive and can lead to small errors. Furthermore, the more the geometry is complex and articulated the more the simplification phase needs to take into account the dimension of the elements. It is

[2] The process was tested on other 3D models obtained both with photogrammetry and projection pattern device and the retopology phase was the same and gave the same results.

well known that where the details are small or the changes in directions are complex, the number of elements in the superficial mesh is higher and their dimension smaller. In this case, the retopology phase has to be done accordingly, without stressing too much the reduction of elements. Another possibility is to segment the mesh in order to perform different type of retopology for the different parts of the artefact.

Future work will consist on a proper study of the properties of the different material composing ancient buildings, study that goes at the same pace with a more accurate analysis of the segmentation of the models.

References

1. Höllig, K. (2003). *Finite element methods with b-splines*. Society for Industrial and Applied Mathematics. Germany: Stuttgart. https://doi.org/10.1137/1.9780898717532
2. Remondino, F. (2011). Heritage recording and 3D modelling with photogrammetry and 3D scanning. *Remote Sensing, 3*, 1104–1138.
3. Remondino, F., & El-Hakim, S. (2006). Image-based 3D modelling: A review. *Photogrammetric Record, 21*(115), 269–291.
4. Vosselman, G., & Maas, H. G. *Airborne and terrestrial laser scanning*. CRC Press
5. Grussenmeyer, P., Landes, T., Voegle, T., & Rongle, K. (2008). Comparison methods of terrestrial laser scanning, photogrammetry and tacheometry data for recording of cultural heritage buildings. *International Archives of Photogrammetry, Remote Sensing and Spatial Information Sciences, 37*(B5), 213–218
6. Guidi, G., Remondino, F., Russo, M., Menna, F., Rizzi, A., & Ercoli, S. (2009). A multi-resolution methodology for the 3D modelling of large and complex archaeological areas. *International Journal of Architectural Computing, 7*(1), 39–55.
7. Fassi, F., Achille, C., & Fregonese, L. (2011). Surveying and modelling the main spire of Milan Cathedral using multiple data sources. *The Photogrammetric Record, 26*, 462–487.
8. Nex, F., & Rinaudo, F. (2011). LiDAR or photogrammetry? Integration is the answer. *Italian Journal of Remote Sensing, 43*(2), 107–121.
9. Gonizzi Barsanti, S., Remondino, F., & Visintini, D. (2013). 3D surveying and modelling of archaeological sites-some critical issues. *ISPRS Annals of the Photogrammetry, Remote Sensing and Spatial Information Sciences, II-5/W1*, 145–150. https://doi.org/10.5194/isprsannals-II-5-W1-145-2013
10. Brune, P., & Perucchio, R. (2012). Roman concrete vaulting in the Great Hall of Trajan's markets: A structural evaluation. *Journal of Architectural Engineering, 18*, 332–340. https://doi.org/10.1061/(ASCE)AE.1943-5568.0000086
11. Erkal, A., & Ozhan, H. O. (2014). Value and vulnerability assessment of a historic tomb for conservation. *The Scientific World Journal*, Article ID 357679. https://doi.org/10.1155/2014/357679
12. Riveiro, B., Caamaño, J. C., Arias, P., & Sanz, E. (2011). Photogrammetric 3D modelling and mechanical analysis of masonry arches: An approach based on a discontinuous model of voussoirs. *Automation in Construction, 20*, 380–388. https://doi.org/10.1016/j.autcon.2010.11.008
13. Zvietcovich, F., Castaneda, B., & Perucchio, R. (2014). 3D solid model updating of complex ancient monumental structures based on local geometrical meshes. *Digital Applications in Archaeology and Cultural Heritage, 2*, 12–27. https://doi.org/10.1016/j.daach.2015.02.001
14. Guarnieri, A., Pirotti, F., Pontin, M., & Vettore, A. (2005). Combined 3D surveying techniques for structural analysis applications. *International archives of photogrammetry, remote sensing and spatial information sciences, 36*, 6.

15. Sadholz, A., Muir, C., & Perucchio, R. (2015) *A 3D kinematic model for assessing the seismic capacity of the Frigidarium of the baths of Diocletian* (pp 89–92)
16. Oliveira, D. V., Ramos, L. F., Lourenço, P. B., & Roque, J. C. A. (2005). Structural monitoring of the Monastery of Jerónimos. In *International Conference on 250th Anniversary of the 1755 Lisbon earthquake* (pp. 466–473), Lisboa, Portugal
17. Castellazzi, G., Altri, A. M. D., Bitelli, G., Selvaggi, I., & Lambertini, A. (2015). From laser scanning to finite element analysis of complex buildings by using a semi-automatic procedure. *Sensors, 15*, 18360–18380. https://doi.org/10.3390/s150818360
18. Bitelli, G., Castellazzi, G., D'Altri, A. M., De Miranda, S., Lambertini, A., & Selvaggi, I. (2016). Automated voxel model from point clouds for structural analysis of cultural heritage. *The International Archives of the Photogrammetry, Remote Sensing and Spatial Information, XLI-B5*, 191–197. https://doi.org/10.5194/isprs-archives-XLI-B5-191-2016
19. Bassier, M., Hadjidemetriou, G., Vergauwen, M., Van Roy, N., & Verstrynge, E.: Implementation of scan-to-BIM and FEM for the documentation and analysis of heritage timber roof structures. In: M. Ioannides, E. Fink, A. Moropoulou, M. Hagedorn-Saupe, A. Fresa, G. Liestøl, V. Rajcic, P. Grussenmeyer (Eds.), *Digital heritage. Progress in cultural heritage: Documentation, Preservation, and Protection: 6th International Conference, EuroMed 2016* (pp. 79–90), Proceedings, Part I, Nicosia, Cyprus, October 31–November 5, 2016. Cham: Springer International Publishing (2016). https://doi.org/10.1007/978-3-319-48496-9_7
20. Bagnéris, M., Cherblanc, F., Bromblet, P., Gattet, E., Gügi, L., Nony, N., Mercurio, V., & Pamart, A. (2017) A complete methodology for the mechanical diagnosis of statue provided by innovative uses of 3D model. Application to the imperial marble statue of Alba-la-Romaine (France). *Journal of Cultural Heritage*. ISSN 1296-2074. https://doi.org/10.1016/j.culher.2017.05.002
21. Shapiro, V., & Tsukanov, I. (1999). Mesh free simulation of deforming domains. *Computer-aided design (CAD), 31*, 459–471. https://doi.org/10.1016/S0010-4485(99)00043-3
22. Freytag, M., Shapiro, V., & Tsukanov, I. (2011). Finite element analysis in situ. In *Finite elements in analysis and design* (Vol. 47, No. 9, pp. 957–972). ISSN 0168-874X. https://doi.org/10.1016/j.finel.2011.03.001
23. Shapiro, V., Tsukanov, I., & Grishin, A. (2011). Geometric issues in computer aided design/computer aided engineering integration. *Journal of Computing and Information Science in Engineering, 11*, 21005. https://doi.org/10.1115/1.3593416
24. Abbate, E., Invernizzi, S., & Spanò, A. (2020). HBIM parametric modelling from clouds to perform structural analyses based on finite elements: A case study on a parabolic concrete vault. *Applied Geomatics*. https://doi.org/10.1007/s12518-020-00341-4
25. Oreni, D., Brumana R., & Cuca, B. (2012). Towards a methodology for 3D content models: The reconstruction of ancient vaults for maintenance and structural behaviour in the logic of BIM management. In *2012 18th International Conference on Virtual Systems and Multimedia* (pp. 475–482), Milan. https://doi.org/10.1109/VSMM.2012.6365961
26. Barazzetti, L., Banfi, F., Brumana, R., Gusmeroli, G., Previtali, M., & Schiantarelli, G. (2015). Cloud-to-BIM-to-FEM: Structural simulation with accurate historic BIM from laser scans. In *Simulation modelling practice and theory* (Vol. 57, pp. 71–87). ISSN 1569-190X. https://doi.org/10.1016/j.simpat.2015.06.004
27. Bommes, D., Zimmer, H., & Kobbelt. L. (2009). Mixed-integer quadrangulation. *ACM Transactions on Graphics, 28*(3), 77:1–77:10. https://doi.org/10.1145/1531326.1531383
28. Benzley, S. E., Perry, E., Merkley, K., & Clarck, B. (1995). A comparison of all hexagonal and all tetrahedral finite element meshes for elastic and elasto-plastic analysis. In *Proceedings of 4th International Meshing Roundtable, Sandia National Laboratories* (Vol. 1, pp. 179–191)
29. Corradi, R., Liberatore, A., & Miccoli, S. (2016). Experimental modal analysis and finite element modelling of a contemporary violin. In *ICSV23 23rd International Congress on Sound & Vibration*, 10–14 July, Athens
30. Jo, Y. H., Hong, S., Jo, S. Y., & Kwon, Y. M. (2020). Noncontact restoration of missing parts of stone Buddha statue based on three-dimensional virtual modeling and assembly simulation. *Heritage Science, 8*, 103. https://doi.org/10.1186/s40494-020-00450-8

31. Hou, M., Yang, S., Hu, Y., Wu, Y., Jiang, L., Zhao, S., & Wei, P. (2018). Novel method for virtual restoration of cultural relics with complex geometric structure based on multiscale spatial geometry. *ISPRS International Journal of Geo-Information.*, *7*(9), 353. https://doi.org/10.3390/ijgi7090353
32. Shakya, S. (2019). Virtual restoration of damaged archaeological artifacts obtained from expeditions using 3D visualization. *Journal of Innovative Image Processing (JIIP)*, *1*(2), 102–110. https://doi.org/10.36548/jiip.2019.2.005
33. Abate, A. F., Barra, S., Galeotafiore, G., Diaz, C., Aura, E., Sánchez, M., Mas, X., & Vendrell, E. (2018). An Augmented reality mobile app for museums: Virtual restoration of a plate of glass. In: M. Ioannides, et al. (Eds.), *Digital heritage. Progress in cultural heritage: Documentation, preservation, and protection* (Vol. 11196). EuroMed Lecture Notes in Computer Science. Cham: Springer. https://doi.org/10.1007/978-3-030-01762-0_47

Digital Enjoyment and Virtual Musification

Roman Theatre Experience the Making of Digital Reconstruction

G. Amoruso and C. Carioni

Abstract A strategic plan for the digital museology of Roman archeological sites in Jordan is proposing an interactive multimedia exhibition on the Roman Theatre at the former forum of Amman. The project intends to represent the lost heritage through an immersive installation to engage visitors in the experience of the Roman Theater. The Roman Theatre in Amman is the largest in Jordan. Between 1967 and 1970, the Department of Antiquities of Jordan restored the building. According to the archaeological data, the Theatre was constructed at the time of Antoninus Pius (138-161 d.C). The project aims to present the scientific study and the 3D reconstruction of the original theatre, based on rigorous analysis using *representative models*. The outcomes will be included in a future exhibition designing site-specific installations and user experience based on digital interaction and tactile artifacts. The paper illustrates the multidisciplinary approach associated with the use of 3D virtual reconstruction to disseminate an innovative proposal for its valorisation. Future investigations will address a workflow for an interactive digital model organized semantically as a cognitive system that shows both the constructive organization of the building and the operating technique of the front of the scene.

Keywords Roman theatre · Scientific reconstruction · Architectural survey · Musealization · Digital interaction

1 Research Gap

During the Roman Empire the majestic and ancient Roman town Philadelphia, part of the network of ten Greek cities, called Decapolis-between present-day Jordan, Syria and Israel- was transformed in an important and complex urban

G. Amoruso (✉) · C. Carioni
Dipartimento di Design, Politecnico di Milano, Milan, Italy
e-mail: giuseppe.amoruso@polimi.it

C. Carioni
e-mail: claudia.carioni@polimi.it

© The Author(s), under exclusive license to Springer Nature Switzerland AG 2023
I. Trizio et al. (eds.), *Digital Restoration and Virtual Reconstructions*, Digital Innovations in Architecture, Engineering and Construction,
https://doi.org/10.1007/978-3-031-15321-1_17

system, acquiring numerous public and entertainment buildings, reaching so glories completely forgotten in contemporary memory; for this reason today the city of Amman could boast one of the largest theaters in the Mediterranean area, for the large part undervalued, underpriced and with potential still underestimated.

The Roman Theatre of Amman, the former town of Philadelphia, part of the *Decapolis* network [8], is the largest in Jordan with its 6000-spectator *cavea.* The monument is only marginally investigated, and the scientific literature is not providing the adequate information and framework for experts and visitors that are not engaged with the Roman Heritage and its influence in the region. The interventions to partially reconstruct the Theatre had a massive impact on the whole touristic area and introduced incoherent additions or modern building technologies after centuries of neglect (Fig. 1).

In addition, there have been significant absences of accurate surveys. Hence, the collection of documentation cannot provide a complete reconstruction in archaeological terms, but it will face a typological approach. After its excavation and restoration, it was partially rebuilt, during the 60s and 70s, today, the Theatre represents one of the Jordanian capital landmarks. Together with the two museums housed in the space initially dedicated to the *parascaenia*, the Folklore Museum, and the Museum of Popular Traditions, it constitutes one of the main attractions of Amman and one of the symbolic monuments of the city. As the site of the Theatre, the two museums,

Fig. 1 Roman Theatre from the top of roman citadel. American Colony. Photo Department, photographer (1932). *Air route to Baghdad* via *Amman and the desert. Amman showing Roman theatre and Philadelphia Hotel.* Jordan Amman, 1932. [?] [Photograph] Retrieved from the Library of Congress, https://www.loc.gov/item/2019706714/

founded in the '70s, are also affected by several critical lacks that are influencing the right understanding of such heritage. In the site there are not available information on the diverse collections and the ancient historical place:

- *General conditions of accessibility*

The site of the Roman Theatre is poorly signposted. There are no communicative artifacts, illustrations, and maps that can make the visit more significant. Even in a context in which the Theatre itself acts as an attractive catalyst, no information is available on the historical evolution of the site and on the Roman Forum and the characteristic buildings such as the *Odeon*, the small Theatre built for musical activities such as singing and poetry competitions. Moreover, there is little integration between the square and the buildings; for example, the *Ninfeo* could be included in a possible Roman-historical itinerary for the visitor's benefit.

Therefore, despite being a site of strategic value for development, local and regional appears underused. The cultural offer is not adapted to its importance and the manufactured ones' environmental, historical, and social value. It is a matter of proposing a model of museum system that also concerns the archaeological area to be considered whole by offering activities and services to the benefit of the local community and the school and university institutions to be involved.

- *Poor attractiveness of the tourist-cultural offer*

Although of extraordinary beauty, the area, in general, is not attractive. The main criticism lies in the scarce documentation available on the Roman site, where it is not possible to recognize and learn of the extraordinary evolution that the Forum has produced in the urban structure of Amman as evidence of the rootedness of Hellenistic-Roman culture.

The installation of an information and multimedia area could significantly improve the quality of the visit by allowing a longer also stay to a broader and intergenerational audience. The presence of the two museums on the Jordanian traditions is poorly reported when visitors arrive in the proximity of the large square. The site is characterized by the absence of a museum system that can network and communicate the historical, architectural, environmental resources present as well as the products of material culture (crafts, textiles, objects of daily life, agricultural products, etc.) and intangible (oral stories, ritual feasts, traditions, etc.) through on-site and off-site digital communication systems.

- *Museums: poor interaction with the exhibition and lack of experiential learning*

Accessibility and reception infrastructures are not adequate. Contents and objects displayed are not effectively described. It is impossible to make a significant experience of the collections and their understanding from the origin, craft, and artisanal traditions. From the environmental point of view, visibility is lacking. In some cases, absent or poorly integrated with the installation component and exhibition and the exhibitors themselves limit the full enjoyment of the collections.

In general, exhibitors are obsolete, inadequate, do not integrate information elements, and do not allow a conscious journey to the visitor because the themes related to the content are not clear.

It is not clear what are the only factors of attraction on which the exhibition can play the role of enhancement and at the same time the definition of the relationships and links between the cultural resources exposed.

Museums do not offer content to read or listen to, not creating an inclusive space or allowing a learning dynamic that could favor the visitor. At the same time, the experience is limited to a short walk inside.

It does not highlight the rich intangible heritage linked to objects, rituals and Bedouin traditions, or other cultures that dominated the city of Amman in past times.

The research encourages a new experiential approach by applying two basic museology principles. Cultural heritage experience is the answer to people's need to connect emotionally to places before dealing with their physical content. Digital experience is a means to approach a new experience mainly for getting close to origins and rediscovering the intangible cultural heritage through the aid of new technologies and new forms of interactivity.

The paper intends to present the methodology as a possible standard, from the data collection to the realization of a dedicated theatre exhibition, considering the relevance of the building for the construction of the city identity (Fig. 2).

Fig. 2 Roman Theatre of Amman nowadays

2 Methodological Framework

Although Roman Archeology generally attracted the scientific community's interest, the Roman Theatre in Amman remained a relatively marginal subject among the international debate. Therefore, the project starts from a different approach focusing on the theatre as a body of knowledge, history, archaeology, and the life and society of ancient Roman *Philadelphia*: a preliminary assessment of the documentation and reconstructions of the theatre is providing the proper framework. Guidelines and recommendations can become significant in research on cultural heritage and virtual archaeology. If the Carta di Venezia in 1964 stated that "the restoration must stop where the hypothesis begins", it is possible to argue that virtual archeology begins when the restoration is finished. In fact, despite this apparent break, the distance between the two subjects is much closer than one would expect, considering that both disciplines share many goals [11].

What is the purpose of digital reconstruction of lost monuments? Who are the recipients of the Roman Theatre Experience?

Together with other multimedia products, the digital replica of buildings is a bearer of incisive, highly participatory news. Many reconstructions of theatres are available online, a broad and diverse user base, including visitors to archaeological sites, museums, and exhibitions, especially in recent years, from more generic online users. Digital images enable people to understand monuments that are no longer or only partially preserved in an immediate, practical, precise, and evocative way. In this situation, designers must rely on and refer to the *London Charter* and the *International Principles of Virtual Archaeology*"[1] to propose methods, contents, and communication issues defining some main aspects:

- Make available all sources used for reconstruction.
- Make clear all the technological tools used.
- Make clear the logical process followed in recomposing the various reconstructive hypotheses.
- Propose more reconstructive hypotheses, especially in particularly controversial cases.
- Propose different reconstructive hypotheses for each period or phase of the ancient object considered.

So far, the *London Charter* has enriched the rules, guiding principles, and guidelines, so it has greater credibility in the field of virtual archaeology. The *London Charter* also includes a recommendation on visualization and the idea of digital reconstruction (computer-based visualization); it aspires to suggest the proper use of digital visualization in academic or professional environments and in these fields as the entertainment industry involves cultural heritage.

[1] International Principles of Virtual Archaeology—The Seville Principles 2011: http://smartheritage.com/sevilleprinciples/seville-principles.

In contrast to the *born-digital* artifacts, some others originated from 3d acquisition and modeling. They are to be considered not as simple virtual copies but as resources to study, know, and tell aspects that before were not grasped of the same finding.

The *Seville Charter* was released as part of this activity [6]; the aims added to those of the *London Charter* are:

- Generate criteria that are easily applicable by the whole community and experts.
- To establish guidelines to give the public a better understanding of the archaeological activity.
- Establish criteria for assessing the quality of projects carried out under Virtual Archaeology.
- To promote the responsible use of new technologies to manage cultural heritage.
- Using the new technologies, research in archaeological heritage, conservation, and dissemination of processes shall be improved.
- To open new perspectives for the applications of digital methods and techniques in archaeological research and its dissemination.
- Raising awareness within the international scientific community of a prevailing need for concerted efforts within the ever-increasing scope of Virtual Archaeology.

In the context of the restoration, the Seville Charter acts as a charter; it is not providing a system of rules but the orientations of a vast scientific community. Scholars from different fields of investigation are operating on virtual archeology as a new and mature discipline, which lives in compliance with these rules persecuting scientifically valid and widely shared methods.

The *London Charter* and the *Seville Charter* have in themselves a great deal of untapped potential, which, if disseminated and applied, would make the debate on the appropriateness of confident choices more exciting and prolific [11].

The first task is to give voice to that creative impulse, which gives rise to interpretations useful for formulating reconstructive hypotheses based on a preliminary study.

Virtual archaeology aims to deal with many targets. The possibility of using digital tools allows innovation in science, increasing the teaching and dissemination of ideas and historical discoveries. It pursues informational and chic purposes because it can effectively communicate the work of archaeologists and provide visitors with a correct and complete reading of the past.

Digital Heritage (DH) is a digital cultural heritage, and UNESCO sanctions such cultural heritage as an integral part of Cultural Heritage (CH) or Intangible Cultural Heritage. Through computer science, cultural heritage has become the subject of new interpretation, understanding, and dissemination forms. In this way, the traditional activities of historical and cultural production and consumption have changed, and a large amount of digital data is available to determine its undisclosed meaning.

Sometimes it happens in fact that in the frenzy of proving to be "in time," one subordinates the contents to the containing and elects the cultural object as a trophy, losing the opportunity to transform it from *exposed object* to *narrated object*" [16].

Due to this in 2003, the ICOM[2] statute defined the concept of intangible heritage as "*the practices, representations, expressions, knowledge, know-how as well as the tools, objects, artifacts and cultural spaces associated with them—which communities, groups and in some cases, individuals recognize as part of their cultural heritage*".

In principle, virtual reconstruction can be helpful to answer questions and complexities such as those just enunciated because it can give a comprehensive and transparent visual compendium and, connecting directly with the sources, the three-dimensional model can become an extraordinary database.

Technological tools must not distract the public in the various phases of museum visits. However, they must be a means of conveying quality cultural content, the only ones able to guarantee a valuable experience.

The European Commission has continued and continues now more than ever to work for the development of the European cultural system, having specific interests concerning digital culture: "*The Commission is working on legislation that supports the creation of an authentic Digital Single Market of content and promotion of media freedom and pluralism in Europe. It makes sure that the proper regulatory framework in terms of copyright is in place for all citizens to fully benefit from the great opportunities offered by a Truly connected digital single market. In social media platforms and networks, the Commission addresses the various challenges relating to fake news and online disinformation, and the dissemination of illegal content. Open data and re-use of public sector information, open government data and encouraged. The digitization of the European cultural heritage of European libraries, archives, museums, and audiovisual archives makes it accessible and accessible to all*".[3]

These programs started with the *European Agenda for Culture*. It is a work plan, updated periodically according to the changes in the Contemporary Society and the needs of Cultural Institutions, also promoting the Cooperation between the various European States, favored by the creation of the OMC Group (*Open Method of Coordination*).[4] More than 150 international leaders in 2015 met at the United Nations to contribute to global development, promote human well-being and protect the environment. The *2030 Agenda for Sustainable Development*, released after the meeting, identifies 17 goals (*Sustainable Development Goals—SDGs*) to end poverty, inequality, and social and economic development.

Inevitably, the challenge has also been promoted by cultural institutions that assume new roles within contemporary society. Michela Rota refers to a particular goal to move from museum to sustainable museum through culture—society—environment:

[2] ICOM—International Council of Museums is the only international organization representing museums and their professionals. Since 1946, ICOM has been assisting the global museum community in its mission to preserve, conserve and share cultural heritage.

[3] https://digital-strategy.ec.europa.eu/en/policies/supporting-media-and-culture.

[4] The *Open Method of Coordination* (OMC) is a European group of cultural experts, active since 2007 to coordinate the various European countries in cultural and museum policies.

- **Goal 11**: Making cities and human settlements inclusive, safe, lasting, and sustainable.
- **Target 11.4**: Strengthen efforts to protect and safeguard the world's cultural and natural heritage.

Sustainability and innovation are the essential combinations for a world that wants to stay up to date and protect its models of digital growth and the future well-being of people. Sustainability means not only a better use of energy resources. If applied to museums, it addresses a paradigm shift that involves, in a multiplicity of integrated domains, the entire life of the museum, the totality of its staff, the building with collections and exhibition, relations with the territory, the public and local communities in a logic of participation and installation [17].

3 *Representation Models* as a Cognitive and Expressive Means of Communication

The Theatre of Philadelphia is mentioned in several volumes regarding the Roman Theatres. No recent research explicitly focused on it, and scholars cited the information provided by previous studies with different information. Worthy of note is the lack of consistency between the authors: the descriptions are partial, and sometimes they even contradict each other, especially concerning dimensions, without giving precise and exhaustive references. At present, the Theatre embodies the cultural life of Amman, and it hosts several cultural manifestations of contemporary and traditional performing arts. Thanks to the restoration, the building has revived its ancient purpose and its relation to the people of Amman. On the contrary, very little information about its history and significance in ancient times is available.

For this reason, the exhibition will bring back to life the profound link between the building and its social relevance, between past and present, material, and intangible memory. To obtain a digital replica of the Theatre, the project of reconstruction proposes the investigation of three different *Representation Models* (Fig. 3).

Cognitive Model

An informative and heterogeneous model is the container of all information, data collection, photographs, papers, documents, and considerations regarding the subject. Such a model fits perfectly within the guidelines of the *London Charter*. The different types of data collected regards:

- Pictures: old pictures that clearly show the area's state at the end of the Eighteenth and beginning of the nineteenth century.
- Architectural drawings: plans and sections made by architects during the old expeditions that give information about measurements of the original building.
- Descriptions: All travelers' notes resulted from the great interest in some structural details, such as the corridors.

MODEL	Cognitive Model	Architectural Model	Reconstructive Model

METHOD	-Data Collection -Photographic collection -Literary review -Documentation -Archive -Archaeological studies -Historical sources	-Typological reconstruction -Relief with specific instrumentation (Laser Scanner) -Field survey -Raw Data Processing -Digital Modeling	-Data Simulation -Virtual reconstruction -Studies of functionality -Game Engine processor -Virtual Tour
GOAL	-**Report** through all channels of information and knowledge. -Lay the scientific, technical, anthropological, archaeological **foundations useful** for including next *model.* -**Understanding gaps**, criticality and missing data	-Understand the **technical - architectural aspect**. -**Detect the object** from the data collected. -Identify **typological elements** and compare with parameters and rules. -Obtain the most **up-to-date technical document**. -Design in a uniform and **consistent manner** with the available data.	-Obtain a reconstructive model useful for **different purposes**. -**Cultural enhancement** of the object. -Get **interactive features** -**Online** Archive -**VR** tour -**AR** fruition

Fig. 3 Representation model—method and goals

- Reports: documentation made by archaeologists who worked in the area.
- Parallels: data related to other theatres in the same area and time that can help integrate the missing parts of the building.

Architectural Model

A technical and specific model, which dialogues with technical drawings and all relevant information. A collector of reconstructions, reliefs, types of equipment used, and analyzer of the state of alterations. This model becomes a perfect technical document in continuous updating.

A documentation campaign included direct analysis, on-site measured drawing survey, and laser scanner data acquisition. Direct measurements provided a high level of accuracy of the dimension of the different parts, including detailed descriptions of the state of the monument.

Finally, comparing old and new images pointed out all the modifications obtained during the restoration works. Ideally, this procedure and a correct stratigraphic excavation constitute the base for the reconstruction of the main historical phases of the building: phase 0 (the area before the construction of the Theatre); phase 1 (first construction); phase 2 (comprehending even more minor structural modification of some parts of the building); phase 3 (abandon, possibly characterized by a superimposed structure with different functions); phase 4 (second abandon and collapse); phase 5 (restoration).

Reconstructive Model

Represents a model of cultural enhancement that enhances the functions and functionality through the Cultural Heritage and Intangible Cultural Heritage valorization.

A scientific reconstruction is a tool with great cognitive value since it facilitates the management and understanding of complex information thanks to 3D simulations [10]. For the Roman Theatre of Amman, the reconstruction of the different phases of the building has multiple purposes:

- It is a tool for collecting information for the correct interpretation of the building history by using all the data available.
- It is the media for the valorization and dissemination of the theatre building in a future exposition.
- It is the metadata container that will be the digital contents available online to the digital public.

In virtual reconstruction, intangible heritage storytelling is the innovative strategy for audience development and people engagement. These unique approaches have become part of the research methods with a cognitive purpose, constituting an update of the traditional studies aimed at reconstructing ancient artifacts. The workflow has led to new documentation methods, restitution, use, and valorization by processing materials available as a basis for dissemination work on various supports and applications [13].

In principle, virtual reconstruction can be helpful to answer questions and complexities such as those just enunciated because it can give an exhaustive and precise visual compendium and, connecting directly with the sources, the three-dimensional model can become an extraordinary database.

The rigorous and scientific digital model includes the 3D metadata of the Theatre building, its components (decorative, epigraphic), and metadata. Moreover, the same model can be used for simulations to test the validity of working hypotheses on statics, capacity, visibility, lighting, use of structures and space, on their functionality [20].

A critical methodological step in this direction was presented at the international symposium of digital sensory archaeology *"Capturing the Senses: digital methods for sensory archaeologies"*[5] organized by Giacomo Landeschi (Lund University) and Eleanor Betts (Open University).

[5] https://sensorystudiesinantiquity.com/2017/07/03/review-of-capturing-the-senses-digital-methods-for-sensory-archaeologies/.

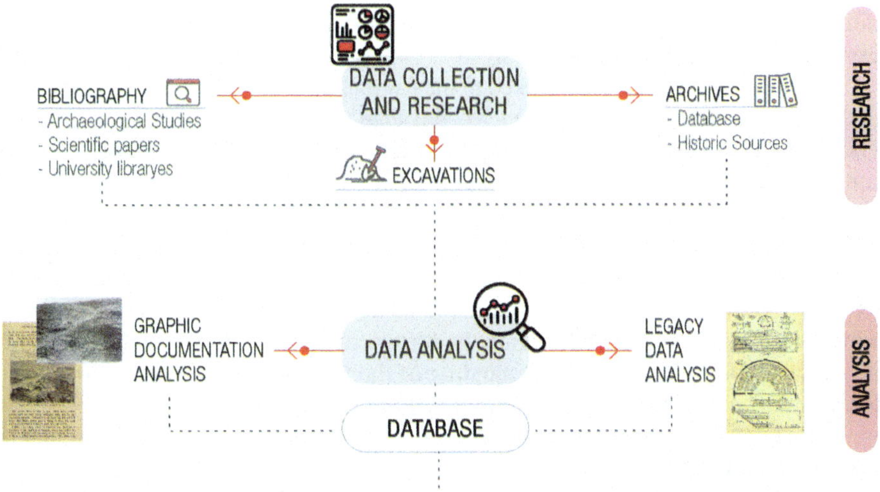

Fig. 4 Workflow Roman Theatre of Amman / Research – Analysis

The Roman Theatre experience investigates the application of innovative digital methodologies to archaeological research to stimulate the sensory perception of ancient, urban, and architectural space. The research challenge is exploring how digital methods can grasp the social meaning of spaces through a multisensory perspective (sight, hearing, smell, touch, taste). In parallel with the composition and completion of the three representation models, a specific workflow was developed in 3 steps: research, analysis, acquisition/simulation. Each step is associated with completing purposes related to models of representation, becoming a reservoir for all information regarding the reconstruction of the Theatre (Fig. 4).

4 Museum Experience

Museology and curatorship evolved from displaying collections or giving emphasis on masterpieces. Heritage is a complex subject embodying intangible issues, and artisans are the messenger of stories and practices. The strategy is to make objects significant through the creation of new knowledge. Heritage cannot be considered simply in its material value but is the product of a cultural process, rooted in some territories and using materials and techniques passed on from generation to generation within the reference communities. The Cultural Heritage elements (CH—Cultural Heritage) make up museums. This family, which once included only physical and cultural assets, is much larger today than in the past and contains ICH (ICH—Intangible Cultural Heritage) with traditions, memories, and testimonies. This new cultural asset as a graft of memories and heritage gives communities a sense of identity and continuity, thus promoting respect for cultural diversity and human creativity. Its dimension transcends the meaning of material good and includes traditions, lifestyles, experiences, and cultural memories.

The aim is to design the CH linked to the Roman Theatre of Amman by displaying dedicated content to interact with tactile physical replicas intended for a multimedia visit, making knowledge of the characters that have inhabited the Forum in ancient times. This activity aims at the practical application of the *Recommendation on the protection and promotion of museums and collections, their diversity, and their role in society* [25], which highlights the importance of technologies for the task of the Cultural Heritage sector and spaces conducive to education and life-long learning. The activities always pursue the social responsibility of carrying out an educational mission by appropriating the historical heritage. Therefore, the installations must favor the correct understanding of the tangible and sensorial dimensions to communicate and share an intangible cultural reality (Fig. 5).

Virtual reconstruction of an Ancient Theatre must embody both intangible and tangible issues [20].

Among the thousands of components and stories that revolve around a spectacular scenic machine such as the Roman Theater in Amman, the main trace of content choices concerns the rediscovery of two fundamental aspects: the architectural structure's story and the level of the scenic dynamics. This exhibition comprises a tangible and intangible sphere, making a clear path. While creating a dialogue between the monument and the public, the symbolic meaning and classical and performing fruition modes of the different spaces: *cavea, orchestra, scaena frons* become the protagonist.

These three elements feature the architectural typology of the Theatre, and the exhibition will create an understanding of such complex structures. One more target is to create a link between the public and actors as they were used to performing in the past. The information from the Roman Theatre's cognitive and architectural models settled some fundamental scientific data that have become the basis and structure in the choice of contents to be displayed, integrating the elements collected with the sphere of the ICH (Intangible Cultural Heritage) at the time of ancient Philadelphia.

The most relevant building element is the *cavea*, partially excavated on the hill slope and partially built to conclude the semicircular shape with new additions. The sector has three-level, the *ima, media*, and *summa cavea*. There are 41 lines, 13 in the *ima* and *media cavea*, and 15 in the *summa cavea*. Such articulation reflects the society where the social status determined the distribution of the public: *nobiles* and *knights* in the *ima cavea, togata* in the *media cavea*, and *plebes*, enslaved people, and women in the *summa cavea* (Fig. 6).

Unconventionally the shape of the orchestra is slightly bigger than a semicircle. The orchestra floor was completely rebuilt after the restoration; some coins and a fragment of inscribed lintel during the excavations dated the Theatre to the second half of the second century AD. Some recent studies on the sensorial perception of Ancient Theatres shine the lights on the scenic space, the actors' movements, the acoustic and sensory perceptions of the public [20]. The *scaena frons* dominates the stage building and the entire Theatre. A fundamental aspect was the versatility in its functions, and it acts as a background and support for the representation of Roman

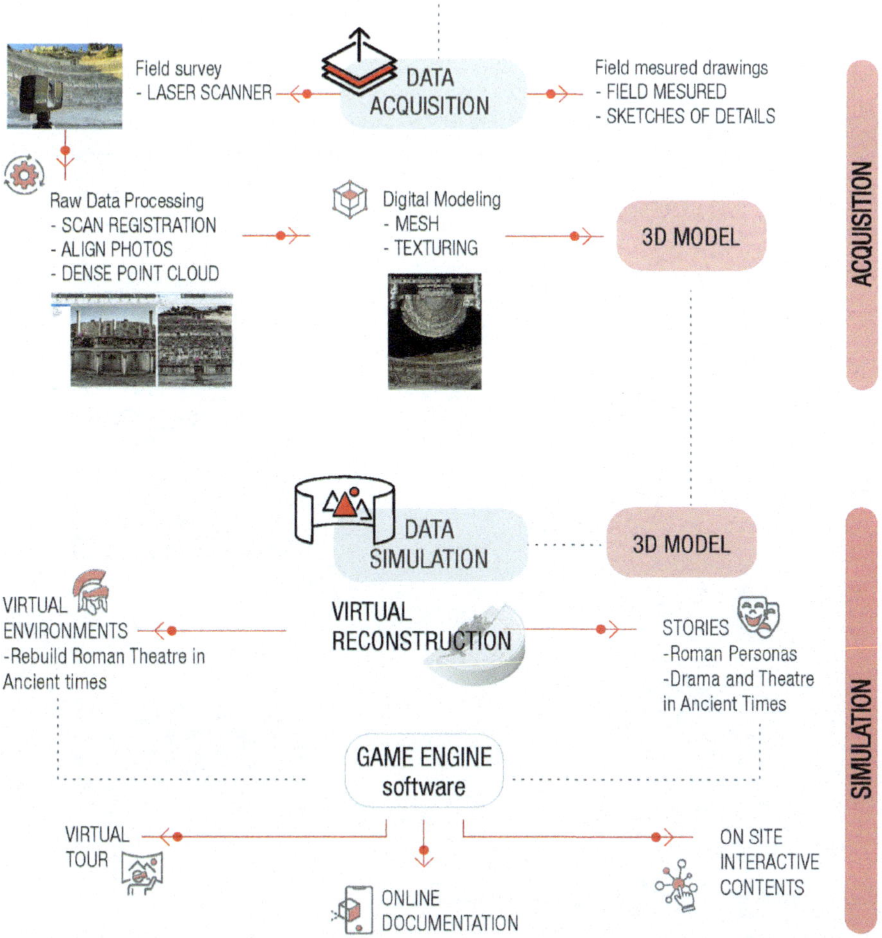

Fig. 5 Workflow Roman Theatre of Amman / Acquisition-Simulation

comedies and tragedies. Its reconstruction will be very complex because of the few archeological remains. It will be based on the comparative studies and parallel with other similar theatres and checking the references from Vitruvius (*De Architectura*, Chaps. 3–9 of Book V) and Heron (*De Mensuris 24*, chapter “Measurements of Theatres”). Vitruvius’ criteria address the *cavea* configuration and its ratio with the theatre stage, the proportions of the theatre stage, the proportions and composition of the wall behind the theatre stage.

Content-driven installations allow reflecting people’s past and make the experience a place to acquire a sense of belonging and involvement. Precisely by enhancing the aspect of dialogue, the museum provides three different communication channels to bring the museum’s visitors closer to a complete knowledge of the extraordinary theatrical machine as it was the Ancient Theatre, a place for listening and seeing.

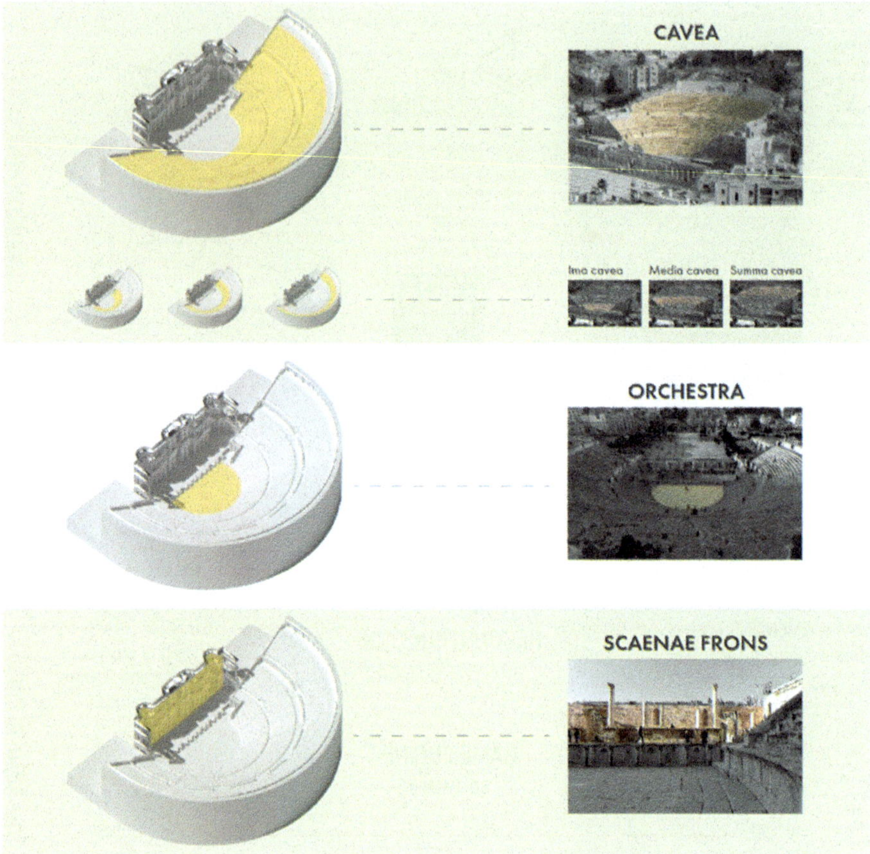

Fig. 6 Visual glossary of the typological elements

In this project's phase, the output of Representation models has represented: restitution of the intangible. The protagonist is the Theater as a messenger of stories, a basin, and a collection of memories; the use of metaphors to describe Roman Theater is a way of representing the model, a logical process for dealing with a collective scenario that can collect ideas, memories, suggestions, and therefore represent an identity. Metaphor is the narrator of the multimedia installation, driving the experience of visitors; it encloses the complete identity by addressing three different themes, starting from different points of view:

- *Theatre as a vibrating body*: The story of the Theatre as an architectural structure, made up of well-defined parts, rules and schemes, architectural orders, and geometric principles but also on those constructive exceptions, common elements with the Theatres of the Decapolis and extraordinary components of originality, an all-round story about architecture, the generating principles, and the complex balance of the parts.

The metaphor of the vibrating body contains all that basic information that makes the Theater like a body, a solid structure made up of parts, a body considered alive.

However, it is done of architectural component but the human sphere, the spectators complete the space and make it dynamic and changeable. The Theatre is animated by the shows, and it changes; it adapts to different situations. It is a body able of being a magic place for spectators and their ability to envision, animated by actors, and transformed in a livable environment by the performances.

- *Theatre as mystery keeper*: Entering the world of theatrical representation made up of the body and gestures, a significant and prominent expression for most shows, an opening to new spaces for the imagination with extreme importance to ancient stories. Theatrical divisions are given by well-defined rules, particular historical moments, and accomplices of political and social choices. Stories full of tales, truths, and scenes; getting involved was the prerogative of the theatre actor, a passion for entering the lives of those who attended the shows, to enrich their minds and upset their beliefs.

The metaphor of the mystery keeper is synonymous with research and a desire to bring attention back to the actors and their way of representing the shows. The attention always remains on the human and emotional level, enhancing not the story and the scenic representations but rather the dynamics, the choreographies, the references, and the study of the gestural and emotional sphere that revolved around the actors.

Mystery is associated with discovering what could have been the representations, what happened behind the curtain, *auleum*; a keeper of mystery tells what the stories of shows were, but before that, understands the dynamics of the protagonists of the staging.

- *Theatre as an exchange of synergies*: The show is overturned, the audience enters the scene within what was a physical and mental space full of rules and prejudices, a "structure" dominated by hierarchies and social classes in which everyone had his place in the space, having access to the shows differently. A structure that wants to hide but, at the same time, reveal truth and reality. The viewer becomes the protagonist of his own story, with different enemies and a distorted identification of roles.

The metaphor of *Theater as an exchange of synergies* tells the aspect of emotional participation and identification of spectators. The Theater told stories as a mirror of reality, a true representation of society, completing and showing the scenario of scandals, situations, and collective identity politics.

Synergy is a complete expression that reveals the exchange of contents, emotions, and reality.

The proposal deals with an innovative design process and new installation at the Museum of Popular Traditions in Amman to enhance the intangible heritage embodied in the site of the Roman Theatre.

Above all, a total absence of a path capable of guaranteeing knowledge also on the aspect of the intangible object concerning the theatrical machine of the Great Roman Theatre in Amman.

The exhibition proposal contextualizes cultural contents into the exhibition. It develops storytelling along the path, connecting in a logical succession all the displayed elements. The strategy builds an educational-cultural message understandable to any visitor category through the integrated use of media, multimedia, technological supports, and edutainment activities. The goal is to conjure up a memory. The final step is to recall the atmosphere of ancestors in the city's ancient Roman civilization and discover the relationship between the origin and the present day.

Over the years, we have gone from the design of purely conservative models, in which the "fixity" of forms and or meanings of the heritage overcomes as a document to be preserved, to transformative interventions, in terms of designing new purposes (context where scene models and innovative fruition paradigms), and new forms of goods [15]. Activating and increasing the value of the cultural asset in its heritage, historical, civil, symbolic, social, and development function aim at the development of platforms and connection systems capable of connecting areas and communities through culture and knowledge.

The metaphor of activation and enhancement for the exhibition and storytelling of the Roman Theatre inside the Museum of Popular Traditions relates to the theatrical world: breaking of the fourth wall. The space replicates the experience of the stage; guests act as both actors and audience, while all the activities within the space make up the show. The interpretative metaphor breaking of the fourth wall tries to unite the world of the scene and the audience with a formula that evokes the tension between the parts through the persistence of a force capable of solidifying itself in action: dialogue. Visitors are engaged in dialogue with the exhibition's contents are no longer just like objects but will be treated as vectors of emotions. They create the visitors' atmosphere, identifying with the exhibition and the environment. The exhibition is composed of four chapters. The first is a space to introduce the typological reconstruction work of the Ancient Roman Theatre of Amman, explaining its phases and all the geometric components. The space will have a double reading level to immediately understand the dualism between actor and spectator. The second part shows the tactile model, with the possibility of implementing the previously learned knowledge and acquiring new ones, enhancing an underlying multidisciplinary approach that also takes inspiration from the Roman art world. The third and fourth parts explain respectively the protagonists' actors and the audience protagonists, with their roles, gestures, and characteristics (Figs. 7, 8 and 9).

The exhibition design is a gradual process like discovering, so the project starts with learning information to finish with imagination and suggestions. A multilayer system with different learning densities, from a more immediate level that includes a visual acquisition of details up to subsequent levels, continues with contents that trigger past knowledge and rediscovery of a primitive memory. A clear understanding of the close relationship between body movement and communication comes from the study of space.

Fig. 7 Visual metaphor to explain the link between elements

The visitor is a body in motion, and contemporary exhibition design can examine movement patterns to explore communication and exhibition experience. The user becomes the leading subject/actor of a special visit that places himself as the protagonist, reconfiguring his relationship with the cultural asset: it is no longer a question of attributing the concept of uniqueness only to the work of art but to the person.

The good museum-visitor relationship necessarily passes through the activation of an empathic process [21, pp. 90–92].

[...] *The theme of borders in the museum space is of fundamental importance and more than ever in continuous transformation especially today that exploration and visits do not take place only through the physicality of the body and movement in current spaces but also in virtual areas* [...] says Raffaella Trocchianesi on new museum proxemics, on the gestures and behaviors of relationships. It is considered proper to specify the assumptions, implications, and logic underlying this theme, making it of clear interest and reason for exploration for research in this field.

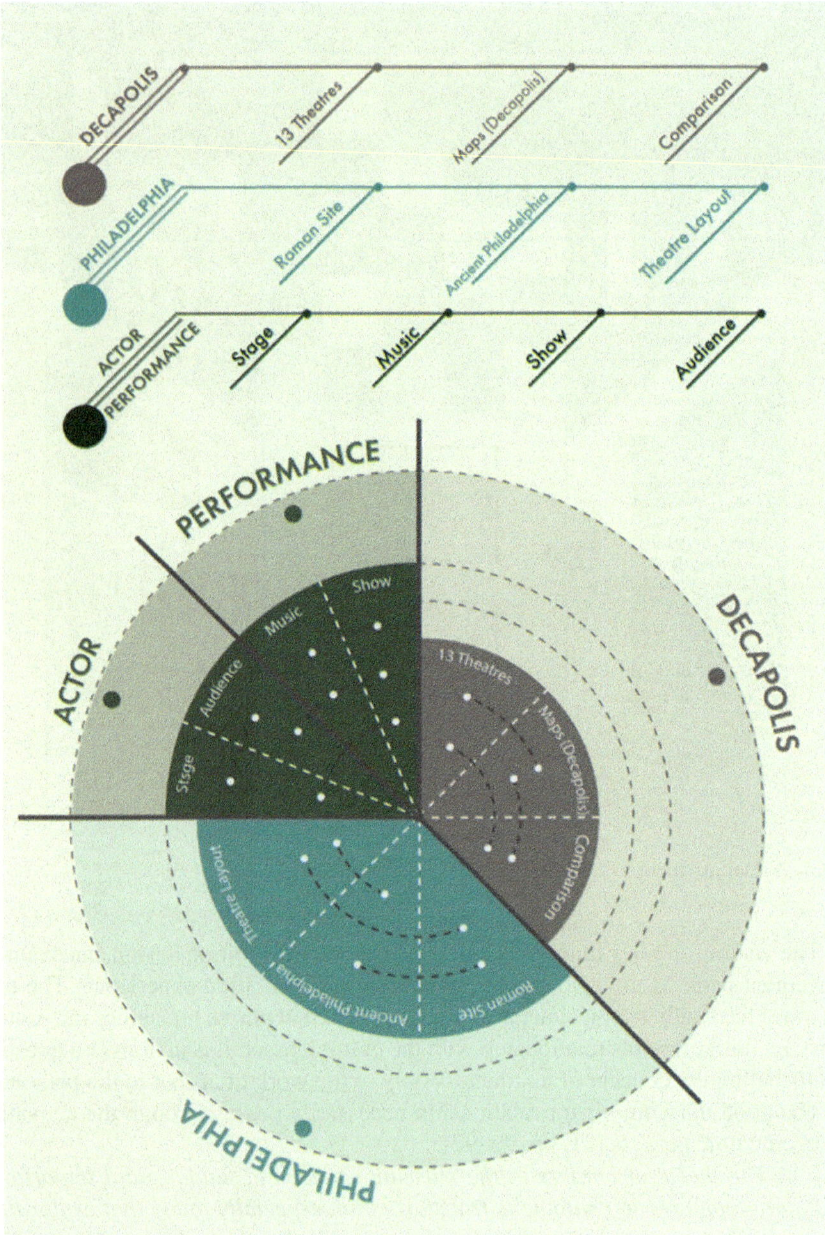

Fig. 8 Visual table of contents

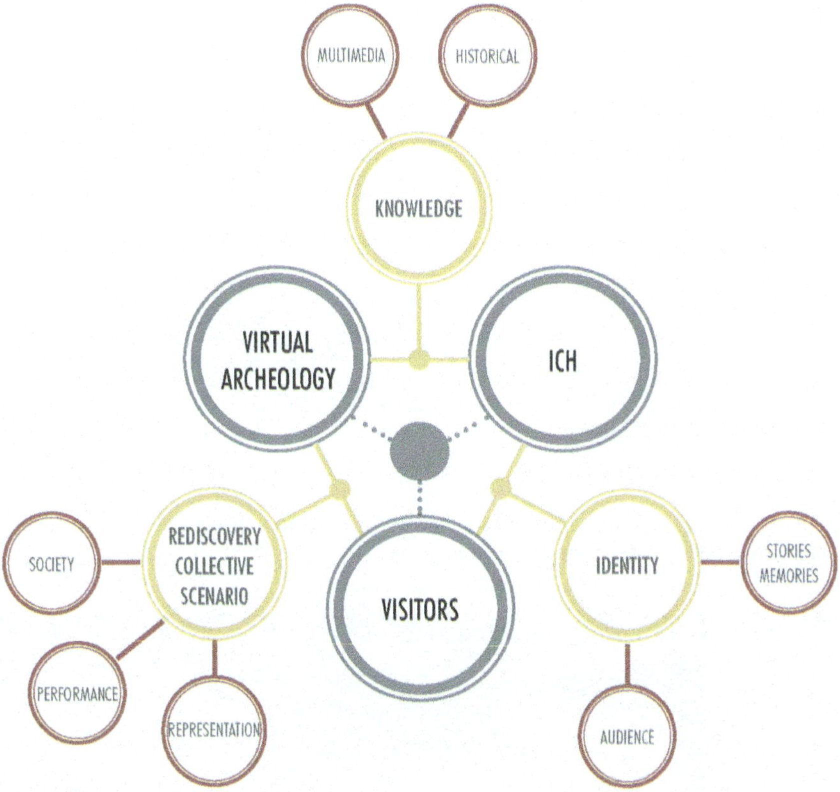

Fig. 9 Exhibition scenario

Placing one's body at the center of the fruition project means addressing the issue in these terms; consider the body as a bearer of social relations (memory and sharing) and consider the body as an instrument of appropriation of texts (places, narratives) (Figs. 10 and 11).

The exhibition starts from a space suitable for understanding and processing information passes to an evocative area, host of multimedia and interactive content. It is a gradual process like discovery, starting with learning information to ending with imagination and suggestions.

A multilayer system with different learning densities starts from a more immediate level that includes a visual acquisition of information. Later levels instead continue with content that triggers past knowledge and rediscovery of a primitive memory. The exhibition aims to make evident the Greek-Roman society and its reflection on places, environment, and way of life. Today, the Roman Forum is the center of today's city of Amman and an essential trace of an ancient culture rich in memories. On a parallel learning stream, both the Popular Traditions and Folklore museums exhibit the vibrant Bedouin culture and its practices.

Fig. 10 *Theatre as a vibrant body*, focus view of the display with the infographic about the architectural element of the Amman Theatre

Fig. 11 *Theatre as an exchange of synergies*, an overall view of the environment with the characters of the speaking audience

So, what will be the future of the Roman Theatre of Amman and Roman culture?

The project also includes multiple fruition options; the strategy provides different visit packages that emphasize specific information, change the mode of narration, or add details to the contents. Visitor's experience and the management of the museum are articulated through different tours, diversified for in-depth analysis and to satisfy as many needs as possible:

1. FIRST TIME IN THE MUSEUM?

Basic package for tourists with a general overview of all content suitable for a broad audience from children to families to move up to groups over 60. Information weighed to be effectively understood by all categories, mediating more digitized systems to others more direct and analog. A path that is left as open as possible, providing an everyday basis for everyone and freedom to deepen if you want 'integration with other packages.

2. DISCOVER

Attractive package, suitable for children and tourists, and visitors who like to write down their impressions, judgments, or appreciations. A diary designed to review all the exhibition steps, with challenging questions and spaces for personal notes.

However, it does not end here. Instead, it includes pages dedicated to other Roman sites in the area (Petra and Jerash, for example) to encourage discovery and stimulate comparisons.

3. EXPLORE

A path suggested by the curators, who emphasized specific contents of the exhibition, adding details sometimes of an artistic nature other time of a technical/scientific ones. A path that adapts to the forum's events, bearer of novelty and originality in content; Tourists, experts, and enthusiasts of the sector will have the opportunity to deepen and investigate unpublished issues under the guidance of curators and experts.

The path will change in timing as the deepening of some issues will expand the time of stay and experience of specific points of the exhibition.

4. LISTEN

A journey guided by the expert's voices will take a new and unprecedented look at the contents through stories in the form of narration. Specific fruition is suitable for children and not only. The narrations will highlight some information and complete the fruition with compelling stories of the Roman protagonists, through stories in the first person. The visit will follow a particular itinerary marked by the narrator's voice that will be activated at some specific points of the exhibition; chapter after chapter, the story will take shape.

The proposal of these visit packages brings a change in the use of exhibition, the times of stay in the museum change and adapt to the level of information proposed. An interesting aspect is the ability to choose and feel free to open the drawers of the curiosity from the beginning of the user experience, a narrative log that changes and adapts to the needs of each visitor.

5 Conclusion

The study illustrates all the potentiality of a multidisciplinary approach associated with using virtual archeology and Theatre reconstruction to make an experience of the site during the Roman times. Reconstruction of ancient archeological sites stimulates the scientific debate about architecture and its re-use as a learning environment to connect people to heritage as promoted by the Faro Convention. For this purpose, inclusion and participation policies are very relevant goals, and the strategy provides a robust user experience and target analysis.

Innovative and original aspects of the study are a new approach through the different types of representation model to set a visual database of sources. The visual glossary is a semantic description of each part of the building to support the virtual reconstruction process.

The aim is to set up an interactive digital model organized semantically: the 3D segmentation will constitute a cognitive system that shows both the constructive organization of the building and the operating technique of the front of the scene. The qualified model favors the multi-scale representation of both the architecture and the scenography with different levels of resolution while keeping the understanding of the different parts logical.

The scientific study frames the theater in Philadelphia's historical and social context. Its reconstruction aims to understand its parts according to a functional and constructive decomposition that is no longer fully visible due to the numerous losses that the building has suffered. The experience will be guided by those who animated it in ancient times, actors, ordinary citizens, members of civil society, representatives of political power. The future installation of the exhibition will be able to offer the visitor, in an experiential form, a visit to an environment renewed with technologies and digital content.

Acknowledgements This work is part of the international research and cooperation project: *Program for the definition of a strategic plan for the improvement and the enhancement of the Folklore Museum, the Museum of Popular Traditions and the site of the Roman Theater in Amman*, launched in 2020 funded by the Italian Agency for Development Cooperation in agreement with the Department of Antiquities of the Ministry of Tourism and Antiquities of the Hashemite Kingdom of Jordan and operated by the Department of Design of the Politecnico di Milano. Giuseppe Amoruso is the scientific director and the curator of the digital museology project for the site; he is also the author of paragraphs 1, 2, and 3 and curated the scientific reviews of all the texts and illustrations; Claudia Carioni wrote paragraph 4.

References

1. Almagro, A. (1983). The survey of the Roman Monuments of Amman by the Italian mission in 1930. *Annual of the Department of Antiquities of Jordan, 27*, 607–639
2. Amoruso, G. (2019). Reconstructing the Roman amphitheatre in Milan. Geometry, drawing and generative modeling. In G. Granado, C. Joaquín, & A. Camacho (Eds.), *Proceedings of*

the XIV International Conference on Graphic Expression applied to Building, APEGA 2019 (pp. 300–303), De la linea a la nube, Universidad de Sevilla. Roquetas de Mar: Almería Editorial Círculo Rojo ISBN 978-84-1317-575-1

3. Amoruso, G. (2019). Drawing as an experience. Knowledge, simulation, and participation in DISEGNARECON (Vol. 12, No. 23). Experiential Design for Heritage and Environmental Representation
4. Amoruso, G. (2020). Figuring out the interiors through the representation of experiential and interactive e*nvironments.* In L. Crespi (Ed.), *Cultural, theoretical, and innovative approaches to contemporary interior design* (pp. 367–386). Hersey, PA: IGI Global
5. Anastasio, S., & Botarelli, L., (2015). The 1927–1938 Italian archaeological expedition to Transjordan in Renato Bartoccini's Archive, Archaeopress Archaeology
6. Bendicho, V. M. L. M. (2013). International guidelines for virtual archaeology: The Seville principles. In C. Corsi, B. Slapšak, & F. Vermeulen (Eds.), *Good practice in archaeological diagnostics.* Natural Science in Archaeology. Cham: Springer. https://doi.org/10.1007/978-3-319-01784-6_16
7. Butler, H. C. (1907). Ancient architecture in Syria
8. Chancey, M.A., & Porter, A.L. (2001) . The Archaeology of Roman Palestine. *The University of Chicago Press Journals, 64*(4), 164–203. https://doi.org/10.2307/3210829
9. El-Fakharani, F. (1975). Das Theater von Amman, *Jordanien, Archäologischer Anzeiger*, 377–403
10. Ferdani, D., & Demetrescu, E. (2021). ATON: An open-source framework for creating immersive, collaborative and liquid web-apps for cultural heritage. *Applied Sciences, 11*(22), 11062. https://doi.org/10.3390/app112211062
11. Gabellone, F. (2012). La trasparenza scientifica in archeologia virtuale: una lettura critica al principio N.7 della Carta di Siviglia. *SCIRES Scientific Research and Information Technology, 2*(2)
12. Hadidi, A. (1992). Amman-Philadelphia: Aspects of Roman Urbanism. *Studies in the History and Archaeology of Jordan, 4*, 295–298.
13. Limoncelli, M. (2012). *Il restauro virtuale in archeologia,* Roma: Carrocci Editore
14. Lupo, E. (2009). Il design per i beni culturale. In E. Lupo, & M. Parente (a cura di), *il sistema Design Italia e la valorizzazione dei beni culturali.* Milano: POLI.design
15. Lupo, E., & Trocchianesi, R. (2013). Progetto e memoria del temporaneo. Milano: Electa.
16. Rosa, P. (2011). Dai musei di collezione ai musei di narrazione in DISEGNARECON (Vol. 4, No. 8, pp. 129–138). Tecnologie per la comunicazione del patrimonio culturale
17. Rota, M. (2019). *Musei per la sostenibilità integrata.* Collana: Geografie Culturali, Editrice Bibliografica Fondazione Fitzcarraldo
18. Sear, F. (2006). Roman Theatres. An architectural study
19. Trocchianesi, R. (2014). Design e narrazioni per il patrimonio culturale. Sant'Arcangelo di Romagna: Maggioli.
20. Viccei, R. (2019). Ricostruzioni digitali per la conoscenza e la valorizzazione dei teatri antichi: Usi, significati, questioni aperte, in Massimo Limoncelli, Furio Sacchi, Raffaella Viccei, Indagini archeologiche e ricostruzioni digitali per lo studio degli spazi teatrali greci e romani. *Visioni e prospettive, Milano, EDUCatt, 2019*, 83–148.
21. Vitale, G. (2013). *Design di sistema per le istituzioni culturali. Il Museo empatico.* Zanichelli.
22. http://smartheritage.com/sevilleprinciples/seville-principles
23. https://digital-strategy.ec.europa.eu/en/policies/supporting-media-and-culture
24. https://www.londoncharter.org/
25. *Recommendation concerning the Protection and Promotion of Museums and Collections, their Diversity and their Role* in Society adopted by the UNESCO General Conference at its 38th Session, Paris, 17 November 2015

Archaeology of the Present: Knowledge as a Strategy for Claiming the Value of Contemporary Authorial Architecture

Paolo Belardi and Valeria Menchetelli

Abstract How can contemporary authorial architecture be protected? The need to define intervention strategies aimed at this objective is the premise of the *Censimento nazionale delle architetture del secondo Novecento* (National Census of the Buildings of the Second Half of the Twentieth Century), undertaken by the Ministry of Culture and aimed at cataloguing the numerous buildings on the Italian territory marked by high-quality characteristics. In the wake of the experience of multimedia cataloguing of the architecture of the second half of the twentieth century scattered throughout the Umbrian territory, the contribution illustrates a research path aimed at reclaiming the value of some specific architectural works built in recent decades, today completely demolished and obliterated by memory. The memory of these works resides solely in knowledge, just as their direct accessibility lies solely in virtual reconstruction and material scale replication. Transferring the title and the cultural approach that distinguishes the iconic collection of essays by Giovanni Urbani, the contribution presents the application to the Umbrian territory of a valorisation methodology that aims at constructing an 'archaeology of the present'. This methodology aims at embodying a cultural oxymoron capable of celebrating the history and the past, even if recent, perpetuating and reaffirming its presence in contemporaneity. The aim is to preserve long the knowledge of a segment of the architectural heritage not yet unanimously recognised as a holder of cultural value for the city, the landscape and society in general. The aim is to build an archaeological museum of the present on a national level.

Keywords Digital reconstruction · Architectural survey · Contemporary authorial architecture · Conservation · Archaeology of the present

P. Belardi · V. Menchetelli (✉)
Department of Civil and Environmental Engineering, University of Perugia, 06100 Perugia, Italy
e-mail: valeria.menchetelli@unipg.it

P. Belardi
e-mail: paolo.belardi@unipg.it

© The Author(s), under exclusive license to Springer Nature Switzerland AG 2023

I. Trizio et al. (eds.), *Digital Restoration and Virtual Reconstructions*, Digital Innovations in Architecture, Engineering and Construction,
https://doi.org/10.1007/978-3-031-15321-1_18

1 The Legal Framework and the Need for Protecting Contemporary Authorial Architecture

The contemporary architecture protection issue occupies a prominent role in the national debate, which highlights a series of problematic questions and consequently outlines a continuously evolving legislative framework. First of all, it is necessary to clarify the exact definition of 'contemporary architecture', specifying the requirements and time limits within which the validity of this identification can occur. In 1902, the so-called "Legge Nasi" was the first establishing the protection applied to monuments, buildings and movable objects having the value of antiquity or art. Excluded are buildings and artworks by living authors, or whose execution does not date back more than fifty years [GU No. 149 of 27 June 1902, Art. 1]. Even in this context, a large segment of architectural heritage remains de facto invisible and excluded from preservation, and exposed to considerable risk of degradation, alteration or significant transformation. Over a century later, a new and updated legislative instrument intervenes, introducing a much more articulated regulatory structure but not substantially modifying the status quo. The Code of Cultural Heritage and Landscape [D.L. 22 January 2004, n. 41] provides the safeguarding of the architectural heritage. The Code establishes the constraint both for publicly owned buildings (if built more than seventy years ago and considered to be of historical-artistic interest) and for privately owned buildings (if built more than fifty years ago and declared to be of particularly important interest through the so-called "constraint decree"). In this sense, there are no consolidated operating methods oriented towards the safeguarding of public buildings less than seventy years old (or private buildings less than fifty years old). This is not even the case when these buildings have represented, or still represent, an important cultural value, not only for the territory they belong to but also for the national and international architectural debate. By extending to 70 years the time frame that must elapse to safeguard a public building, the architectural heritage built since the second half of the twentieth century cannot be protected. So how can contemporary authorial architecture be protected?

A further instrument, still used in practice, is represented by the Law on Copyright (L. 633/1941). Although not comparable to a 'direct' protection measure executed through the affixing of a constraint, this instrument allows applying a form of 'indirect' protection because it concerns the author rather than the work. Although a helpful expedient to preserve part of an otherwise unprotected heritage, this type of protection shows limits both from a normative and an applicative point of view. In fact, "the protection measure is born on a private initiative" and "has a destiny strictly connected to the private subject" who applies for it [20]. This request can only be formulated by the author of the artwork, while the authors' heirs can only request and obtain the recognition of the "moral copyright", which effectively excludes the possibility of issuing a declaration for the works of no longer living authors. The outcome is not a constraining decree but a "Declaration of important artistic character" (aap.beniculturali.it/dirittoautore.html).

At the international level, numerous initiatives stand out for their approach to the protection and enhancement of the contemporary. These activities have their roots in the *Carta di Venezia* (1964), which is the reference document for ICOMOS (International Council on Monuments and Sites) established in 1965 to promote the conservation, protection and enhancement of monuments and sites of cultural interest. DoCoMoMo International (International Committee for Documentation and Conservation of Buildings, Sites and Neighbourhoods of the Modern Movement) was born in 1988 out of ICOMOS. Its mission is the documentation, conservation and valorisation of 20th-century architecture through the promotion of knowledge activities and the study of methodologies and intervention criteria based on the historical-documentary value of buildings. The national group DoCoMoMo Italia was founded in 1990 and is a reference point in Italy for 20th-century architecture's preservation. The theme of safeguarding contemporary architecture has long been the subject of attention by UNESCO, which in the 2005 update of its Operational Guidelines on the Implementation of the World Heritage Convention (https://whc.unesco.org/archive/opguide05-en.pdf) further specifies the concept of "outstanding universal value" that "groups of buildings" and "sites" must possess to be included in the World Heritage List. This definition has been the subject of debate for some time and, it is evident that in the face of heritage in continuous evolution and transformation, "the notion [...] is a modern concept, and the justification needs to be in some way 'constructed'" [22, p. 46], opening the way towards a redefinition of classification standards. Another international initiative oriented towards the objective of safeguarding twentieth-century buildings is the international programme Conserving Modern Architecture Initiative (CMAI) undertaken by the Getty Conservation Institute and aimed "to advance the practice of conserving twentieth-century heritage, with a focus on modern architecture, through research and investigation, the development of practical conservation solutions, and the creation and distribution of information through training programmes and publications" (www.getty.edu/conservation/our_projects/field_projects/cmai). The initiatives described above certainly contribute to the knowledge, documentation, preservation and enhancement of works of universally shared and recognised artistic value; however, a more inclusive approach to protection seems necessary, based on regulatory provisions applicable to a wider variety of cases, given the fundamental testimonial value of the works considered.

Based on these premises, the need to define effective intervention strategies for the protection of contemporary Italian authorial architecture, together with the need to identify a system of objective criteria for the classification of buildings belonging to the recent architectural heritage, was the starting point for the *Censimento nazionale delle architetture del secondo Novecento* (National Census of Buildings of the Second Half of the Twentieth Century, Fig. 1), undertaken by the Ministry of Culture and aimed at cataloguing the numerous buildings scattered throughout the Italian territory and marked by high-quality characteristics, with a view to 'certifying' their value. In the absence of an actual legal protection system, the path traced by the Ministry of Culture appears to be a viable way to at least recognise a sort of 'right to protection'

Fig. 1 National census of the buildings of the second half of the twentieth century, homepage. http://www.architetturecontemporanee.beniculturali.it/architetture/

of contemporary architecture. All the more so in a scenario in which, sometimes even after a widespread acquisition of knowledge (and awareness) of the intrinsic value of significant architectural works, the instruments currently available and in force are unable to ensure their survival.

2 Umbrian Architecture: Not Only Middle Ages

That is the context for the research activity carried out by an interdisciplinary team within the Department of Civil and Environmental Engineering of the University of Perugia, long been engaged [31] in identifying, cataloguing and claiming the value of several architectural masterpieces built in the region over the last few decades [10]. The ups and downs of these buildings have led both to the dismantling of two interior designs (in Bastia Umbra the Sala Consiliare Comunale, built in 1975 and designed by Gian Carlo Leoncilli Massi, Fig. 2, and in Perugia the Infopoint of the Rocca Paolina, designed by Italo Rota in 2003, Fig. 3) and the demolition of three buildings (in Perugia, the Civic Slaughterhouse, designed by Claudio Longo in 1967; in Corciano, the EH System housing module, designed by Renzo Piano and Peter Rice in 1976; and in Todi, the Exhibition Pavilion of the Antiquarian Fair, designed by Carlo Aymonino and Paolo Portoghesi in 1982), risking to obliterate them forever from the collective memory. That might definitively compromise the already complex relationship between Umbria and contemporary architecture. In fact, although Umbria is not lacking in recent architecture, both late-modern (from Marco Zanuso's Senatore Borletti nursery school in Gubbio, Fig. 4, to Julio Lafuente's Santuario dell'Amore Misericordioso in Collevalenza, Giancarlo De Carlo's Matteotti village in Terni and Arrigo Rudi's municipal library in Foligno) and post-modern (from Tomaso Buzzi's Scarzuola acropolis in Montegabbione to Aldo Rossi's business centre in Perugia,

Piero Sartogo's San Lorenzo church in Bevagna, Fig. 5, to Francesco Cellini's Rowing Club in Baschi), Umbria is not fully aware of either the small quantity or the great quality of its contemporary architectural heritage [5]. Above all, this is because Umbria continues to be afflicted by an embarrassing 'Mulino Bianco syndrome'[1] [1, 7], expressed in grotesque neo vernacular revivals (both in terms of composition and construction). Umbria displays a prejudicial aversion to contemporary architecture that, at a close look, has its roots in the suppression of the School of Architecture of the Perugia Academy of Fine Arts imposed by the Fascist regime [2]. In fact, in 1938, the rationalist wave of international ascendancy fuelled by teachers at the Academy committed to freeing themselves from the historicist legacies of Ugo Tarchi, such as Pietro Angelini and Manfredi Franco, was abruptly halted by a provision of the Ministry of National Education. By imposing the transfer of teaching skills to universities and polytechnics, the Ministry stifled the creativity of an entire generation of modern Umbrian architects at birth [3]. That produced a cultural impasse which, by anachronistically restoring the image of Umbria as a purely mediaeval region, launched by the Savoy commissioners the day after the unification of Italy, concentrated the many eclectic architectures in the historic cities and relegated the rare modern expressions to the periphery (such as the Casa della GIL built in Narni in 1939 and designed by Agnoldomenico Pica). But the road was already marked out. Neither the splendid modern architecture scattered around Terni and Perugia by two 'clerics of the Modern' such as Mario Ridolfi and Giuseppe Nicolosi [25] nor the valuable late-modern architecture signed by high-level Umbrian designers such as Franco Antonelli (Fig. 6) [23] and Bruno Signorini (Fig. 7) [29] are worth anything. Umbria remains anchored in a medieval image that is as artificial as it is sterile. Hence the deeper sense of the reasons why the presence of the contemporary in Umbria, due to its exceptionality, assumes even greater value and invokes in itself a broad and widespread knowledge.

Fig. 2 Gian Carlo Leoncilli Massi, Sala Consiliare Comunale in Bastia Umbra (PG), construction 1975—demolition 2018. https://terrenostre.info/2021/07/laula-consiliare-dellarch-leoncilli-cancellata-dopo-45-anni/

Fig. 3 Italo Rota, Infopoint and Museum of the Rocca Paolina in Perugia, construction 2003—demolition 2017 [27, pp. 298, 301]

Fig. 4 Marco Zanuso, Nursery School Senatore Borletti in Gubbio (PG), 1959. http://www.architetturecontemporanee.beniculturali.it/architetture, © Università Iuav di Venezia—Archivio Progetti, Fondo Giorgio Casali

3 Three Lost Authorial Buildings

In the context of these works, the article analyses three case studies, which have been the subject of a dedicated survey and have given rise to three different levels of documentation, valorisation and fruition. From reconstructive architectural survey to virtual reconstruction with restoration hypotheses, up to replication through 3D printing.

Fig. 5 Piero Sartogo, San Lorenzo Church in Bevagna (PG), 1973. http://www.architetturecontemporanee.beniculturali.it/architetture, © Julien Proietti Peparelli

Fig. 6 Franco Antonelli, Monastery of Santa Maria di Betlem in Foligno (PG), 1995. http://www.architetturecontemporanee.beniculturali.it/architetture, © Luca Tesei

The research carried out is part of the wide applicative path that for decades now has seen ICT acting synergistically within the knowledge processes and enhancement of cultural heritage in general and architectural heritage in particular (see for example [15, 21]. ICT are nowadays assimilated as an integral part of the processes of analysis and valorisation and find application at a multiplicity of levels: that of

Fig. 7 Bruno Signorini, Monte dei Paschi di Siena Bank Office in Perugia, 1979. http://www.architetturecontemporanee.beniculturali.it/architetture, © Bruno Signorini

documentation, through which the numerous pieces of information that characterise a building (geometric, historical, constructive, etc.) are acquired, stored and stored in a database, and then stored in a database. They are applicable at many levels: documentation, through which the numerous pieces of information that characterise a building (geometric, historical, constructive, etc.) are acquired, memorised and archived; modelling, through which a knowledge model of the object of study is elaborated, in which the information is organised according to a precise structure; representation, through which images are returned that respond to specific purposes, each time established concerning the definite purpose, but all aimed at greater accessibility of the information; interpretation, through which the information is made available in a manner dedicated to different types of users, whether they are of a technical or divulgation nature. Applications such as 3D scanning, VR/AR, GIS for cultural heritage, digital photogrammetry in the broadest sense, 3D printing are now widely used in the cultural heritage sector and seem fundamental strategies for concrete enhancement measures implementation.

The case studies here presented propose a diversified use of ICT, integrated into the different stages of knowledge-documentary research and extension of the accessibility of the heritage also in terms of dissemination.

3.1 Perugia, Civic Slaughterhouse (Claudio Longo, 1967)

The 1958 city plan for Perugia, drawn up by Mario Coppa, Giuseppe Grossi, Francesco Zannetti and Bruno Zevi, included a plan for the city and neighbouring

Fig. 8 Claudio Longo, Civic Slaughterhouse in Perugia, 1969. Picture of the project model [26, p. 113]

municipalities to reorganise their Slaughterhouses, including the construction of the city's Civic Slaughterhouse. A design and technical team composed of architects Claudio Longo and Ugo Laurenti and engineers Marcello Indiati, Mario Aquilino and Sante Aureli developed the project. Highly technologically advanced for its time, the project aims to fit into the urban context while respecting the landscape and the historic environment. The building firm CREA spa awarded the construction contract and completed the work in 1968. The building began his life in 1969 (Fig. 8) [8].

Longo had already shown interest in this type of building, as evidenced by a proposal developed for the new Civic Slaughterhouse in Rieti, never built. In Perugia, he adopted a horizontal functional model for the building, characterised by a large main ground floor accessible from street level and a basement extended below part of the ground floor. The complex plan is composed of the repetition of a square module with a side of 6 m, diagonally arranged to the area's perimeter. Of the two distinct proposals prepared by the design group (in 1962 and 1964, respectively), the work realisation is more closely related to the first (Fig. 9). In addition to the articulated volumetry of the Slaughterhouse, the plan includes a four-storey building destined to house the officers: panorama-faced, it is separated from the complex by a green strip of trees. On the one hand, plastic forms give the complex an unmistakably contemporary character; on the other hand, the use of traditional Umbrian materials makes the building strongly integrated into the city context. Alternating bands of Assisi stone and solid brick characterise the wall façades. Together with the rough cement of the walls, that give the complex high uniformity and regularity.

When the Slaughterhouse ceased its activity, the necessity of regeneration and redevelopment of the area led up to some project hypotheses. In the 1990s, the Controstudio architectural studio drew up a renovation project to reuse the building as municipal offices, which was not approved. In 2015, a new proposal came from the Roman firm 0 spazio architects: the project envisaged maintaining the existing buildings and converting them to commercial, office and residential use, with a view to overall sustainability. Shelved also that proposal, the complex was abandoned to gradual decline, culminating in its demolition in 2018, to make way for

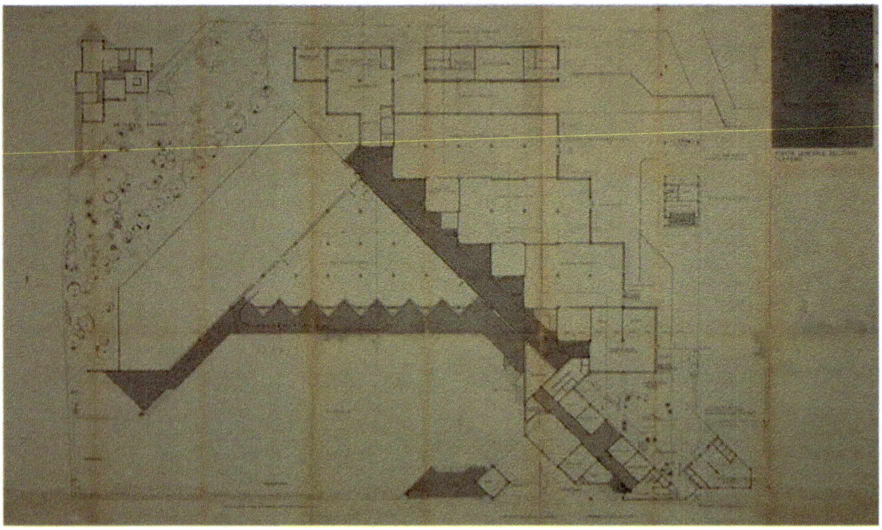

Fig. 9 Claudio Longo, Civic Slaughterhouse in Perugia, 1969. Project plan. © Archivio Centrale dello Stato

a nearby supermarket construction. Today, minimal traces of the building remain, which we can interpret as contemporary ruins, valuable for documenting the plastic and constructive quality of the work. These include the privately-owned apartment building and sporadic portions of the lot's perimeter wall.

The architectural survey of the former Civic Slaughterhouse in Perugia was carried out in 2020 [26], based on the previous didactic activity of direct architectural survey[2] held before the demolition. The survey outcome was two-dimensional drawings (plans, elevations, sections at different scales of representation) and a three-dimensional model (Fig. 10a–c).

3.2 *Corciano, EH System Housing Module (Renzo Piano and Peter Rice, 1978)*

The aftermath of the dramatic earthquake that hit Friuli Venezia Giulia in May 1976 led two years later to the decision to launch a competition to select a project for the reconstruction of housing modules for displaced Friulians. On the occasion of this competition, the associated studio of Renzo Piano and Peter Rice (active from 1977 to 1981) presents an emergency housing prototype [19] based on evolutionary principles and conceived in collaboration with the technical department of an Umbrian company, Vibrocemento Perugia, then a leader in the field of prefabricated construction. The idea [9], which aims at a low-cost product that is easy to assemble and

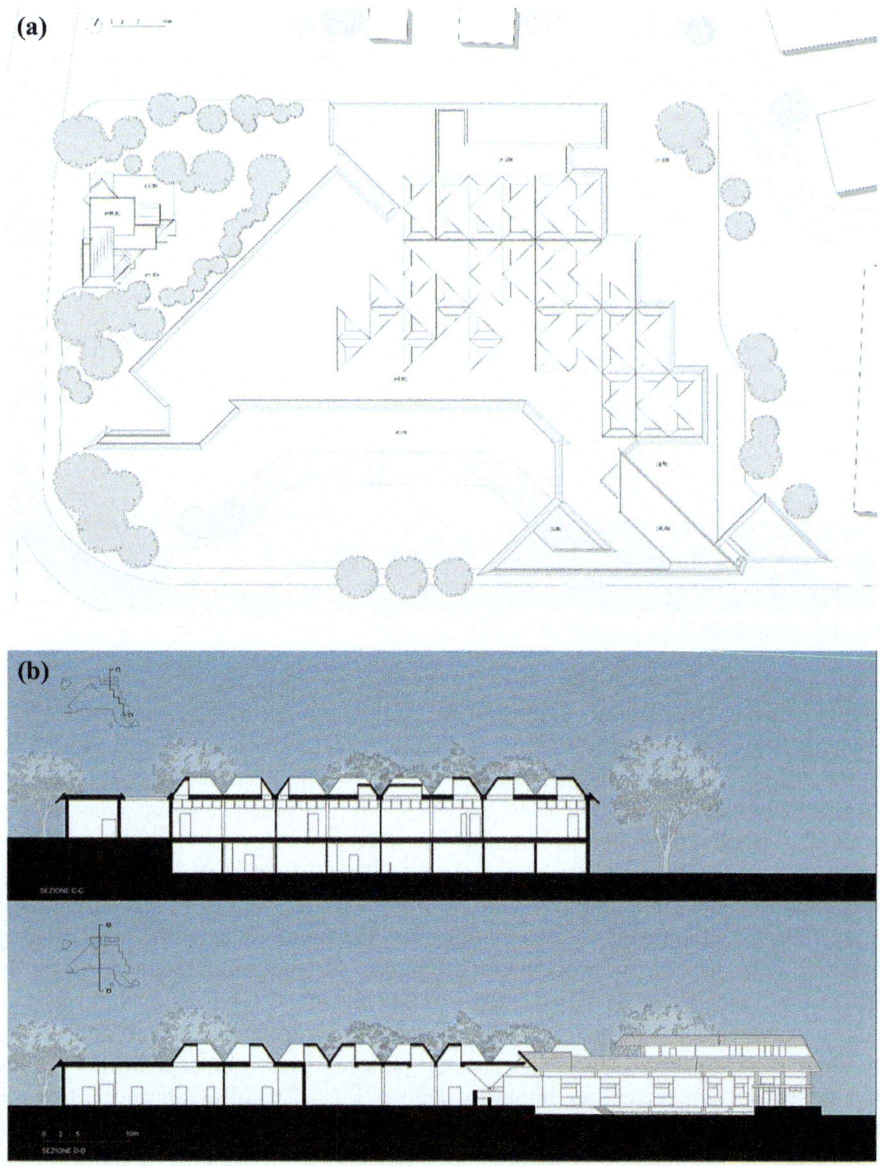

Fig. 10 **a** Claudio Longo, Civic Slaughterhouse in Perugia, 1969. Plan [26]. **b** Reconstructive architectural survey of the Civic Slaughterhouse in Perugia, 2021. Sections [26]. **c** Reconstructive architectural survey of the Civic Slaughterhouse in Perugia, 2021. Ruins of contemporary [26]

Fig. 10 (continued)

highly flexible, is as simple as innovative. But, above all, it is in line with the do-it-yourself principles Renzo Piano has long held dear since the Centre Pompidou in Paris (Fig. 11). The system, although rigorously anti-seismic, relies on elementary assembly techniques. The basic unit, made of reinforced concrete, consists of a three-dimensional C-shaped structure which, once superimposed on another, defines a square (6×6 m) repeatable spatial module. That module acts as a floor, ceiling and sidewall at the same time. Two sliding window walls on rails arranged longitudinally close the module on the two main fronts. The apparent technological banality balances with the utopian propensity that marks (and qualifies) the design idea. In fact, according to Piano and Rice, each user has the right to a basic house, which can be expanded to a maximum of 120 m^2 or reduced to a minimum of 50 m^2 concerning contingent family needs. One can take the missing pieces or deposit the surplus in a collectively managed neighbourhood store (Fig. 12). Although rejected by the competition jury, the prototype took shape in Solomeo in 1978; the industrial partners so well appreciate it that it went into production anyway: redesigned in the spring of 1979 as a social housing module for the Centre for Mental Health in Bastia Umbra (where it still risks demolition despite the constraint imposed by the Superintendency of Cultural Heritage of Umbria in May 2021), it was also used (and, unfortunately, trivialised) in Corciano, as part of the construction of the Il Rigo residential district.

Fig. 11 Renzo Piano e Peter Rice, Evolutive Housing System in Solomeo (PG), 1978. Overview [11, 18]

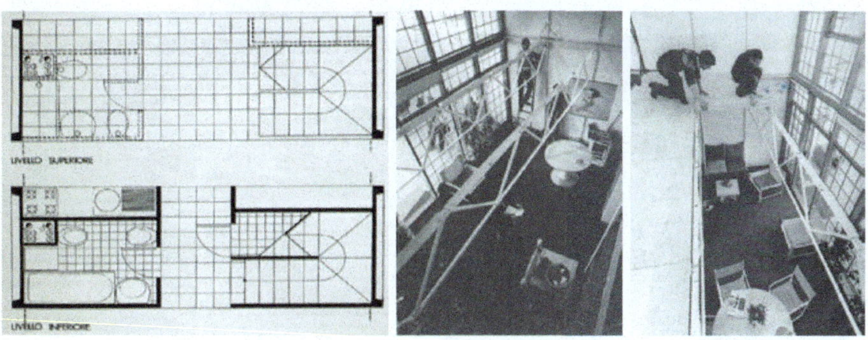

Fig. 12 Renzo Piano and Peter Rice, Evolutive Housing System in Solomeo (PG), 1978. Project drawings and pictures of the prototype under construction [11, 18]

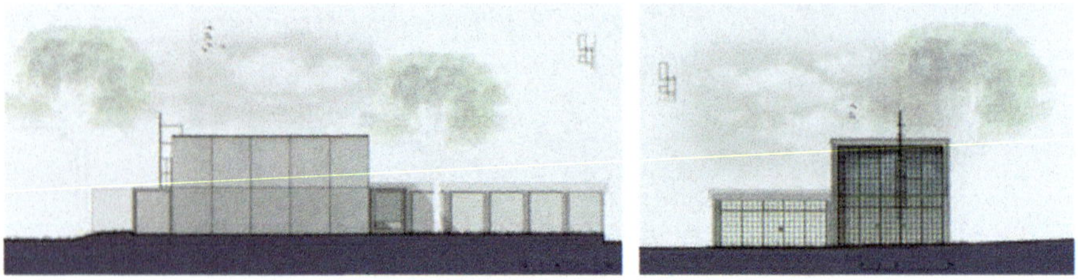

Fig. 13 Reconstructive architectural survey of the Evolutive Housing System in Solomeo (PG), 2019. Elevations [13]

The architectural survey of the EH System housing module was carried out in 2011 [11, 13, 18] before its demolition in 2016. The survey outcome, documented through two-dimensional graphical drawings (plans, elevations, sections at different scales of representation) and a three-dimensional model (Fig. 13), formed the basis for the creation of a 1:33 scale reproduction obtained through 3D printing,[3] currently in production (Fig. 14) and soon to be exhibited.

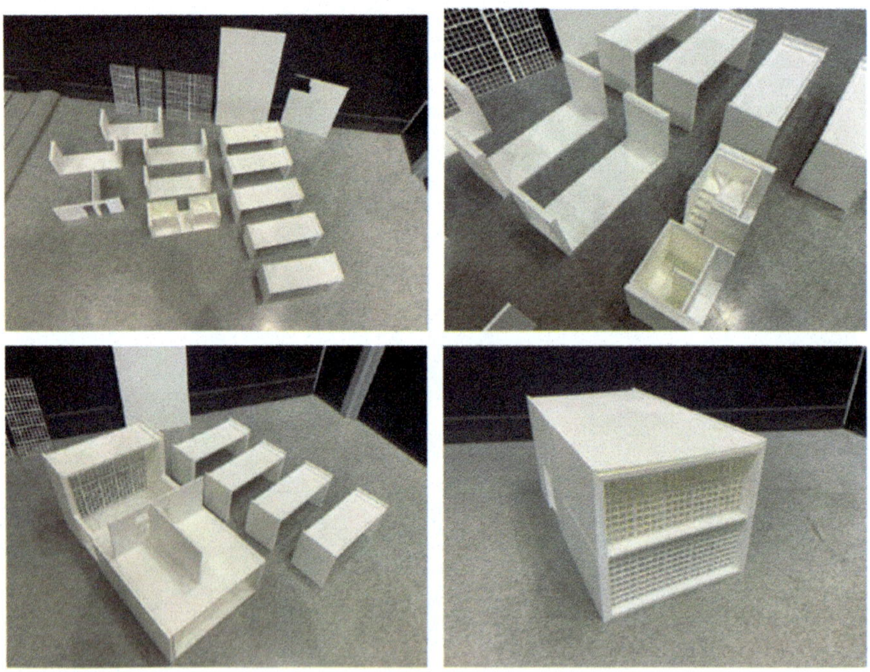

Fig. 14 Reconstructive architectural survey of the Evolutive Housing System in Solomeo (PG), 2019. Replication by 3D printing, © FABLAB Perugia

3.3 *Todi, Exhibition Pavilion of the Antiquarian Fair (Carlo Aymonino and Paolo Portoghesi 1982)*

Since 1969, the town of Todi hosts the famous Antiquarian Fair, traditionally held in Palazzo Landi Corradi, known as Palazzo del Vignola, to which we owe the main door. In 1982, however, a disastrous fire broke out in the palazzo on the final day of the exhibition: that led to the decision to build a temporary exhibition structure to host the event. The city administration, quickly acting with the help of a citizen's committee set up for the purpose, decided to implement the project by architects Carlo Aymonino and Paolo Portoghesi. The project consists of a mobile multifunctional pavilion made up of an elongated removable tensile structure with a metal structure covered by a translucent PVC roof and provided with a dense series of perimeter openings. At both ends of the structure, two small wooden pavilions designed as a tribute to Aldo Rossi's *Teatro del mondo* (World Theatre) mark the entrance and exit (Fig. 15). Originally destined to stand in the central Piazza del Popolo (Fig. 16), the work is the object of bitter controversy (whose main animator is Antonio Cederna [17]) aimed to preserve the 'historical' image of the city, even on the occasion of its temporary placement. The decision, obliged by a monumental constraint expressly imposed on the square, was, therefore, to build the pavilion at Porta Fratta, at a tangential location compared to the historic centre: the project building, modified from a tensile structure to a tent with an independent supporting structure, thus became less temporary and more stable, and could also serve as a cinema and for other initiatives. Inaugurated in August 1982 on the occasion of the 5th Mostra Mercato dell'Artigianato, the Todi exhibition pavilion received considerable attention from the major Italian architectural publications [28, 30, 32]. After the first public event, the pavilion was reused for several minor exhibitions and events; then, it was abandoned, first in the municipal warehouses of Todi and then in those of Spoleto, and moved to the shed of a Spoleto-based transport company. Here it remained unused for a long time until its components finally deteriorated.

The architectural survey of the exhibition pavilion of the Antiquarian Fair was carried out in 2007 [14] and aimed to obtain a virtual reconstruction of the work. The survey resulted in the construction of a three-dimensional model and the collection of extensive documentary material (Fig. 17), in addition to the traditional two-dimensional drawings (plans, elevations, sections, axonometric exploded views).

4 From the Survey of Lost Architecture to the Archaeological Museum of the Present

Based on the experience of multimedia cataloguing of the architecture of the second half of the twentieth century scattered in the Umbrian territory, conducted in 2018 by the internal research team of the Department of Civil and Environmental Engineering

Fig. 15 Carlo Aymonino and Paolo Portoghesi, Exhibition Pavilion of the Antiquarian Fair inTodi (PG), 1982. Overview [14]

of the University of Perugia, it was possible to develop a knowledge enhancement methodology that took and applied a properly archaeological approach. The translation of the title and the cultural approach that distinguish the emblematic collection of essays by Giovanni Urbani published in 2012 [33], give the name to a valorisation strategy designed to build an "archaeology of the present". This strategy aims at embodying a cultural oxymoron capable of celebrating the history and the past, even if recent, perpetuating and reaffirming its presence in contemporaneity. The aim is to preserve long the knowledge of a segment of the architectural heritage not yet unanimously recognised as a holder of cultural value for the city, the landscape and society in general.

The proposed methodology was tested and applied based on the outcomes of the survey of contemporary architecture carried out as part of the *Censimento nazionale delle architetture del secondo Novecento* (National Census of Buildings of the Second Half of the Twentieth Century).[4] The Census revealed a significant number of cases in which valuable works requiring protection underwent substantial transformation (according to procedures which, despite the absolute regularity of the authorisation path, entailed the distortion and impoverishment of the work's original conception). But the case of those works which have seen their existence drastically and irreversibly ended by demolition despite their intrinsic value is even more striking. The memory of these works resides solely in knowledge, just as their direct accessibility, no longer materially possible, lies solely in virtual reconstruction and replication in the form of a physical model obtained by 3D printing.

Fig. 16 Carlo Aymonino and Paolo Portoghesi, Exhibition Pavilion of the Antiquarian Fair inTodi (PG), 1982. Perspective [14]

Faced with the material obliteration and the consequent impossibility of valorising buildings that no longer exist in their concrete form, the only instrument that has the power to rehabilitate their memory is thus the architectural survey, which in this context takes on a properly vindictive role [4]. The cognitive activity undertaken, inevitably conducted in the absence of the concrete object [24], identifies its scientific foundation in the documentary evidence produced over time concerning the work, and especially in the drawing, the preferred place in which the identity of an architectural work thickens. Sketches and project drawings, drawings resulting from previous surveys, documentary material conserved in public and private archives, oral sources that can still be consulted on the works and their construction and functional history: by putting together the historical information and the graphic characteristics deduced from this complex information structure, it is possible to compose a reconstructive knowledge model of the analysed work, which can embody its simulacrum. The digital survey of the contemporary architectural heritage that no longer exists thus

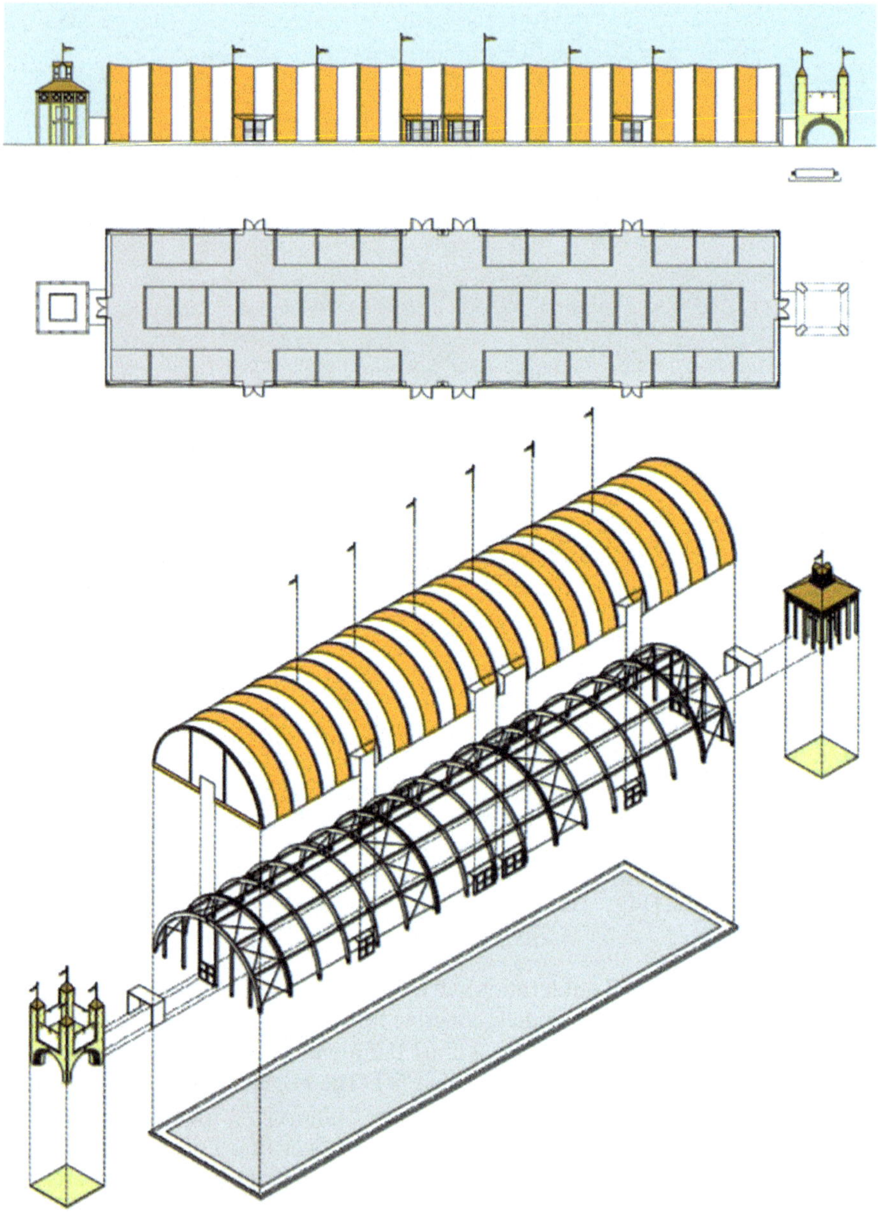

Fig. 17 Reconstructive architectural survey of the Exhibition Pavilion of the Antiquarian Fair inTodi (PG), 2006. Elevation, plan, exploded axonometry [14]

configures a multiplicity of valorisation actions, ranging from virtual reconstruction to scale reproduction through 3D printing. These actions allow the recovery of lost works experience, bringing them closer to a large public and spreading their knowledge. But, even more in keeping with the techniques, methodologies and tools of archaeological investigation, the reconstructive processes employed in the models' creation are based on the survey of the 'ruins of the contemporary'. These kinds of ruins are those surviving traces of a near (but no longer accessible) past that can embody the starting point for undertaking processes of virtual anastylosis applied to recent heritage.

This is the starting point for the idea of setting up, beginning with the pilot experiment currently underway in Umbria, an 'Archaeological Museum of the Present', in which virtual reconstructions of lost architecture can coexist with their three-dimensional reproductions and with digital models that users can interrogate in a personalised manner. A 'knowledge institution' that can build an overall view of the architectural heritage that the territory has had the opportunity to use and enhance, but which ineffective management has led to being erased. A museum conceived as a constantly updated digital database that can be freely consulted by a plurality of users, but at the same time organised in a network of physical locations throughout the country, where the digital archive and material reproductions of the catalogued works can be set up and made directly accessible to the public.

The Museum's structure relies on ICT use, which will be employed in all phases of the knowledge-documentary process: from the acquisition of all the information available on the heritage object of interest through scanning and digitisation procedures to the construction of digital models consultable and semantically implementable. But also from the representation of these models through VR, AR and 3D printing applications to the creation of interpretative documents to support the technical figures involved in the protection and enhancement processes.

The database conceived as a national digital archive and thus continuously updated is to be consulted through a device of a typological nature and thus adaptable to the various physical locations hypothetically selected for the exhibition. In the Umbrian case, this location is in a representative place of the Perugian *Studium*, underlining the reference role played by the University of Perugia in the promotion of a culture of valorisation applied to a continuously updated cultural heritage. The database, constantly updated through the cognitive models emerging from a series of dedicated educational and research initiatives, will be made available to users through a black box designed as a modular multimedia totem, which acts as both a symbolic container and a user interface, ideally replicable and adaptable for setting up in all the Italian locations chosen to host the Museum. The exhibition will be completed by 3D-printed scale models that, built based on contemporary ruins, will once again concretely display the lost architecture, keeping its memory alive.

Notes

[1]For a General Overview of Twentieth Century Umbrian Architecture See [6, 10, 12, 16].
[2]The direct architectural survey activity was carried out during the 2012–2013 academic year as part of the Rilievo dell'architettura course (professors: Paolo Belardi, Valeria Menchetelli; tutor: Maria Elena Lascaro) by students Francesca Romana Gabrielli, Francesca Nafissi, Andrea Pacchierotti, Elena Pierbattisti, Maria Beatrice Sardegna and Chiara Settimi.
[3]Reproduction is being carried out as part of the project *MUSAE—Museums, Social Use and Accessibility to contrast Marginalisation*, coordinated by Key&Key Communications and co-financed by the Fondazione Cassa di Risparmio di Perugia as part of the public notice 2018 *Insieme per un Welfare di Comunità*; in the partnership, the scientific contribution by the Department of Civil and Environmental Engineering of the University of Perugia is aimed at developing knowledge-gathering models required to create corresponding tactile models.
[4]The *Censimento nazionale delle architetture italiane del secondo Novecento nella Regione Umbria* (National Census of Italian Buildings of the Second Twentieth Century in the Umbria Region), an integral part of the initiative undertaken by the Ministry of Culture on the national territory, was promoted by the then Direzione Generale Arte e Architettura Contemporanee e Periferie Urbane (General Directorate for Contemporary Art and Architecture and Urban Peripheries), now Direzione Generale Creatività Contemporanea (General Directorate for Contemporary Creativity). The research agreement was signed in February 2018 between the Ministero dei Beni e delle Attività Culturali e del Turismo per l'Umbria (Regional Secretariat of the Ministry of Cultural Heritage and Activities and Tourism for Umbria) and the Department of Civil and Environmental Engineering of the University of Perugia. The Research Coordination Committee included: Aurora Raniolo, Regional Secretary of the Regional Secretariat of the Ministry of Cultural Heritage and Activities and Tourism for Umbria (Chair); Camilla Mannocci, Scientific Responsible for the Regional Secretariat of the Ministry of Cultural Heritage and Activities and Tourism for Umbria; Florian Castiglione, Delegate for the Archaeological, Fine Arts and Landscape Superintendence of Umbria; Paolo Belardi, Scientific Responsible for the Department of Civil and Environmental Engineering of the University of Perugia. Other members of the research group were: Valeria Menchetelli, Operations Coordinator for the Department of Civil and Environmental Engineering of the University of Perugia; Luca Tesei, holder of the grant "Scheduling of architecture from the second half of the twentieth century in Umbria"; Veronica Zoccolini, holder of the grant "Cataloguing of architecture from the second half of the twentieth century in Umbria".

References

1. Belardi, P. (2009, May 25). Contro la sindrome da Mulino Bianco. Corriere dell'Umbria, p. 6
2. Belardi, P., Boco, F. (2010). Il disegno della nuova architettura. La svolta moderna dell'Accademia di Belle Arti di Perugia. In: M. Docci, & M.G. Turco (Eds.), L'Architettura dell'"Altra" Modernità (pp. 313–317). Roma: Gangemi Editore
3. Belardi, P. (2010). «Povera architettura nostra!» La ventata razionalista dell'Accademia del Disegno di Perugia. In: Ikhnos. Annale di Analisi grafica e Storia della Rappresentazione (pp. 119–128). Lombardi editori, Siracusa
4. Belardi, P. (2011). SOS architettura contemporanea. Il disegno rivendicativo. In G. De Fiore (Ed.). *Disegno dunque sono* (pp. 15–18). Università degli Studi di Genova, Genova
5. Belardi, P. (2012a). L'amnesia del presente. In M, Pisani (Ed.), *Italy now. Architecture 2000–2012* (p. 128). Roma: Edilstampa
6. Belardi, P. (2012b). Profilo storico dell'architettura umbra del Novecento. Dal lessico quotidiano delle case operaie di Terni all'astrazione metafisica delle piazze pubbliche di Perugia. In P.

Belardi, & V. Menchetelli (Eds.), *Da case popolari a case sperimentali. Un secolo di architettura nell'edilizia residenziale pubblica della provincia di Perugia* (pp. 49–68). Perugia: EFFE Fabrizio Fabbri Editore

7. Belardi, P. (2014a). L'architettura del Novecento. Dall'architettura-arte all'arte-architettura. In: Tosti, M. (Ed.), *Storia dell'Umbria dall'Unità a oggi. Uomini e risorse* (pp. 331–361). Marsilio, Venezia
8. Belardi, P. (2014b). Il mattatoio pubblico di Claudio Longo a Perugia. In: P. Belardi (Ed.), *NAU Novecento Architettura Umbria* (pp. 166–169). Il Formichiere, Foligno.
9. Belardi, P. (2014c). Il modulo abitativo di Renzo Piano e Peter Rice a Corciano. In: P. Belardi (Ed.), *NAU Novecento Architettura Umbria* (pp. 206–209). Il Formichiere, Foligno
10. Belardi, P. (Ed.). (2014d). NAU Novecento Architettura Umbria. Il Formichiere, Foligno
11. Bellucci, A. (2011–2012). Evolutive Housing System di Renzo Piano e Peter Rice. Dal rilievo architettonico integrale alla proposta di riuso in chiave energetica [Unpublished degree thesis]. Università degli Studi di Perugia, Perugia.
12. Biscottini, G. (2010–2011). L'architettura contemporanea in Umbria attraverso le riviste (1945–2009) [Unpublished degree thesis]. Università degli Studi di Firenze, Firenze
13. Bonelli, F. (2018–2019). Evolutive Housing System di Renzo Piano e Peter Rice. Rilievo architettonico di un'architettura perduta [Unpublished degree thesis]. Università degli Studi di Perugia, Perugia
14. Briganti, N. (2006–2007). Il Padiglione per esposizioni di Carlo Aymonino e Paolo Portoghesi. Dall'idea progettuale all'ipotesi ricostruttiva [Unpublished degree thesis]. Università degli Studi di Perugia, Perugia
15. Brusaporci, S., & Trizio, I. (2013). La "Carta di Londra" e il patrimonio architettonico: riflessioni. circa una possibile implementazione. *Scires.it 3*(2), 55–68
16. Canali, F. (2010). L'architettura nel Novecento. In R. Avellino (Ed.), Percorsi d'architettura in Umbria (pp. 531–607). Editrice Centro Italia, Foligno.
17. Cederna, A. (1982, July 4). Lo scempio antincendio. L'espresso, 46–48
18. De Matteis, L. (2011–2012). Evolutive Housing System di Renzo Piano e Peter Rice. Dal rilievo architettonico integrale alla proposta di riuso in chiave strutturale [Unpublished degree thesis]. Università degli Studi di Perugia, Perugia
19. Dini, M. (1983). Renzo Piano. Progetti e architetture 1964–1983. Electa, Milano (1983). Sistema costruttivo industrializzato per abitazioni a tipologia evolutiva (pp. 104–113)
20. Iovane, G. (2020, November 17). La tutela dell'architettura contemporanea: spunti di riflessione a partire dal caso "Geco". *e-lex.it*. Retrieved December 28, 2021, from https://www.e-lex.it/it/la-tutela-dellarchitettura-contemporanea-spunti-di-riflessione-a-partire-dal-caso-geco/
21. Ippolito, A., & Cigola, M. (Eds.) (2017). *Handbook of research on emerging technologies for digital preservation and information modeling*. Hershey, PA: IGI Global.
22. Jokilehto, J. (2007). Modern Built Heritage as World Heritage. https://www.icomos.org/risk/2007/pdf/Soviet_Heritage_10_I-4_Jokilehto.pdf
23. Lenci, S., & Antonelli, M. (Eds.). (1996). Franco Antonelli architetto. Cassa di Risparmio di Foligno, Foligno
24. Maggio, F., & Villa, M. (2008). Architettura demolita. Modelli abitativi alla V Triennale di Milano. Ridisegno e analisi grafica. Caracol, Palermo
25. Manzo, E. (2010). I "chierici" del Moderno. L'Italia della continuità in provincia: l'Umbria e Perugia. In A. Giannetti, & L. Molinari (Eds.), Continuità e crisi. Ernesto Nathan Rogers e la cultura architettonica italiana del secondo dopoguerra (pp. 153–172). Alinea editrice, Firenze
26. Massotti, D. M. (2019–2020). Rilievo architettonico di un'architettura perduta: l'ex mattatoio di Claudio Longo in via Palermo a Perugia [Unpublished degree thesis]. Università degli Studi di Perugia, Perugia
27. Molinari, L., & Alebbi, V. (2008). Italo Rota. Projects, works, visions 1997–2007. Skira, Milano
28. Monumentale, L. (1982). *Domus, 634*, 28–29.
29. Pisani, M. (2004). Signorini: Architetture. Libria, Melfi

30. Priori, G. (1983). Un'architettura "dentro" la città progettata da Carlo Aymonino e Paolo Portoghesi. *Eupalino, 1*, 18–19.
31. Soletti, A., & Belardi, P. (Eds.). (1996). *Architettura contemporanea in Umbria*. Nuove tendenze. Università degli Studi di Perugia.
32. Un serpentone per la rinascita di Todi. (1982). *Casa Vogue, 133*, 56.
33. Urbani, G. (2012). Per una archeologia del presente. Saggi sull'arte contemporanea (B. Zanardi, Ed.). Milano: Skira

Visual Journalism Applications: 3D Modeling for Cultural Heritage Sites Visualization

Amedeo Ganciu, Marta Pileri, Andrea Sias, and Michele Valentino

Abstract The purpose of digital cultural heritage dissemination requires multiple communication tactics to reach as broad an audience as possible. The relationship between high-tech and low-tech knowledge for visual communication requires an overall reflection on the role of expert knowledge to reach a wide audience. In this respect, Visual Journalism is a practical methodology for communicating cultural heritage and facilitating public access to complex content. The elaboration of this kind of visualisation, with particular reference to the use of 3D models of heritage sites, becomes an opportunity for a reflection on the role of heritage communication to achieve a broad strategy for its accessibility in a general sense.

Keywords Visualisation · Communication · Visual journalism · 3D modeling · Low-tech knowledge

1 Introduction

The concept of *Digital Twin* was used for the first time during the *Product Lifecycle Management* course held in 2001 at the *Florida Institute of Technology* by Michael Grieves. It is intended as a virtual and digital equivalent of a physical product, based on a dynamic *Mirrored Spaces Model*. The dimensions, real and virtual, remain linked during the entire lifecycle of the artefact: creation, production (manufacturing), and disposal. The evolution of the production paradigm towards Digital Twins could also be considered inevitable considering other determining factors such as the technological development of the *Internet of Things*, the cloud, mobile technologies, Artificial Intelligence and Blockchain.

The fields of application are many and varied, from agriculture to biomedical sciences to the design and planning of *Smart Cities*. One example is the digital model of entire metropolises, as in Singapore. In these cases, digital copying helps urban

A. Ganciu (✉) · M. Pileri · A. Sias · M. Valentino
Dipartimento di Architettura, Design e Urbanistica, Università degli Studi di Sassari, Alghero, Italia
e-mail: aganciu@uniss.it

© The Author(s), under exclusive license to Springer Nature Switzerland AG 2023

I. Trizio et al. (eds.), *Digital Restoration and Virtual Reconstructions*, Digital Innovations in Architecture, Engineering and Construction,
https://doi.org/10.1007/978-3-031-15321-1_19

Fig. 1 VR-Studio Zoan (2020), *Photo Virtual Helsinki*. <https://www.myhelsinki.fi/en/see-and-do/sights/virtual-helsinki>

planners and decision-makers understand how to improve energy efficiency, manage traffic and public services more efficiently, and identify new applications that can improve citizens' lives. In the cultural sector, *Virtual Helsinki* is a pioneering digital model implemented by the City of Helsinki to make the metropolis a significant capital of cyberspace. The project's release is the culmination of a long collaboration with the *Zoan* studio, which used the *Unreal Engine* platform for three-dimensional modelling. The model merges drawings and images to create one of the world's most realistic and advanced virtual reality experiences. The *Virtual Helsinki* digital model recreates a simulated experience of the city's most famous landmarks, giving people who cannot physically get to Helsinki the opportunity to experience it (Fig. 1).

If VR and 360° video are used as a marketing tool by many destinations, Virtual Helsinki represents an evolution. In this case, visitors are allowed to move freely in the digital simulation of the city to explore at their own pace and create their own experiences in time and space, virtually 'experiencing' life in the capital of Flanders. They can participate in concerts, exhibitions, meetings, and events. While the contemporary trend is to invest mainly in the most technologically advanced digital tools, it must also be considered that, according to the OECD, only 21% of the population has a sufficient level of digital literacy.

The essay illustrates the role of three-dimensional models in cultural heritage sites' communication and fruition processes, particularly monuments of the past and present, based on low-tech communication media. The investigation aims to identify applications of three-dimensional modelling techniques in this new field of study.

The article's focus on this type of communication media derives from the need to facilitate accessibility to multiple audiences with additional skills, abilities and literacy—including digital literacy—prescribed by international guidelines and conventions such as the *ICOMOS Charter for the Interpretation and Presentation of Cultural Heritage Sites*.

In particular, it is intended to highlight how enhancing the accessibility and usability of sites requires integrating different communication strategies linked to the various learning capacities of the different audiences using cultural heritage. In addition to digitally literate audiences, those excluded from high-tech modes of communication must also be considered. However, for generational, economic, or socio-cultural reasons, these can be reached through low-tech strategies based on graphic-visual languages, which are thus essential tools for inclusion and integration.

The methodological rigour of computer visualisation as a research and communication tool for cultural heritage is set out in the *London Charter*, whose aim is to create a debate in the scientific community on the need to apply the new technologies critically. In particular, the document has six principles concerning the application of digital visualisation to Cultural Heritage, which make scientific research in the field of heritage communication compliant [1].

2 Hi-tech and Low-tech Knowledge for Visual Communication

In point one of the sixth principle of the *London Charter*, the opportunity to make assets accessible through digital dissemination plans that are as inclusive as possible is set out. As mentioned above, the level of digital literacy is one of the issues to be overcome to best meet this need [2].

The illustration is one of the most effective transmission techniques as it uses a visual channel, which is easily intuitive and conveniently accessible to a broad audience. As it is a method of communication that does not require any particular digital skills, it can be considered a valid output of 3D modelling applied to cultural heritage.

Among the different visual communication techniques that can be used for the transmission of goods, Visual Journalism is an increasingly practised and widespread reality. At the beginning of the 2000s, the term Visual Journalism meant drawings exclusively to support journalistic information in its various visual forms. More recently, Visual Journalism has been identified as combining architectural representation, illustration, and infographics to interactively visualise data through immediate graphics. 3D modelling becomes an opportunity to elaborate functional elements for the transmission of cultural heritage to non-expert users [3].

Although Visual Journalism is considered a relatively recent practice [4], it can be said that the interaction between images and information has a history rooted in time. It must be remembered that this practice has absorbed aspects of visual disclosure and has taken journalism to a new frontier, changing some models of traditional journalism [5] (Figs. 2 and 3).

Several authors have attempted to define it, also using specific parts of the different points of view that make up this practice. One thinks of *photojournalism* accompanying journalistic reporting with photos and moving images [6] or *comics journalism*

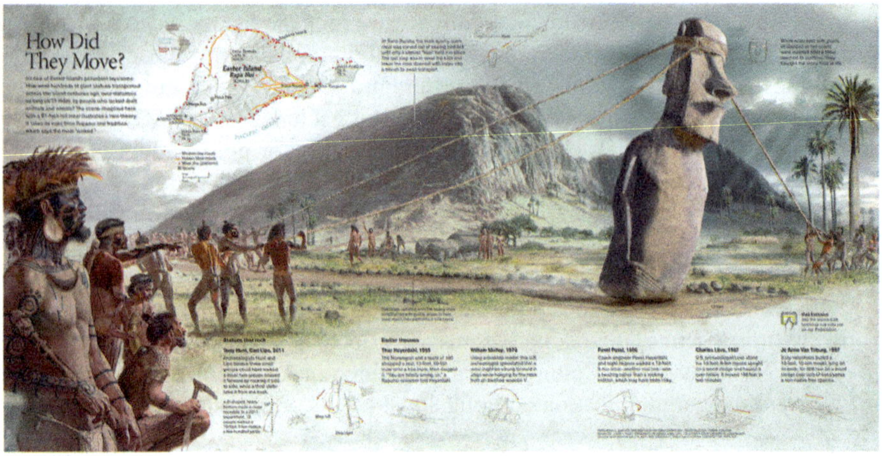

Fig. 2 Fernando G. Baptista (2012), *Easter Island.* National Geographic

Fig. 3 Fernando G. Baptista (2014), *Inside a Wari Tomb.* National Geographic. <https://issuu.com/projectbeagle/docs/national_geographic_usa_2014-06/82>, pp. 82–83

through illustrations and comics [7]. However, the position that captures the potential of Visual Journalism much more broadly is the one that highlights the strong interaction between image and text that makes use of different visualisation techniques [8].

The term 'Visual Journalism' refers to the set of information activities in which the priority is taken away from text and given to images so that the user, thanks to the visual channel, can learn the information immediately and pleasantly. This reverses the traditional order of journalistic communication where the reader's gaze is first drawn to images and then to headlines and text boxes.

Nowadays, society is strongly influenced by the development of technologies and journalism itself is currently experiencing a strong push, not so much in the profession in the strict sense of the word, but in the methods and models of transmitting the information. One of the innovative modes that journalism is confronted with is certainly immersive journalism or *VR journalism* that provides a different level of the sensory experience of reality [9]. The emergence of virtual reality and 360° videos has opened up new narrative perspectives that are strongly influenced by these technologies, opening up new modes of communication within the context of large-scale news reporting [10]. Among the first newspapers to go through this practice, we can include the New York Times, which is 2015, in collaboration with Google and with the involvement of many users, launched a virtual reality project called *NYT VR*. Hundreds of 360° videos are available on its YouTube channel, which allows an immersive experience through VR visors. One example is the video made for the newspaper's web platform by multimedia journalist Guglielmo Mattioli, who trained as an architect and urban planner and filmed in VR and 360° the chapels that the Vatican State built for the 16th *International Architecture Exhibition—Biennale di Venezia* in 2018 [11] (Figs. 4 and 5).

In this respect, an emblematic and apparent case is the use of the 3D model of Notre-Dame Cathedral after the dramatic fire that undermined its entire structure. The images broadcast, almost in real-time, in all the world's media showed the impossibility of recovering the ancient building using traditional restoration techniques. In the article *Why Notre-Dame Was a Tinderbox* [12], published on 17 April 2019, just two days after the fire, the use of the 3D model clarified how modern visualisation techniques in a journalistic key allowed complex elements to be highlighted

Fig. 4 Guglielmo Mattioli (2018), *Norman Foster's Chapel*, 360° image, <https://www.nytimes.com/2018/05/25/arts/design/venices-architecture-biennale-the-vatican.html>, video ID: 100000005919023

Fig. 5 Guglielmo Mattioli (2018), *Terunobu Fujimori's Chapel*, 360° image, <https://www.nytimes.com/2018/05/25/arts/design/venices-architecture-biennale-the-vatican.html>, video ID: 100000005919024

to a non-expert audience (Fig. 6). The 3D model allowed for the parallel visualisation of non-communal structures visible to the public and the possible causes for which the fire spread. Visual journalists could uniquely convey complex elements that belong to a highly specialized niche of users to a broad audience. Visual journalists [13] could uniquely communicate intricate details belonging to a highly specialized place of users to a broad audience. Naturally, this process was made possible by combining the expressive language of 3D graphics with simple explanatory sentences and relying on the university lecturers' sources—Jack J. Murphy [14], Jim Lygate [15] and Guillermo Rein [16] (Figs. 7, 8 and 9).

The 'visuality' of information becomes crucial in a situation where the image is pervasive in our society [17–19]. As Marco Gasperetti points out in his book *Giornalismo 3D: la metamorfosi di una professione* (2015), modern digital techniques have challenged the nature of written communication towards an integration of different tools—such as writing, photos, videos, 3D models—that allow the user to have an increasingly precise and clear picture of the information. The author identifies a possible shift from the 3D to the 4D dimension of information, in which the user can become an active part and agent of this journalistic process [20].

3D by modelling reality from within becomes a fundamental tool for understanding spaces. Applying this investigation methodology to journalism makes it possible to reveal what is hidden from view.

Fig. 6 Jack J. Murphy, Jim Lygate & Guillermo Rein, *Why Notre-Dame Was a Tinderbox?* (2019). <https://www.nytimes.com/interactive/2019/04/17/world/europe/notre-dame-cathedral-fire-spread.html>

Fig. 7 Jack J. Murphy, Jim Lygate & Guillermo Rein, *Why Notre-Dame Was a Tinderbox?* (2019). <https://www.nytimes.com/interactive/2019/04/17/world/europe/notre-dame-cathedral-fire-spread.html>, details

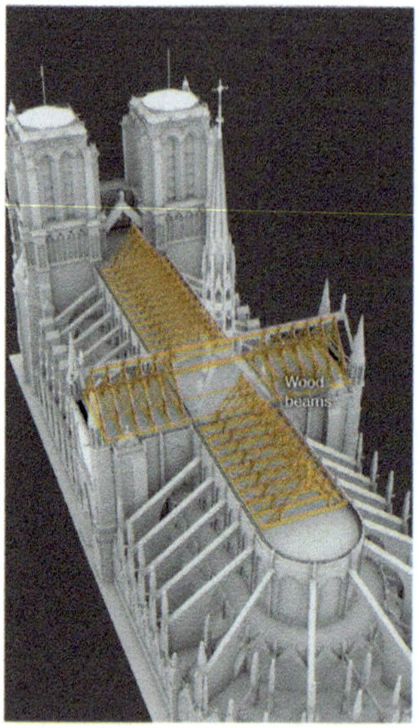

Fig. 8 Jack J. Murphy, Jim Lygate & Guillermo Rein, *Why Notre-Dame Was a Tinderbox?* (2019). <https://www.nytimes.com/interactive/2019/04/17/world/europe/notre-dame-cathedral-fire-spread.html>, details

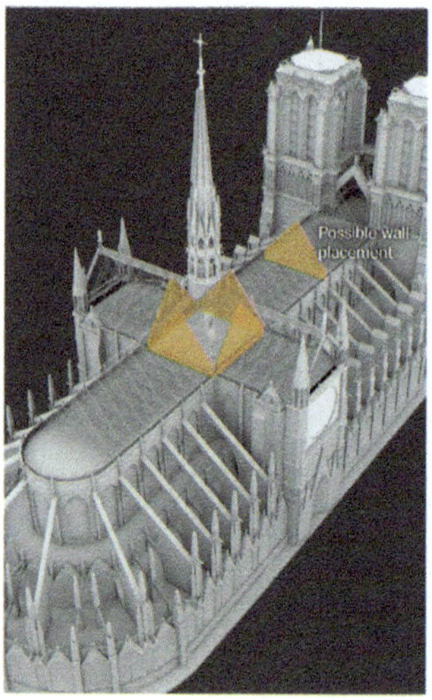

Fig. 9 Jack J. Murphy, Jim Lygate & Guillermo Rein, *Why Notre-Dame Was a Tinderbox?* (2019). <https://www.nytimes.com/interactive/2019/04/17/world/europe/notre-dame-cathedral-fire-spread.html>, details

3 Visual Journalism as Low-tech Tool Communication for Cultural Heritage

In contrast to the above experiences, it can be seen that the creation of a Visual Journalism board can require teamwork involving journalists, graphic designers and designers experimenting with different combinations of text, images and infographics that recreate a different multimedia experience.

Considering that Visual Journalism originates from newspapers, the pioneer of this technique is undoubtedly National Geographic. His team includes Fernando Baptista, senior infographics editor at the magazine's headquarters in Washington, D.C. The illustrator defines his work as narrating different stories through visual communication, constantly exposing different themes.

Architectural works produced based on 3D modelling of a building or object are of enormous importance. The most famous include Antoni Gaudí's *Sagrada Familia* and Filippo Brunelleschi's *Dome of Florence Cathedral* (Figs. 10 and 11).

In these visual articles, the author uses several devices to make the analysis of the cultural heritage communicable. Among these devices, the following are particularly relevant: chronological and typological timelines; construction descriptions and detailed studies of graphic and symbolic elements; axonometric cutaways examined in detail. All these elements transform virtual and abstract concepts into concrete and visible content.

The application of three-dimensional modelling techniques in Visual Journalism has become an expanding field of transdisciplinary investigation. The three-dimensional model is configured as the fundamental element on which the entire communication process is based. Information and contents are disseminated through graphic-visual strategies of various types and technologies and supported with different technological levels. In particular, the article focused on the use of Visual Journalism in the communication of architectural cultural heritage.

The production of images for Visual Journalism requires the construction of three-dimensional supports to base the illustration of the contents. These supports are often physical-material, as in the case of Fernando Baptista's work for *National Geographic*, since the professionals working in the sector often come from artistic rather than technical-designing backgrounds.

Even behind the production of low-tech communication media, such as illustrations on printed publishing products intended to be used and understood by the general public, there is, therefore, the use of three-dimensional models that can be produced even though the most advanced digital modelling technologies. In this case, the possession of skills in digital three-dimensional modelling allows the creation of models on which to base post-production aimed at the illustration for the communication of cultural heritage and, in particular, the architectural heritage. Furthermore, it is evident how, in addition to supporting the elaboration of traditional visual communication products, the three-dimensional digital model can help the most advanced

Fig. 10 Fernando G. Baptista, *Barcelona's Natural Wonder* (2010). <https://www.metalocus.es/en/news/infographic-sagrada-familia-barcelona-gaudi>

Fig. 11 Fernando G. Baptista, *Redefining The Dome* (2014). <https://www.nationalgeographic.org/hires/domed-buildings/>

applications of VR Visual Journalism for the construction of immersive learning and information environments, outlining a multiplicity of applications capable of defining an area of potential development of graphic sciences within the field of information and dissemination (Fig. 12).

The interaction between expert knowledge—using the most advanced digital modelling technologies—and the need to communicate cultural heritage to a public unfamiliar with these artefacts raises relatively open questions about the role that visual communication of the heritage must and can play. Its dissemination, as a

Fig. 12 Bridget Poetker, *What Is Virtual Reality?* (2019). <https://learn.g2.com/virtual-reality>

strategy for its accessibility, becomes a vital issue for cultural heritage management. The planning and implementation of sustainable solutions to preserve the visual documentation of cultural heritage need multiple communication tactics to reach as broad an audience as possible.

References

1. Brusaporci, S., & Trizio, I. (2013). La "Carta di Londra" e il patrimonio architettonico: riflessioni circa una possibile implementazione. *SCIentific RESearch and Information Technology. Ricerca Scientifica e Tecnologie dell'Informazione, 3*(2), 55–68
2. Denard, H. (2016). A new introduction to the London Charter. *In Paradata and transparency in virtual heritage* (pp. 83–98). Routledge
3. Gynnild, A. (2019). Visual journalism. In P. V. Tim, & H. Folker (Eds.), *The international encyclopedia of journalism studies*. Hoboken: Wiley
4. Luigini, A., & Moretti, M. (2018). L'attualità di Otto Neurath. Da ISOTYPE al visual journalism per un racconto visuale della società. XY, 3(6), 74–93
5. Schianchi, P. (2018). Visual Journalist: L'immagine è la notizia. FrancoAngeli
6. Newton, J. (2013). *The burden of visual truth: The role of photojournalism in mediating reality*. Routledge
7. Koçak, K. (2017). Comics journalism: Towards a definition. *International Journal of Humanities and Cultural Studies, 4*(3), 173–199. https://www.ijhcs.com/index.php/ijhcs/issue/view/30
8. Machin, D., & Polzer, L. (2015). Visual journalism. *Macmillan International Higher Education*
9. Sirkkunen, E., Väätäjä, H., Uskali, T., & Rezaei, P. P. (2016). Journalism In virtual reality: Opportunities and future research challenges. In *Proceedings of the 20th International Academic Mind Trek Conference* (pp. 297–303)
10. Paino Ambrosio, A., & Rodríguez Fidalgo, M. I. (2019). Proposal for a new communicative model in immersive journalism. *Journalism*, 1464884919869710
11. Retrieved January 20, 2022, from https://www.nytimes.com/2018/05/25/arts/design/venices-architecture-biennale-the-vatican.html
12. Retrieved January 20, 2022, fromhttps://www.nytimes.com/interactive/2019/04/17/world/europe/notre-dame-cathedral-fire-spread.html
13. Buchanan, L., Glanz, J., Grothjan, E., Rebecca Lai, K. K., McCann, A., Patanjali, K., & Parshina-Kottas, Y. Jeremy White e Graham Roberts
14. Chairman of the Fire Safety Directors Association of Greater New York
15. Visiting professor of fire investigation at University of Edinburgh
16. Professor of fire science at Imperial College London
17. Eribon, D., & Gombrich, E. H. J. (1994) Il linguaggio delle immagini: intervista. Einaudi
18. Barthes, R. (2003). *La camera chiara: nota sulla fotografia*. Torino: Einaudi.
19. Eco, U., Augé, M., & Didi-Huberman, G. (2015). La forza delle immagini. FrancoAngeli
20. Gasperetti, M. (2015). Giornalismo 3D: la metamorfosi di una professione. Edizioni ETS

The manufacturer's authorised representative in the EU is Springer Nature Customer Service Centre GmbH, Europaplatz 3, 69115 Heidelberg, Germany. If you have any concerns regarding our products, please contact ProductSafety@springernature.com

Printed and bound by CPI Group (UK) Ltd, Croydon, CR0 4YY
28/09/2026
02224078-0005